Studies in Classification, Data Analysis, and Knowledge Organization

Springer
Berlin
Heidelberg
New York
Barcelona
Hong Kong
London
Milan
Paris
Singapore
Tokyo

Titles in the Series

H.-H. Bock and P. Ihm (Eds.)
Classification, Data Analysis, and Knowledge Organization. 1991
(out of print)

M. Schader (Ed.)
Analyzing and Modeling Data and Knowledge. 1992

O. Opitz, B. Lausen, and R. Klar (Eds.)
Information and Classification. 1993
(out of print)

H.-H. Bock, W. Lenski, and M.M. Richter (Eds.)
Information Systems and Data Analysis. 1994
(out of print)

E. Diday, Y. Lechevallier, M. Schader, P. Bertrand, and B. Burtschy (Eds.)
New Approaches in Classification and Data Analysis. 1994
(out of print)

W. Gaul and D. Pfeifer (Eds.)
From Data to Knowledge. 1995

H.-H. Bock and W. Polasek (Eds.)
Data Analysis and Information Systems. 1996

E. Diday, Y. Lechevallier and O. Opitz (Eds.)
Ordinal and Symbolic Data Analysis. 1996

R. Klar and O. Opitz (Eds.)
Classification and Knowledge Organization. 1997

C. Hayashi, N. Ohsumi, K. Yajima, Y. Tanaka, H.-H. Bock, and Y. Baba (Eds.)
Data Science, Classification, and Related Methods. 1998

I. Balderjahn, R. Mathar, and M. Schader (Eds.)
Classification, Data Analysis, and Data Highways. 1998

A. Rizzi, M. Vichi, and H.-H. Bock (Eds.)
Advances in Data Science and Classification. 1998

Maurizio Vichi · Otto Opitz
Editors

Classification and Data Analysis

Theory and Application

Proceedings of the Biannual Meeting of the
Classification Group of Società Italiana di Statistica (SIS)
Pescara, July 3–4, 1997

With 97 Figures
and 78 Tables

Springer

Prof. Maurizio Vichi
University "G. D'Annunzio" di Chieti
Dipartimento Metodi Quantitativi e
Teoria Economica
Viale Pindaro 42
I-65127 Pescara, Italy

Prof. Dr. Otto Opitz
University of Augsburg
Lehrstuhl für Mathematische Methoden
der Wirtschaftswissenschaften
D-86135 Augsburg
Germany

ISBN 3-540-65633-2 Springer-Verlag Berlin Heidelberg New York

Cataloging-in-Publication Data applied for
Die Deutsche Bibliothek – CIP-Einheitsaufnahme
Classification and data analysis: theory and application; proceedings of the biannual meeting of the Classification Group of Società Italiana di Statistica (SIS), Pescara, July 3–4, 1997; with 78 tables/Maurizio Vichi; Otto Opitz ed. – Berlin; Heidelberg; New York; Barcelona; Hong Kong; London; Milan; Paris; Singapore; Tokyo: Springer, 1999
(Studies in classification, data analysis, and knowledge organization)
ISBN 3-540-65633-2

Printed in Germany

Softcover-Design: Erich Kirchner, Heidelberg

SPIN 10700416 32/2202-5 4 3 2 1 0 – Printed on acid-free paper

PREFACE

International Federation of Classification Societies

The International Federation of Classification Societies (IFCS) is an agency for the dissemination of technical and scientific information concerning classification and data analysis in the broad sense and in as wide a range of applications as possible; founded in 1985 in Cambridge (UK) from the following Scientific Societies and Groups: British Classification Society - BCS; Classification Society of North America -CSNA; Gesellschaft für Klassifikation - GfKl; Japanese Classification Society - JCS; Classification Group of Italian Statistical Society - CGSIS; Sociétè Francophone de Classification - SFC. Now the IFCS includes the following Societies: Dutch-Belgian Classification Society - VOC; Polish Classification Section - SKAD; Portuguese Classification Association - CLAD; Group-at-Large; Korean Classification Society - KCS.

Biannual Meeting of the Classification and Data Analysis Group of SIS

The biannual meeting of the Classification and Data Analysis Group of Società Italiana di Statistica (SIS) was held in Pescara, July 3 - 4, 1997.

The 69 papers presented were divided in 17 sessions. Each session was organized by a chairperson with two invited speakers and two contributed papers from a call for papers. All the works were referred. Furthermore, during the meeting a discussant was provided for each session. A short version of the papers (4 pages) was published before the conference.

The scientific program covered the following topics:

- ***Classification Theory***

Fuzzy Methods - Hierarchical Classification - Non Hierarchical Classification - Optimisation approach in Classification. - Classification of Multiway Data - Probabilistic Methods for Clustering - Consensus and Comparison Theories in Classification - Spatial data and Clustering - Validity of Clustering - Neural Networks and Classification - Genetic Algorithms - Classification with Constraints

- ***Multivariate Data Analysis***

Categorical Data Analysis - Factor Analysis and Related Methods - Discrimination and Classification - Visual Treatment in Data Analysis Symbolic Data Analysis - Non Linear Data Analysis

- ***Multiway Data Analysis***

Analysis of Multiway Data - Panel Data Analysis

- ***Proximity Structure Analysis***

Multidimensional Scaling - Similarities and Dissimilarities -

- ***Software Developments for Classification and Data Analysis***

Algorithms for Hierarchical and Non Hierarchical Classification - Computer Data Visualization. Statistical Algorithms for Multivariate Analysis

- ***Applied Classification and Data Analysis in Social, Economic, Medical, and other Sciences***

Classification and Data Analysis of Textual Data - Data Analysis in Economics - Classification and Discrimination Approaches in Medical Science

The present volume contains 45 referred papers presented in four chapters as follows:

Classification
- Methodologies in Classification
- Fuzzy clustering and fuzzy methods

Other Approaches for Classification
- Discrimination and Classification
- Regression Tree and Neural Networks

Multivariate and Multidimensional Data Analysis
- Proximity Methodologies in Classification
- Factorial methods
- Spatial Analysis
- Multiway Data Analysis
- Multivariate analysis

Case Studies

Acknowledgements

The Editors would like to extend their sincere thanks to the authors whose enthusiastic participation made the present meeting possible. We are very grateful to the reviewers of the short papers and the time spent in their professional reviewing work. We are also grateful to the chairpersons and discussants of the sessions that also provided very useful suggestions to the authors.

Special thanks are due to the Local Organizing Committee of Pescara: Mauro Coli, Tonio Di Battista, Stella Iezzi, Eugenia Nissi, Antonio Pacinelli, Tonino Sclocco, Maurizio Vichi (*chairman*)

Finally, thanks are extended to Springer-Verlag, Heidelberg.

Maurizio Vichi

TABLE OF CONTENTS

Preface V

PART I: Classification

Methodologies in Classification

A. Cerioli (Università di Parma)
Measuring the influence of individual observations and variables in cluster analysis 3
P. D'Urso, M. G. Pittau (Università "La Sapienza" di Roma)
Consensus classification for a set of multiple time series 11
T. Di Battista, D. Di Spalatro (Università di Chieti)
A bootstrap method for adaptive cluster sampling 19
D. Iezzi, M. Vichi (Università di Chieti)
Forecasting a classification 27

Fuzzy Clustering and Fuzzy Methods

A. Bellacicco (Università di Teramo)
Neural networks as a fuzzy semantic network of events 35
L. Cerbara (IRP-CNR)
Hierarchical fuzzy clustering: an example of spatio-temporal analysis 43
G. Iacovacci (ISTAT)
A new algorithm for semi-fuzzy clustering 49
A. Maturo, B. Ferri (Università di Chieti)
Fuzzy classification and hyperstructures: an application to evaluation of urban projects 55
M. A. Milioli (Università di Parma)
Variable selection in fuzzy clustering 63

PART II : Other Approaches for Classification

Discrimination and Classification

M. Alfò, P. Postiglione (Università di Chieti)
Discriminant analysis using markovian automodels 73
F. Esposito, D. Malerba, G. Semeraro, S. Caggese (Università di Bari)
Discretization of continuous-valued data in symbolic classification learning 81

S. Ingrassia (Università di Catania)
Logistic discrimination by Kullback-Leibler type distance measures 89
A. Montanari, D. Calò (Università di Bologna)
An empirical discrimination algorithm based on projection pursuit density estimation 97

Regression Tree and Neural Networks

R. Miglio, M. Pillati (Università di Bologna)
Notes on methods for improving unstable classifiers 105
F. Mola (Università "Federico II" di Napoli)
Selection of cut points in generalized additive models 113
R. Siciliano (Università "Federico II", Napoli)
Latent budget trees for multiple classification 121

PART III: Multivariate and Multidimensional Data Analysis

Proximity Analysis and Multidimensional Scaling

G. Bove, R. Rocci (Università di Roma)
Methods for asymmetric three-way scaling 131
S. Camiz (Università "La Sapienza" di Roma)
Comparison of Euclidean approximations of non-Euclidean distances 139
A. Montanari, G. Soffritti (Università di Bologna)
Analysing dissimilarities through multigraphs 147
C. Quintano (Istituto Universitario Navale di Napoli)
Professional positioning based on dominant eigenvalue scores (DES), dimensional scaling (DS) and multidimensional scaling (MDS) synthesis of binary evaluations matrix of experts 155
M. Vichi (Università di Chieti)
Non-metric full-multidimensional scaling 163

Factorial Methods

I. Corazziari (Università "Federico II" di Napoli)
Dynamic factor analysis 171
V. Esposito, G. Scepi (Università "Federico II" di Napoli)
A non symmetrical generalised co-structure analysis for inspecting quality control data 179
R. Lombardo, G. Tessitore (II Università di Napoli - Università "Federico II" di Napoli)
Principal surfaces constrained analysis 187

R. Verde (Università "Federico II" di Napoli)
Generalised canonical analysis on symbolic objects 195
G. Vittadini (Università di Milano)
Analysis of qualitative variables in structural models with unique solutions. 203

Spatial Analysis

A. Capobianchi , G. Jona-Lasinio (Università di Roma)
Exploring multivariate spatial data: line transect data.. 211
A. Giusti, A. Petrucci (Università di Firenze)
On the assessment of geographical survey units using
constrained classification 221
L. Romagnoli (Università di Chieti)
Kalman filter applied to non-causal models for spatial data 229

Multiway Data Analysis

S. Bolasco,A. Morrone, F. Baiocchi (Università "La Sapienza" di Roma)
A paradigmatic path for statistical content analysis using an
integrated package of textual data treatment 237
M. Chiodi, A.M. Mineo (Università di Palermo)
The analysis of auxological data by means of nonlinear
multivariate growth curves 247
M. Coli, L. Ippoliti, E. Nissi (Università di Chieti)
The Kalman filter on three-way data matrix for missing data:
a case study on sea water pollution.. 255
P. A. Cornillon, P. Amenta, R. Sabatier, (Laboratoire de physique moléculaire et structurale, Università "Federico II"di Napoli)
Three-way data arrays with double neighborhood relations as
a tool to analyze a contiguity structure 263
A. Lemmi , D. Stefano Gazzei (Università di Siena, Università di Firenze)
Firm performance analysis with panel data 271

Multivariate Data Analysis

M. R. D'Esposito, G. Ragozini (Università di Salerno, Università "Federico II" di Napoli)
Detection of multivariate outliers by convex hulls 279
M. Di Marzio, G. Lafratta (Università di Chieti)
Reducing dimensionality effects on kernel density estimation:
the bivariate gaussian case 287
M. Giacalone (Università di Napoli)
Shewhart's control chart: some observations 295

A. Laghi, L. Lizzani (Università di Bologna)
Projection pursuit regression with mixed variables 303
P. Mantovan, A. Pastore, S. Tonellato (Università di Venezia)
Recursive estimation of system parameter in environmental time series models 311
A. Pallini (Università di Bologna)
Kernel methods for estimating covariance functions from curves 319
G. Porzio (Università "Federico II", Napoli)
Detection of subsamples in link-free regression analysis 327
A. Roverato (Università di Modena)
Asymptotic prior to posterior analysis for graphical gaussian models 335

PART IV: Case Studies

Applied Classification and Data Analysis

S. Borra, A. Di Ciaccio (Università di Urbino)
Using qualitative information and neural networks for forecasting purposes in financial time series 345
A. M. Mineo, A. Plaia (Università di Palermo) ....
A new approach to the stock location assignment problem by multidimensional scaling and seriation 353
A. Turrini (Istituto Nazionale della Nutrizione)
Food coding in nutritional surveys 361
C. Capiluppi, L. Fabbris, M. Scarabello (Università di Padova)
UNAIDED: a PC system for binary and ternary segmentation analysis 367

Author Index 375
Key Words Index 377

PART I

Classification

- **Methodologies in Classification**
- **Fuzzy Clustering and Fuzzy Methods**

Measuring the Influence of Individual Observations and Variables in Cluster Analysis

Andrea Cerioli
Istituto di Statistica, Università di Parma
Via Kennedy 6, 43100 Parma, Italy, email: Statec1@ipruniv.cce.unipr.it

Abstract: In this paper we address some issues in the field of cluster stability. In particular, we study the effect of deleting individual cases and variables on the results of a (nonhierarchical) cluster analysis. We do not restrict to computation of a single influence measure for each data point, or variable, but we analyze how individual influence varies when the number of clusters changes. For this purpose we suggest the use of simple deletion diagnostics computed by cross-validation. The suggested approach is applied to real data and results are displayed by means of a simple tool of modern multivariate-data visualization. Furthermore, the performance of our diagnostics is assessed through Monte Carlo simulations both under the null hypothesis of well-behaved data and the alternative hypothesis of isolated contamination.

Keywords: cluster stability; deletion diagnostic; k-means; outlier; stalactite plot.

1. Introduction

In a hierarchy of problems, influence detection has become a major issue in the field of *classification stability*, which in turn falls within the broad framework of *clustering validation* (see Milligan, 1996, and Gordon, 1996, Section 7, for general reviews of these topics). Specifically, the purpose of influence detection is to identify those data points that have a large impact on the partitions obtained from a clustering method. Influence detection usually proceeds by comparing a reference partition, computed on the complete data set by means of a specified clustering algorithm, and a modified partition, obtained from a reduced data set using the same algorithm. The reduced data set is produced by simply deleting either a single case or a single variable from the complete one. So far, research efforts have mainly focused on measuring the influence of individual *observations* on results from *hierarchical* clustering (see, e.g., Cheng and Milligan, 1996, and the references therein), although occasional interest has also arisen in the related area of identifying influential variables (Gnanadesikan *et al.*, 1977).

The purpose of this paper is to extend previous work in the field of influence detection in a number of ways. Firstly, we analyze how single-case influence varies in (nonhierarchical) clustering when the number of cluster changes.

This task is accomplished by computing cross-validation deletion diagnostics at successive steps of the clustering process. The resulting pattern of influential observations is then displayed by means of a stalactite plot (Atkinson, 1994). Secondly, we adopt a similar approach to study the influence of individual variables across a varying number of clusters. Finally, we perform a Monte Carlo experiment to assess the behaviour of our deletion diagnostics both under the null hypothesis of well-behaved data and the alternative hypothesis of isolated contamination. In addition, we suggest an extension of the seed selection technique adopted in the FASTCLUS procedure of SAS (1990), in order to overcome the possible effect of the observation order on results from the k-means algorithm.

2. Deletion Statistics in Cluster Analysis

Recent contributions to the identification of influential observations in cluster analysis include Jolliffe *et al.* (1995), and Cheng and Milligan (1996). Both papers explicitly consider only hierarchical methods and, more importantly, quantify the effect of each data unit through a single number. However, we believe that observing individual influence at successive steps of the clustering process can lead to a better insight into the data. As our simulations show, this approach can also have beneficial consequences on the choice of the final number of clusters.

Let n be the total number of objects to be classified and let P_k denote the k-clusters reference partition. In our applications P_k is obtained by clustering the complete data set of n elements. On the contrary, we do not address the related problem of cluster recovery, where P_k is the true, but usually unknown, underlying cluster structure. Furthermore, let P_k^i be the k-clusters partition which is computed after deletion of object i $(i = 1, \dots, n)$. A measure of influence of object i on the k-clusters solution, say θ_k^i, is then defined as a disagreement measure between P_k and P_k^i, provided that information about i is removed from the reference partition. Any index for partition comparison belonging to the class proposed by Hubert and Arabie (1985) can be used at this step.

Examination of values θ_k^i, $2 \leq k \leq n-1$, provides the basis for identifying the pattern of outliers or other highly influential observations across different clustering solutions. Clearly, a detailed graphical presentation of all such values becomes infeasible in many practical applications, where n is reasonably large. Therefore we choose to display only units for which at least one θ_k^i is sufficiently "high". In principle, each θ_k^i could be judged with respect to the distribution of the corresponding random variable under the null hypothesis that P_k and P_k^i are independent partitions of the same set of objects (Hubert and Arabie, 1985; Cerioli, 1997). However, such an extreme hypothesis does not seem to be adequate in the present context, where P_k and P_k^i are computed from data sets differing only by one observation. As it is difficult to devise what alternative distribution might be relevant, we take a more practical approach and standardize θ_k^i by cross-validation.

Let $\theta_k = \sum_i \theta_k^i / n$ be the average disagreement measure at the k-clusters level, and

$$\sigma_k^2 = \frac{\sum_i (\theta_k^i - \theta_k)^2}{n}.$$

The cross-validated standardized value of θ_k^i is then defined as

$$z_k^i = \frac{\theta_k^i - \theta_k}{\sigma_k} \qquad \text{if} \qquad \sigma_k > 0, \tag{1}$$

and $z_k^i = 0$ otherwise.

We regard as potentially influent on the k-clusters solution those objects for which $z_k^i > z^*$, a fixed threshold. A graphical display of potentially influent data units is then produced through a stalactite plot (Atkinson, 1994). In the examples that follow, we take $z^* = 2$ and $z^* = 3$ as useful cutoff points. The adequacy of these thresholds is assessed through Monte Carlo simulations in section 3. Furthermore, a number of measures of cluster cohesion can be computed at each step, as a supplement to z_k^i, to see whether any influential object is either an inhibitor or a facilitator in the clustering process (Cheng and Milligan, 1996).

The approach outlined above is easily extended to the detection of influential variables, in the spirit of the seminal paper by Gnanadesikan *et al.* (1977). Let p denote the total number of variables used in the clustering process. A measure of influence of variable j ($j = 1, \ldots, p$) on the k-clusters solution can be obtained by simply comparing P_k and P_k^j, where P_k^j now denotes the partition computed after deletion of variable j. In this case, we also suggest to monitor how distances from centroids change when passing from P_k to P_k^j. Let d_{ik} denote the Mahalanobis distance of object i from its cluster centroid in the reference k-clusters partition, and let d_{ik}^j be the corrisponding distance in P_k^j. Compute average distances $d_k = \sum_i d_{ik}/n$ and $d_k^j = \sum_i d_{ik}^j/n$. Then, for each $j = 1, \ldots, p$, examination of quantities

$$\Delta_k^j = d_k/p - d_k^j/(p-1) \qquad k = 2, \ldots, n-1 \tag{2}$$

provides further information on the influence of variable j at successive steps of the classification procedure.

3. Simulation Study

A small Monte Carlo experiment is performed in order to assess the behaviour of our standardized deletion statistics z_k^i and the adequacy of threshold values z^* to be used for the identification of influential observations. In particular, we repeatedly generate independent realizations from a eight-variate normal distribution through a slightly modified version of the algorithm in Jolliffe *et al.* (1995). In the present study, $1,000$ data sets are simulated under the null hypothesis of no outliers, and further $1,000$ data sets under the alternative hypothesis of isolated contamination.

In each of the outlier-free simulated data sets, there are 80 observations generated to fall into eight clusters of equal size. Clusters are well separated and have the same dispersion. All cluster centroids lie within the unit hypercube. A k-clusters partition is then obtained for each data set through the convergent k-means algorithm, for several values of k. The clustering algorithm is started from carefully selected seed points. For this purpose, we suggest an extension of the seed selection technique adopted in the FASTCLUS procedure of SAS

Figure 1: *Estimated Type-1 error rates under the null hypothesis of no outliers, for the test based on* $z^* = 3$ *and for different values of* k*. FASTCLUS seed selection algorithm.*

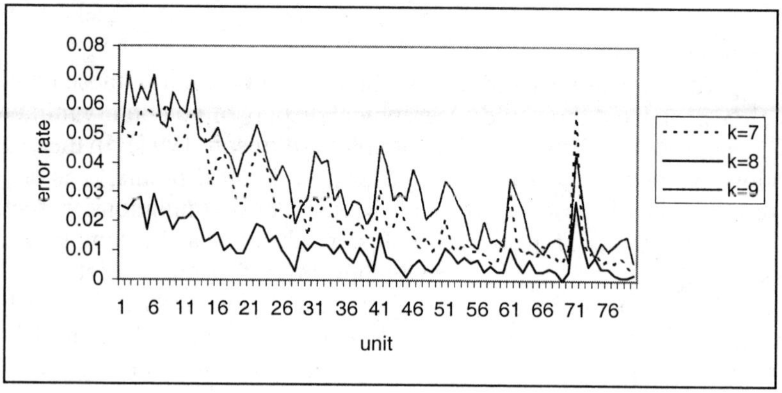

(1990). Our method, which is detailed in the Appendix, is motivated by the large effect that the order of the observations in the data set may have on results for single-case deletion diagnostics when the standard algorithm is applied.

The disagreement measure is defined as $\theta_k^i = 1 - R_k^i$, where R_k^i is the corrected-for-chance Rand index for comparing partitions P_k and P_k^i. Therefore $\theta_k^i = 0$ if a perfect match exists between P_k and P_k^i, while values of θ_k^i near 1 show a chance-level agreement between the two partitions.

For each data set, cross-validation standardized values z_k^i are computed as in (1) and then compared with several thresholds. Given a specific cutoff z^*, the Type-1 error rate for unit i in the k-clusters partition is estimated as the proportion of simulations in which z_k^i exceeds z^*. Figure 1 displays estimated error rates under the null hypothesis of no outliers in the data, for $z^* = 3$ and a few values of k, when the FASTCLUS seed selection technique is adopted. Results from the standard procedure are clearly not satisfactory, due to the decreasing trend in all estimated Type-1 error rates. On the contrary, as Figure 2 shows, our method proves to be largely insensitive to the observation order for all reported values of k.

For our seed selection procedure, Monte Carlo estimates of average error rates under the null hypothesis are given in Table 1 for both cutoffs $z^* = 2$ and $z^* = 3$, and for $6 \leq k \leq 11$. With well-separated clusters, all (estimated) average rates for the test based on $z^* = 2$ are less than 7.5%, while the corresponding rates for the highest threshold do not exceed 3.0%. Indeed, average error rates of these tests are as small as 1.1% and 0.6%, respectively, when k is set equal to the true number of clusters. Thus we conclude that the diagnostic technique proposed in the present paper does not seem to produce spuriously large numbers of outliers with well-behaved clustered data. Furthermore, computation of cross-validation standardized statistics can convey useful information also on the related problem of choosing the appropriate number of clusters.

To assess the performance of our deletion statistics under the alternative hy-

Figure 2: *Estimated Type-1 error rates under the null hypothesis of no outliers, for the test based on $z^* = 3$ and for different values of k. Seed selection algorithm as in the Appendix.*

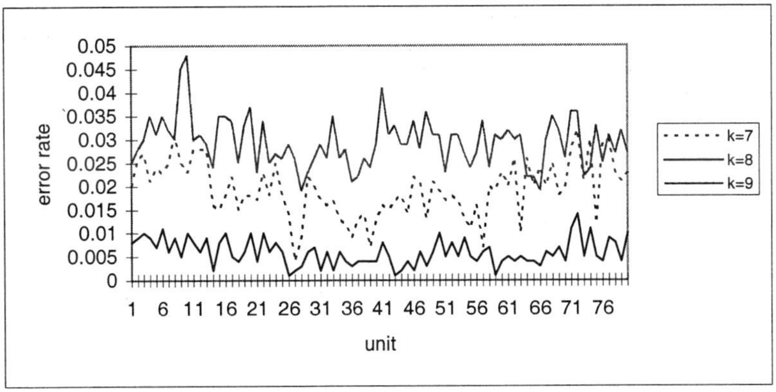

Table 1: *Average error rates for cutoffs $z^* = 2$ and $z^* = 3$, and for different values of k. Seed selection algorithm as in the Appendix.*

k	6	7	8	9	10	11
$z^* = 2$	0.056	0.048	0.011	0.058	0.073	0.074
$z^* = 3$	0.023	0.019	0.006	0.029	0.028	0.024

pothesis of isolated contamination, $1,000$ additional data sets are simulated by adding one outlier to 80 well-behaved clustered data (generated as before). In all contaminated data sets, the outlier is generated to fall near (or slightly outside) the upper boundaries of the unit hypercube. Figure 3 displays boxplots of estimated Type-1 error rates for the test based on $z^* = 3$. For ease of presentation, we restrict to $7 \leq k \leq 10$. The outlying unit is clearly revealed irrespective of the chosen number of clusters, as it exhibits a disproportionately large error rate for all values of k. Also note that in this case the distribution of estimated error rates correctly supports the existence of nine clusters with one outlier.

4. Example

As an expository example, we classify the municipalities (*comuni*) in the province of Parma, Italy, according to 5 demographic indicators taken from the 1991 Italian Census. In the present example $n = 47$. A k-clusters partition is obtained through the convergent k-means algorithm, with the seed selection procedure described in the Appendix. The clustering process is repeated for several values of k. For simplicity, we restrict the analysis to $3 \leq k \leq 8$, which is the range of practical interest in this application.

Figure 3: *Boxplots of estimated Type-1 error rates under the alternative hypothesis of isolated contamination, for the test based on* $z^* = 3$ *and for different values of* k. *Seed selection algorithm as in the Appendix.*

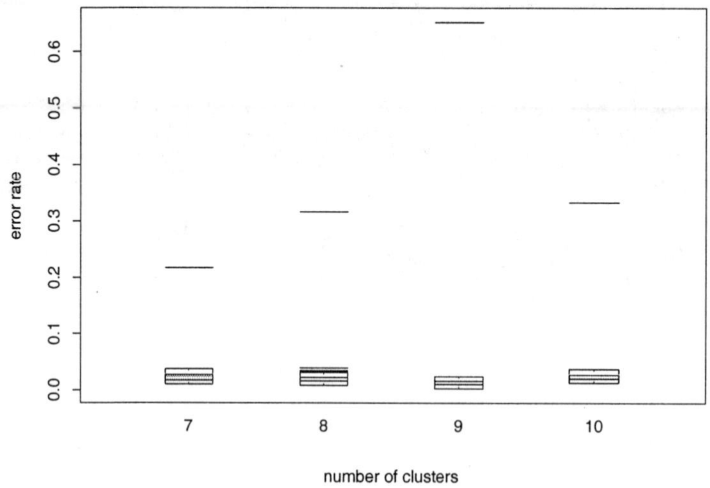

At each step, the disagreement measure is again defined as $\theta_k^i = 1 - R_k^i$. Cross-validation standardized values z_k^i are then computed as in (1). The resulting stalactite plot is given in Figure 4, where columns represent units and rows correspond to values of k. For ease of presentation, we only display units for which at least one $z_k^i > 2$. Note that the actual definition of θ_k^i is not crucial, as very similar results are reached with a number of alternative choices for θ_k^i. This appears to be an advantage of standardization over the use of raw disagreement measures, which can lead to discordant information using different criteria (Jolliffe *et al.*, 1995).

The plot in Figure 4 shows the influential nature of observation 5 across several clustering solutions. At a closer inspection of the data, it can be seen that this unit is an outlier on the first variable, measuring the ratio between the number of elders over 65 and the number of boys aged 14 or less. It is also apparent from Figure 4 that individual influence can vary markedly as the number of cluster changes, and other municipalities can be occasionally influent for some values of k.

Results concerning the identification of influential variables are given in Table 2, where we report the values of Δ_k^j for all variables and $3 \leq k \leq 8$. The figures in Table 2 clearly reveal the large influence of the first variable. In particular, it is seen that deletion of this variable leads to a considerable reduction in average (adjusted) Mahalanobis distance for all values of k in the range of interest. Therefore, clusters become more compact after removing Variable 1. For $k \geq 6$, deletion of Variable 3 also provides a slight improvement in the computed classification.

Figure 4: *Example. Stalactite plot of standardized Rand index. Columns refer to units and rows to values of* k. $**$ *denotes* $z_k^i > 3$, $*$ *denotes* $z_k^i > 2$.

	unit							
k	1	5	13	17	19	24	27	41
3		**						
4		**						**
5								
6	*	**						
7			**		**			
8		*	*	*	*	**	*	

Table 2: *Example. Average change in adjusted Mahalanobis distance* Δ_k^j.

k	Var. 1	Var. 2	Var. 3	Var. 4	Var. 5
3	5.57	-2.00	-0.93	-2.09	-2.10
4	3.18	-1.32	-0.39	-1.42	-1.42
5	3.07	-1.16	-0.32	-1.26	-1.26
6	2.39	-0.91	0.19	-1.05	-1.05
7	2.15	-0.61	0.12	-0.96	-0.96
8	1.83	-0.69	0.51	-0.84	-0.84

5. Discussion

In this paper we propose simple diagnostic tools for the purpose of detecting influential units and variables in nonhierarchical cluster analysis across different clustering solutions. The effectiveness of the suggested method is illustrated both by means of real and simulated data sets. Furthermore, we show how to overcome the possible effect of the order of the observations on results from the k-means algorithm.

However, a well known problem with single-case deletion diagnostics is that they can suffer from the problems of *masking* and *swamping* when multiple outliers are present in the data (Atkinson, 1994). This is true also for quantities like (1) and (2). Therefore, further research must be devoted towards the definition of truly robust methods for the detection of masked multiple outliers in cluster analysis.

Appendix: Seed Selection Algorithm for Nonhierarchical Clustering

Step 0. Fix the number of clusters, say k.

Step 1. Scan the data in natural order. Compute k preliminary seeds as in the

FASTCLUS procedure of SAS (1990). Let S_1 be the set of indexes of the observations selected as preliminary seeds at this step.

Step 2. Scan the data in reverse order. Compute k preliminary seeds as in the FASTCLUS procedure of SAS (1990). Let S_2 be the set of indexes of the observations selected as preliminary seeds at this step.

Step 3. Define $S_3 \equiv S_1 \bigcap S_2$. Let k' be the cardinality of S_3. Take the observations indexed in S_3 as initial seeds for cluster analysis. Stop if $k' = k$; otherwise put $j = k'$ and go to Step 4.

Step 4. Increase j by 1. Let S'_1 and S'_2 be the sets of indexes of preliminary seeds not already selected in S_1 and S_2, respectively. Define $S' \equiv S'_1 \bigcup S'_2$, and $S \equiv (S_1 \bigcup S_2) - S'$. Choose the unit in S' which has the largest distance from the nearest seed in S as the j-th initial seed for cluster analysis. Iterate this step until $j = k$.

References

Atkinson A. C. (1994). Fast Very Robust Methods for the Detection of Multiple Outliers, *Journal of the American Statistical Association*, 89, 1329-1339.

Cerioli A. (1997). Comparing Three Partitions: An Inferential Approach Based on Multi-Way Contingency Tables, *Communications in Statistics, Part A: Theory and Methods*, 26, 2457-2471.

Cheng R. and Milligan G. W. (1996). Measuring the Influence of Individual Data Points in a Cluster Analysis, *Journal of Classification*, 13, 315-335.

Gnanadesikan R., Kettenring J. R. and Landwehr J. M. (1977). Interpreting and Assessing the Results of Cluster Analyses, *Bulletin of the International Statistical Institute*, 47, 451-463.

Gordon A. D. (1996). Hierarchical Classification, in: *Clustering and Classification*, P. Arabie, L. J. Hubert and G. De Soete (eds.), World Scientific, Singapore, 65-121.

Hubert L. J. and Arabie P. (1985). Comparing Partitions, *Journal of Classification*, 2, 193-218.

Jolliffe I. T., Jones B. and Morgan B. J. T. (1995). Identifying Influential Observations in Hierarchical Cluster Analysis, *Journal of Applied Statistics*, 22, 61-80.

Milligan G. W. (1996). Clustering Validation: Results and Implications for Applied Analyses, in: *Clustering and Classification*, P. Arabie, L. J. Hubert and G. De Soete (eds.), World Scientific, Singapore, 341-375.

SAS (1990). *SAS/STAT User's Guide. Ver. 6. 4th Edition*, SAS Institute, Cary, NC.

Consensus Classification for A Set of Multiple Time Series

Pierpaolo D'Urso, Maria Grazia Pittau
University of Rome "La Sapienza", P.le A. Moro, 5, 00185 Rome, Italy.
e-mail: durso@pow2.sta.uniroma1.it; pittau@axrma.uniroma1.it

Abstract: In multiple time series analysis, when there are a very large number of series, a classification into homogeneous clusters might be useful to reduce the problem's complexity and eliminate possible redundancies (Zani, 1983). Furthermore, when we have different classifications, one for each statistical unit (e. g. spatial units), a consensus classification allows one to obtain a classification which summarizes the given classifications. The present paper focuses on the problem of identifying consensus classifications in a set of multiple time series (panel data), using a consensus method (Vichi, 1993, 1994). First, a distance among time series is defined and a hierarchical classification among time series, for each temporal lag and for each unit, is performed. Then, a consensus classification among different units for the same temporal lag is carried out. Finally, a hierarchical classification among the different consensus classifications, with the same temporal lag, is carried out.

Keywords: Distance between Time Series, Hierarchical Clustering of Time Series, Consensus Classification.

1. Introduction

Consensus analysis applied to time series allows us to summarize the information about the dynamic structure of phenomena concerning different units. This implies that the data structure associated to these phenomena is a three-way array. In the present paper we suggest an original procedure based on some new tools of multivariate statistics, generally applied to cross-sectional data sets. Different approaches and applications about time series classification are present in the literature (Bohte *et al.*, 1980; Piccolo, 1990; Zani, 1983). In particular, the approach by Bohte *et al.* (1980), which takes into account the dimension of time through specific functions of time series (autocorrelation and cross-correlation), has been considered in this paper. Then, the consensus classification proposed by Vichi (1993, 1994) is briefly described (section 3) and we suggest applying the consensus method, usually utilized for cross-sectional data sets, to stationary time series (section 4). The suggested procedure allows us to obtain an informative synthesis of structural and dynamic changes of a phenomenon measured in different units.

2. Clustering of a Multiple Time Series

In the literature on clustering time series, among the different approaches mentioned above, we follow the procedure suggested by Bohte *et al.* (1980), based on different distances between time series. Let $\mathbf{x}_1,\ldots,\mathbf{x}_j,\ldots,\mathbf{x}_J$ be a multiplc timc series where $\mathbf{x}_j=\left[x_{j1},\ldots,x_{jt},\ldots,x_{jT}\right]'$ and x_{jt} denotes the observed value at time t of the *j-th* time series, $j=1,\ldots,J$. Assume that the time series are stationary.
In order to analyze the proximity relationships among a pair of stationary time series $\mathbf{x}_j$ and $\mathbf{x}_m$ $j\neq m=1,\ldots,J$, considering the temporal lag, the dissimilarity coefficient with lag k, is utilized:

$$d_{jm}(k)=\frac{\left|r_{jj}(k)r_{mm}(k)-r_{jm}(k)r_{mj}(k)\right|}{1+\left|r_{jj}(k)r_{mm}(k)-r_{jm}(k)r_{mj}(k)\right|},\qquad k\geq 0 \tag{1}$$

where $r_{jj}(k)$ is the autocorrelation coefficient of time series $\mathbf{x}_j$ $j=1,\ldots,J$ at lag k and $r_{jm}(k)$ is the cross-correlation coefficient between the time series $\mathbf{x}_j$ and $\mathbf{x}_m$ $j\neq m=1,\ldots,J$ at lag k. It is easy to prove that (1) satisfies the axioms: $d(\mathbf{x}_j,\mathbf{x}_m)\geq 0$; $d(\mathbf{x}_j,\mathbf{x}_m)=d(\mathbf{x}_m,\mathbf{x}_j)$; $\mathbf{x}_j=\mathbf{x}_m \Rightarrow d(\mathbf{x}_m,\mathbf{x}_j)=0$ so that (1) can be defined as a dissimilarity coefficient but not as a distance function. After the dissimilarities d_{jm} $j\neq m=1,\ldots,J$ between pairs of time series have been defined, it is possible to identify groups of time series which show links in their temporal trend, with possible lag. In order to identify groups of series, which show a high similarity in their temporal trend, a hierarchical classification method can be used, as proposed by Zani (1983). As a basis for the clustering of time series the single linkage method has been chosen. It is well known that in this case only the above axioms, a subset of the full set of axioms satisfied by a distance function, have to be satisfied by the dissimilarity coefficient chosen for the analysis. This clustering of time series is different from a usual classification for the presence of the temporal lag which marks the linkage among the groups already obtained.

3. Consensus Classification

The methods for determine a consensus classification allows us to synthesize several different hierarchical classifications into a single one which detects the information in the given classifications. Among the approaches proposed in the literature, we consider the *Average Consensus* or *Least Squares Consensus Dendrogram*, proposed by Vichi (1993, 1994). This method achieves the consensus classification solving the following quadratic problem:

$$\min \sum_{i=1}^{I} \left\| \mathbf{U}_i - \mathbf{U}^* \right\|^2 \tag{2}$$

over $\mathbf{U}^* = \left[u^*_{jm}\right]$ such that u^*_{jj}=*0* and under the constraint that $\mathbf{U}^*$ must be an ultrametric matrix. $\mathbf{U}_i$ is the ultrametric matrix associated to the *i-th* unit and $\mathbf{U}^*$ is the closest least squares matrix subject to the ultrametric conditions (ultrametric consensus matrix) (Simeone & Vichi, 1996). In order to achieve a consensus classification for a set of multiple time series we consider the method proposed by Vichi (1993, 1994).

4. Classification of Multiple Time Series Observed in Several Units

Suppose we have to classify a set of time series referring to the same phenomenon but measured in different units $i = 1,\dots,I$. Thus, the collected data set may be organized as a three way array:

$$\mathbf{X} \equiv \left[\mathbf{X}_1,\dots,\mathbf{X}_i,\dots,\mathbf{X}_I\right]' \text{ where } \mathbf{X}_i \equiv \left[\mathbf{x}_{i1.},\dots,\mathbf{x}_{ij.},\dots,\mathbf{x}_{iJ.}\right] \text{ and } \mathbf{x}_{ij.} \equiv \left[x_{ij1},\dots,x_{ijt},\dots,x_{ijT}\right]'. \tag{3}$$

The procedure proposed may be formalized as follows.

1. First, a hierarchical clustering method (single linkage) has to be applied to the multiple time series $\mathbf{X}_i \equiv \left[x_{ijt}\right]$, for each unit *i* ($i = 1,\dots,I$) and for each temporal lag *k* ($k = 0,\dots,K-1$). In particular, for each unit *i*, a classification of the time series' elements, associated with the same temporal lag, must be determined. Thus, for each unit *i*, we have *K* hierarchical classifications and then *K* ultrametric matrices. Referring to the *i-th* unit we have the following *K* ultrametric matrices: $\mathbf{U}_i(0),\dots,\mathbf{U}_i(k),\dots,\mathbf{U}_i(K-1)$ for a total number of $I \times K$ ultrametric matrices. The above matrices have been obtained using a generalization of (1) to *I* units:

$$d_{ijm}(k) = \frac{\left|r_{ijj}(k)r_{imm}(k) - r_{ijm}(k)r_{imj}(k)\right|}{1 + \left|r_{ijj}(k)r_{imm}(k) - r_{ijm}(k)r_{imj}(k)\right|}, \qquad k \geq 0, \qquad i = 1,\dots,I \tag{4}$$

where $r_{ijj}(k)$ is the autocorrelation coefficient of time series $\mathbf{x}_j$ $j=1,\dots,J$ at lag k, referring to the i-th unit, while $r_{ijm}(k)$ is the cross-correlation coefficient between the time series $\mathbf{x}_j$ and $\mathbf{x}_m$ $j\neq m=1,\dots,J$ at lag k referring to the same unit.

2. Regarding the same temporal lag a least square consensus classification (section 3) among the ultrametrics, associated to the different units, must be determined. Thus, we have K ultrametric consensus matrices, one for each temporal lag:

$$\mathbf{U}^*(0),\dots,\mathbf{U}^*(k),\dots,\mathbf{U}^*(K-1) \quad (5)$$

obtained solving (2), under the same constraint, for each lag k ($k=0,\dots,K-1$):

$$\min \sum_{i=1}^{I} \left\| \mathbf{U}_i(k) - \mathbf{U}^*(k) \right\|^2 . \quad (6)$$

3. This part can be formalized as follows.

3.1 The different ultrametric matrices (5) are summarized in a three way matrix:

$$\mathbf{U}^* \equiv \left[\mathbf{U}^*(0),\dots,\mathbf{U}^*(k),\dots,\mathbf{U}^*(K-1)\right]', \ \mathbf{U}^*(k) = \left[u^*_{jm}(k)\right], \ (j,m=1,\dots,J), \ (k=0,\dots,K-1) \quad (7)$$

and $u^*_{jm}(k)$ is the ultrametric between $\mathbf{x}_j$ and $\mathbf{x}_m$ calculated for each lag k.

3.2 In the $\mathbf{U}^*$ we detect $d^*_{jm}(k^*) = \inf_k \left(u^*_{jm}(k)\right)$, where k^* is the temporal lag referring to the minimum distance between $\mathbf{x}_j$ and $\mathbf{x}_m$. Then, we have a dissimilarity matrix $\mathbf{D}^*(k^*)=[d^*_{jm}(k^*)]$ obtained by considering the minimum distance among the same pairs of time series for the temporal lag k^*.

3.3 We apply a hierarchical clustering algorithm (single linkage) to $\mathbf{D}^*(k^*)$.

Finally, we have a single hierarchical classification $\mathbf{U}^{**}(k^*)=[u_{jm}(k^*)]$ characterized by the association into homogeneous groups of multiple time series referring to different units. This classification takes into account the distance among the pairs of time series (or groups of time series), as well as the temporal lag k at which the series have been associated.

The ultrametric matrix $\mathbf{U}^{**}(k^*)$ is in bijection with a "labeled" dendrogram in which one can find, for each association step, the temporal lag k^*.

The obtained classification allows us to consider the information about a data set referring to different units, as well as the evolutive behavior of the studied phenomenon. This dynamic behavior is measured by the temporal lag at which the time series have been clustered. With the proposed procedure it is thus possible to analyze complex phenomena considering both structural and dynamic aspects.

5. An Example of the Proposed Procedure

In order to illustrate the procedure described in this paper, we will consider a three way data set regarding the economic performance of the most industrialized countries.

5.1. Time Series Classification Referring to Countries for Each Temporal Lag

The three way data set is made up of 7 countries × 8 economic variables × 27 years, from 1970 to 1996. The 7 countries considered are the G7 countries: Canada, France, Germany, Great Britain, Italy, Japan and USA. The 8 economic variables (source: Datastream) are: Gross National Product per capita in US dollar terms (v_1), Gross Domestic Product at price and exchange rate of 1985 (v_2), Unemployment Rate (v_3), Consumer Price Index (v_4), Exports in US dollar terms (v_5), Imports in US dollar terms (v_6), Industrial Production Index (v_7), Current Account Balance in US dollar terms (v_8). In order to equalize the size and the variability of the input variables, a standardization procedure has been performed, as suggested by Milligan & Cooper (1988). Furthermore, the time series already standardized, have been made stationary by taking the 1'*th* differences (Box & Jenkins, 1976). Considering the dissimilarity coefficient $\left\{d_{ijm}(k),\ i=1,\ldots 7;\ j \neq m=1,\ldots,8;\ k=0,\ldots,4\right\}$ 35 dissimilarity matrices between time series have been calculated. The 35 matrices refer to the 7 countries multiply by the 5 temporal lags. The single linkage clustering algorithm has been applied in order to classify the 8 time series for each country and for each temporal lag. Thus, we have 35 ultrametric matrices and hence 35 dendrograms.

5.2. Consensus Classification for each Temporal Lag

Regarding the same temporal lag, a least square consensus classification among the 35 ultrametrics has been obtained in order to obtain 5 synthesized classifications. Specifically, each consensus matrix has been obtained by synthesizing the dissimilarity matrices referring to the different G7 countries at the same temporal lag. The ultrametric consensus matrices and the consensus dendrograms at lag *k*=0, *k*=1, *k*=2, *k*=3, *k*=4, are respectively shown in Figure 1. As shown in the figure, for the lags k=0,1,3,4, in correspondence of small dissimilarity levels, it may identify two groups. In the first one, the Gross Domestic Product (v_2) is amalgamated with the Industrial Production Index (v_7). In the second one, the Exports (v_5) and the Imports (v_6) are jointed together. The other variables are aggregated at higher dissimilarity levels. At the lag $k=2$ one can find again two groups: the first one is the same (v_2,v_7), but the second one is different: the Unemployment Rate is connected with the Imports (v_3,v_6).

Figure 1: *Consensus Classifications for each Temporal Lag (each distance value is multiply by 100)*

	v1	v2	v3	v4	v5	v6	v7	v8
v1	0	41.94	41.94	45.99	41.94	41.94	41.94	43.68
v2		0	38.82	45.99	37.02	37.02	26.64	43.68
v3			0	45.99	38.82	38.82	38.82	43.68
v4				0	45.99	45.99	45.99	45.99
v5					0	24.48	37.02	43.68
v6						0	37.02	43.68
v7							0	43.68
v8								0

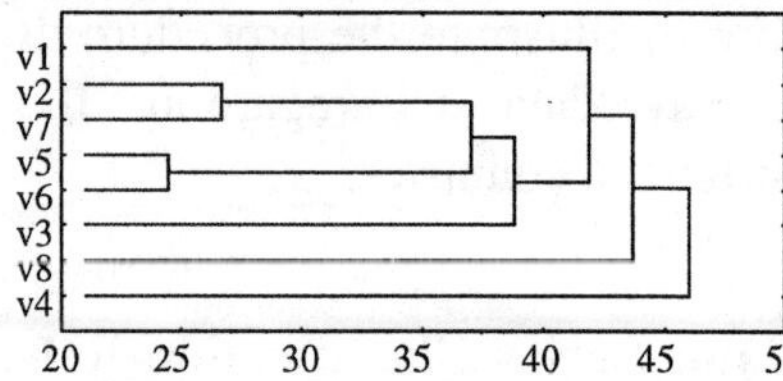

	v1	v2	v3	v4	v5	v6	v7	v8
v1	0	1.91	1.91	2.93	1.91	1.91	1.91	2.21
v2		0	1.85	2.93	1.82	1.82	0.85	2.21
v3			0	2.93	1.85	1.85	1.85	2.21
v4				0	2.93	2.93	2.93	2.93
v5					0	1.14	1.82	2.21
v6						0	1.82	2.21
v7							0	2.21
v8								0

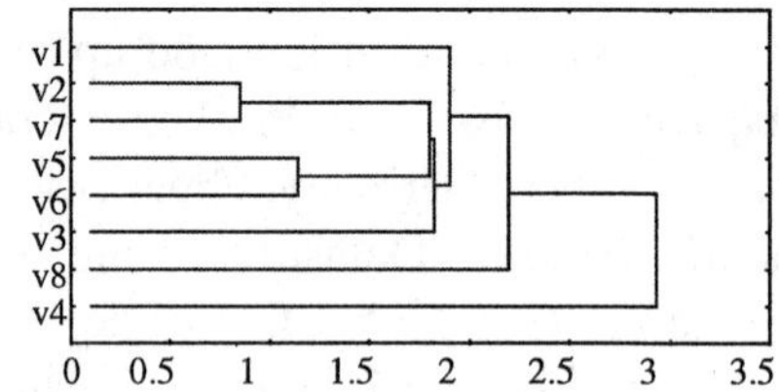

	v1	v2	v3	v4	v5	v6	v7	v8
v1	0	0.75	0.77	0.91	1.03	0.77	0.75	1.19
v2		0	0.77	0.91	1.03	0.77	0.55	1.19
v3			0	0.91	1.03	0.65	0.77	1.19
v4				0	1.03	0.91	0.91	1.19
v5					0	1.03	1.03	1.19
v6						0	0.77	1.19
v7							0	1.19
v8								0

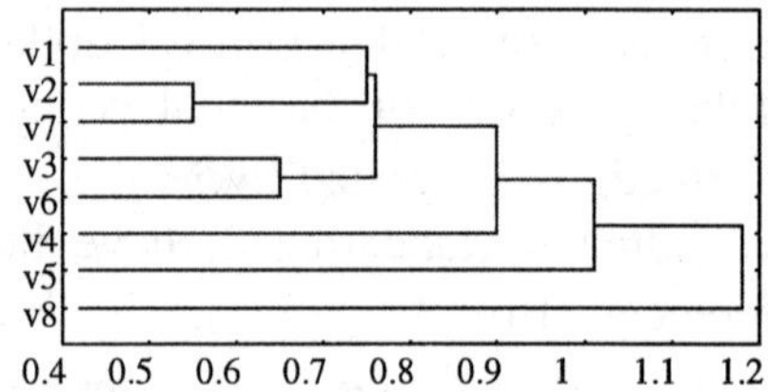

	v1	v2	v3	v4	v5	v6	v7	v8
v1	0	1.98	1.98	1.98	1.98	1.98	1.98	1.98
v2		0	0.75	1.17	0.71	0.71	0.38	1.33
v3			0	1.17	0.75	0.75	0.75	1.33
v4				0	1.17	1.17	1.17	1.33
v5					0	0.44	0.71	1.33
v6						0	0.71	1.33
v7							0	1.33
v8								0

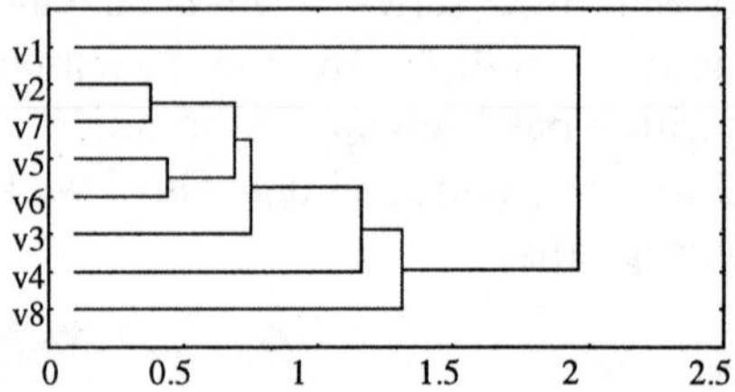

	v1	v2	v3	v4	v5	v6	v7	v8
v1	0	1.34	1.34	1.34	1.34	1.34	1.34	1.34
v2		0	1.07	0.82	0.82	0.82	0.59	0.82
v3			0	1.07	1.07	1.07	1.07	1.07
v4				0	0.78	0.78	0.82	0.78
v5					0	0.52	0.82	0.73
v6						0	0.82	0.73
v7							0	0.82
v8								0

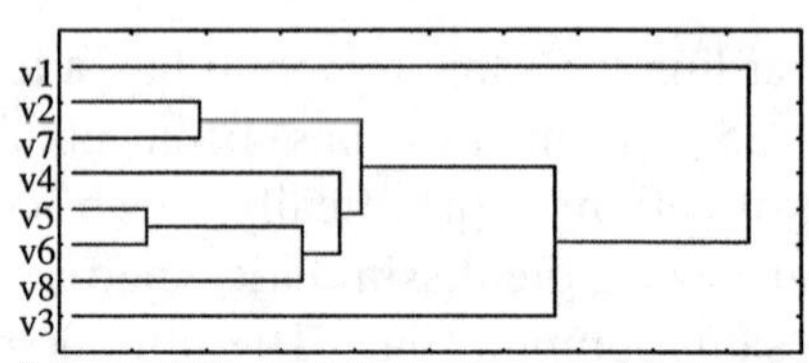

5.3. Final Classification

Considering the 5 consensus ultrametrics, the minimum distance between each pair of time series, $d^*_{jm}(k^*) = \inf_k \left(u^*_{jm}(k)\right)$, $k = 0,1,2,3,4$, has been calculated. k^* is the temporal lag referring to the minimum distance between the pairs of time series. Thus we have a dissimilarity matrix $\mathbf{D}^*(k^*)$.

Finally, the single linkage clustering method has been applied to $\mathbf{D}^*(k^*)$. A final classification has thus been obtained. This classification allows us to identify the temporal lag at which the variables are amalgamated. The ultrametric matrix and the associated dendrogram, referring to this final classification, are displayed in Table 1 and in Figure 2, respectively.

Table 1: *Final Labeled Ultrametric Matrix (each distance value is multiply by 100)*

	v1	v2	v3	v4	v5	v6	v7	v8
v1	0	0.75 **(2)**	0.75 **(2)**	0.78 **(4)**	0.75 **(2)**	0.75 **(2)**	0.75 **(2)**	0.75 **(2)**
v2		0	0.71 **(3)**	0.78 **(4)**	0.71 **(3)**	0.71 **(3)**	0.38 **(3)**	0.73 **(4)**
v3			0	0.78 **(4)**	0.65 **(2)**	0.65 **(2)**	0.71 **(3)**	0.73 **(4)**
v4				0	0.78 **(4)**	0.78 **(4)**	0.78 **(4)**	0.78 **(4)**
v5					0	0.44 **(3)**	0.71 **(3)**	0.73 **(4)**
v6						0	0.71 **(3)**	0.73 **(4)**
v7							0	0.73 **(4)**
v8								0

In parentheses there is the temporal lag k^*.

Figure 2: *Final Labeled Dendrogram (each distance value is multiply by 100)*

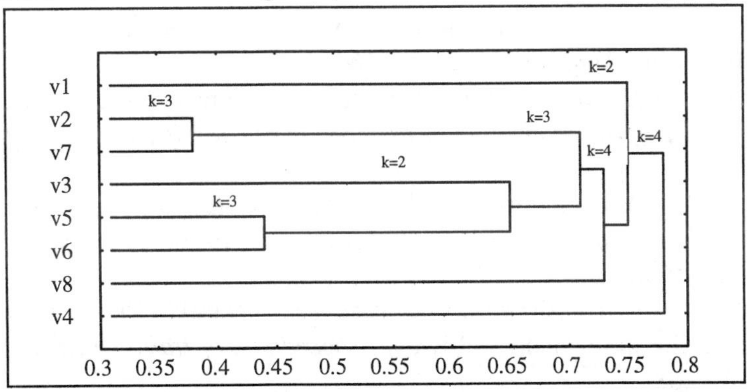

As shown in Figure 2, the following results may be observed.

The variables v_2 and v_7 and the variables v_5 and v_6 are amalgamated at lag 3 even if the aggregation distance levels are different. Then, v_3 is connected with the group (v_5,v_6) at lag 2, the group (v_2,v_7) with (v_3,v_5,v_6) at lag 3, v_8 with (v_2,v_7,v_3,v_5,v_6) at lag 4, v_1 with (v_2,v_7,v_3,v_5,v_6,v_8) at lag 2, finally v_4 is jointed to (v_1,v_2,v_7,v_3,v_5,v_6,v_8) at lag 4.

6. Final Remarks

Multiple time series classification has been used so far for series of data containing single units (Bohte *et al.*, 1980; Piccolo, 1990; Zani, 1983). In the present paper, a generalization of time series classification to the situation in which there are several units is proposed. This original procedure has been obtained using the consensus method (Vichi, 1994). Furthermore, the suggested procedure allows us to consider the temporal lag at which the series are associated. Finally, an example of the proposed procedure has been presented.

Acknowledgments

The Authors share the responsibility of this paper. However Sections 4 and 5.2 are due to Pierpaolo D'Urso, Sections 2, 3, 5.1, and 5.3 are due to Maria Grazia Pittau, while all the other sections are due to both the authors.

References

Box, G. E. P. & Jenkins, G. M. (1976). *Time Series Analysis: Forecasting and Control*, Holden-Day, San Francisco.

Bohte, Z., Cepar, D, & Kosmelj, K. (1980). Clustering of time series, *Compstat 80*, 587-593.

Milligan, G. W. & Cooper M. C. (1988). A study of standardization of variables in cluster analysis, *Journal of Classification*, 5, 181-204.

Piccolo, D. (1990). A distance measure for classifying ARIMA models, *Journal of Time Series Analysis*, 11, 2, 153-164.

Simeone, B. & Vichi, M. (1996). Consensus of hierarchical classifications, *5th Conference of the International Federation of Classification Societies*, Kobe, Japan.

Vichi, M. (1993). Un algoritmo dei minimi quadrati per interpolare un insieme di classificazioni gerarchiche con una classificazione consenso, *Metron*, 51, 3-4, 139-163.

Vichi, M. (1994). Un algoritmo per il consenso tra classificazioni gerarchiche con l'ausilio di tecniche multiway, *Proceedings of Italian Statistical Society*, 37, 261-268.

Zani, S. (1983). Osservazioni sulle serie storiche multiple e l'analisi dei gruppi, in: *Analisi Moderna delle Serie Storiche*, Piccolo, D. (Ed.), F. Angeli, 263-274.

A Bootstrap Method for Adaptive Cluster Sampling

Tonio Di Battista, Domenico Di Spalatro*
Dipartimento di Metodi Quantitativi e Teoria Economica
Università degli Studi "G. d'Annunzio" di Chieti
Viale Pindaro 42, 65127 Pescara

Abstract: In adaptive designs defined by selecting an initial simple random sample with or without replacement, the sample mean estimator is unbiased only if the initial sample is used, whereas, it is biased when a sample obtained at the end of the adaptive procedure is considered. In the last situation the estimator has been opportunely modified (Thompson and Seber 1996). However, for several estimators different from mean, such as the variance, the construction, in adaptive design, of a corresponding unbiased estimator has not been solved.
In this paper the *BACS* (Bootstrap for Adaptive Cluster Sampling) procedure based on a resampling is proposed to estimate the bias of an estimator.

Keywords
Adaptive Cluster Sampling, Resampling, Bootstrap, Bias

1. Introduction

In this paper we consider a sampling design where the n units are selected with simple random sampling with replacement and where the probability selecting the i-th unit in a draw is known. It is common knowledge that we may obtain unbiased estimators of the parameters of the population. In particular, regarding the estimator mean of the population we have an unbiased estimator of the type where every y_i observed in $(i=1,2,....,n)$ sample units is divided by the probability of associate selection and it is multiplied by the number of times that the unit is selected. The Hansen-Hurwitz estimator is unbiased, but in adaptive designs, are not applicable because the probabilities of selection are unknown for every unit of the sample.
Using network Thompson (1990) has proposed for the mean a modified unbiased Hansen-Hurwitz estimator also if it is applied on the final sample.

---

*This work, though it was the result of a close collaboration of the two authors, has been specifically elaborated as follows: the sections 1, 3 and 5 by T. Di Battista and the remainder by D. Di Spalatro.

However, if $T_n = t(X_1, X_2, \ldots X_n)$ is an estimator that differs from the sample mean, such as the variance estimator, it may be difficult to construct a correspondent unbiased adaptive one. Nevertheless the use of biased estimators is in same situations the unique alternative. For example, if the statistical data have a point pattern and the aim of research is to study the spatial dispersion; a function of the variance, such as the ratio between sample mean and variance, is used. In this case it may be useful at least to estimate the bias of the variance estimator obtained at the end of an adaptive procedure. The aim of this paper is to give a useful estimate procedure of the bias of an estimator T_n that is applied at the end of the adaptive procedure.

In section 2 we propose a method based on a resampling technique, called *BACS* (Bootstrap for Adaptive Cluster Sampling). In this context, we consider a sampling design where the *n* units are drawn with replacement and have the same drawn probability; in section 3 we prove the consistency of the method; in section 4 through a simulation, we evaluate the behaviour of the method for small samples. Finally, section 5 gives a discussion and possible future developments.

2. The BACS method

The aim of this paper is therefore to estimate,

$$bias(T_\Omega) = E_F[T_\Omega - \theta] = E_F[T_\Omega] - \theta, \tag{1}$$

where Ω indicates that the T_n estimator is computed on the final sample produced by an adaptive procedure. In particular, since we work with a finite population, (1) is given by

$$bias(T_\Omega) = \left(\sum_{k=1}^{M} [T_{\Omega,k} - \theta]\right) \Big/ M = \left(\sum_{k=1}^{M} T_{\Omega,k} \Big/ M\right) - \theta, \tag{2}$$

where *M* is the number of samples of size *n* that we may draw from a finite population of size *N*.

The estimation procedure of (2), may be obtained with a resampling procedure (Efron and Tibshirani, 1993), where after drawing with replacement from a population composed of *N* units a sample of size *n* $X = (X_1, \ldots X_j, \ldots, X_n)$, we deduce the empirical distribution function $\hat{F}$ from the sample *X* and we calculate the estimator,

$$T_n = t\left(X_1, X_2, \ldots . X_n\right). \tag{3}$$

Then, we draw from $\hat{F}$ a bootstrap sample

$$X^* = \left(X_{11}^*, \ldots, X_{j1}^*, \ldots, X_{n1}^*\right). \tag{4}$$

By applying the adaptive procedure, for each unit draw, we get additional units obtaining a bootstrap final sample of the type

$$X_\Omega^* = \left(X_{11}^*, \ldots, X_{1i}^*, \ldots, X_{1A_1}^*; \ldots, X_{j1}^*, \ldots, X_{ji}^*, \ldots, X_{jA_j}^*; \ldots, X_{n1}^*, \ldots, X_{ni}^*, \ldots, X_{nA_n}^*\right) \tag{5}$$

where A_j $\left(j = 1, \ldots, n\right)$ is the number of the units that are aggregated to i-th initial unit included the i-th unit itself.
Sample (5) includes every initial unit comprised in the bootstrap initial sample and the units produced at the end of the adaptive procedure. From the sample (5) we may obtain the estimator

$$T_\Omega^* = t\left(X_\Omega^*\right). \tag{6}$$

An estimate of $bias\left(T_\Omega\right)$ is given by:

$$bias\left(T_\Omega^*\right) = \left(\sum_{k=1}^{m^*} T_{\Omega,k}^* \Big/ m^*\right) - T_n \tag{7}$$

where m^* is the number of bootstrap samples drawn from $\hat{F}$.
The main difficulty of this procedure is to compute the following expression

$$\left(\sum_{k=1}^{m^*} T_{\Omega,k}^* \Big/ m^*\right). \tag{8}$$

Actually, this computation requires the use of all m^* bootstrap samples.
The problem is numerically solved by drawing a large number $B \leq m^*$ of bootstrap samples from which we obtain

$$T_{\Omega,1}^*, \ldots . T_{\Omega,k}^* \ldots \ldots T_{\Omega,B}^*$$

from which an estimate of (7) is given by:

$$bias_B\left(T_{\Omega}^{*}\right)=\left(\sum_{k=1}^{B} T_{\Omega,k}^{*} \Big/ B\right)-T_n. \tag{9}$$

3. Consistency of the BACS method

The BACS method is consistent in the sense that when the sample size increase the bias estimated by the BACS method converges to the bias generated by adaptive design.

Let $\{\overline{X}_{i,\Omega}A_i\}$ denote a sequence of random variables, for i=1,2,...,∞; and $\{S_n\}$ be a sequence of random variables where $S_n=\sum_{i=1}^{n}\overline{X}_{i,\Omega}A_i$ (n=1,2,...,∞), such that $E\left(\overline{X}_{i,\Omega}A_i\right)=\mu_{\Omega}$. Let $\{\overline{X}_{\Omega}\}$ denote the sequence of sample means $\overline{X}_{\Omega}=S_n\Big/\left(\sum_{i=1}^{n}A_i\right)$. $\overline{X}_{i,\Omega}$ is the mean of the cluster obtained applying the adaptive procedure to i-th unit of the population.

Thus we can write:

$$\overline{X}_{\Omega}\xrightarrow{p}\mu_{\Omega}.$$

If we can define a function of the estimator $\overline{X}_{\Omega}$, $T_{\Omega}=g\left(\overline{X}_{\Omega}\right)$, twice differentiable, applying the delta method to the T_{Ω} estimator we can write (Pace and Salvan, 1996):

$$T_{\Omega}=g(\mu_{\Omega})+g'(\mu_{\Omega})\cdot\left(\overline{X}_{\Omega}-\mu_{\Omega}\right)+\frac{1}{2}g''(\mu_{\Omega})\cdot\left(\overline{X}_{\Omega}-\mu_{\Omega}\right)^2+O_P\left(\overline{X}_{\Omega}-\mu_{\Omega}\right)^3.$$

As known $\left(\overline{X}_{\Omega}-\mu_{\Omega}\right)=O_p\left(n^{-1/2}\right)$, so

$$T_{\Omega}=g(\mu_{\Omega})+g'(\mu_{\Omega})\cdot\left(\overline{X}_{\Omega}-\mu_{\Omega}\right)+\frac{1}{2}g''(\mu_{\Omega})\cdot\left(\overline{X}_{\Omega}-\mu_{\Omega}\right)^2+O_P\left(n^{-3/2}\right);$$

$$E(T_{\Omega})=g(\mu_{\Omega})+\frac{1}{2}g''(\mu_{\Omega})\cdot E\left[\left(\overline{X}_{\Omega}-\mu_{\Omega}\right)^2\right]+O_P\left(n^{-3/2}\right);$$

$$Bias(T_\Omega) = g(\mu_\Omega) + \frac{1}{2} g''(\mu_\Omega) \cdot E\left[\left(X_\Omega - \mu_\Omega\right)^2\right] + O_P\left(n^{-3/2}\right) - \theta .$$

Finally, let $\overline{X}^*_\Omega$ be the sample mean of a bootstrap sample and $T^*_\Omega = g\left(\overline{X}^*_\Omega\right)$ then we have:

$$T^*_\Omega = g(\mu_\Omega) + g'(\mu_\Omega) \cdot \left(\overline{X}^*_\Omega - \mu_\Omega\right) + \frac{1}{2} g''(\mu_\Omega) \cdot \left(\overline{X}^*_\Omega - \mu_\Omega\right)^2 + O_P\left(n^{-3/2}\right);$$

$$E\left(T^*_\Omega\right) = g(\mu_\Omega) + \frac{1}{2} g''(\mu_\Omega) \cdot E\left[\left(\overline{X}^*_\Omega - \mu_\Omega\right)^2\right] + O_P\left(n^{-3/2}\right);$$

$$Bias\left(T^*_\Omega\right) = g(\mu_\Omega) + \frac{1}{2} g''(\mu_\Omega) \cdot E\left[\left(X^*_\Omega - \mu_\Omega\right)^2\right] + O_P\left(n^{-3/2}\right) - T_n .$$

Following (Shao and Tu, 1995), we have:

$$Bias(T_\Omega) - Bias\left(T^*_\Omega\right) = \frac{1}{2} g''(\mu_\Omega) \left\{ E\left[\left(X_\Omega - \mu_\Omega\right)^2\right] - E\left[\left(X^*_\Omega - \mu^*_\Omega\right)^2\right] \right\} + \\ + T_n - \theta + O_P\left(n^{-3/2}\right). \quad (10)$$

Therefore, the limit for $n \to \infty$ of equation (10) proves the convergence and the consistency of BACS procedure.

4. Some simulation results

In order to evaluate the behaviour of the *BACS* method we have simulated a finite population *P* of 100 units and we have considered a contiguity criterion north-south, east-west as shown in Figure 1.

Figure1. *The population P and the type of neighbourhood for unit i*

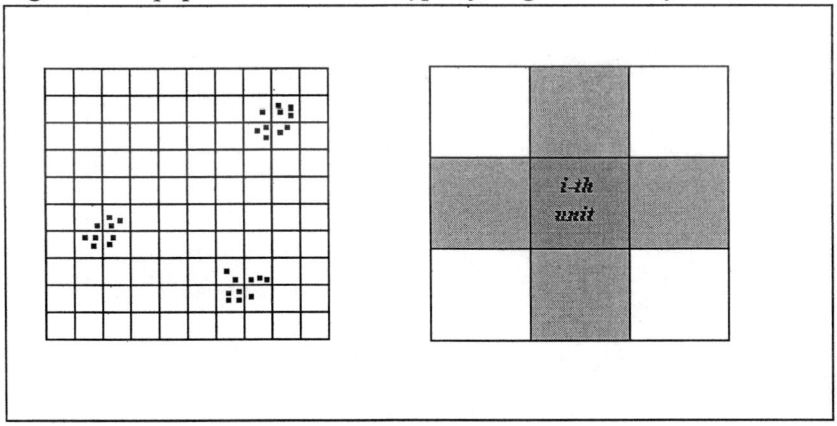

Applying the adaptive procedure we have considered the aggregation condition C of the type $y_i > c$ for c =1,2 (Thompson and Seber 1996).
In this context, we have drawn with replacement from the population P a sample size of order: 2,3,4,5,6 respectively. We have computed the real bias of the sample mean and variance mean applied to final sample with reference to sample size n and C condition.
Then we have drawn B=100 bootstrap samples from initial samples and we have estimated the bias of the sample mean and the sample variance.

Figure 2. *Real Bias and Estimate Bias of sample mean with c > 1.*

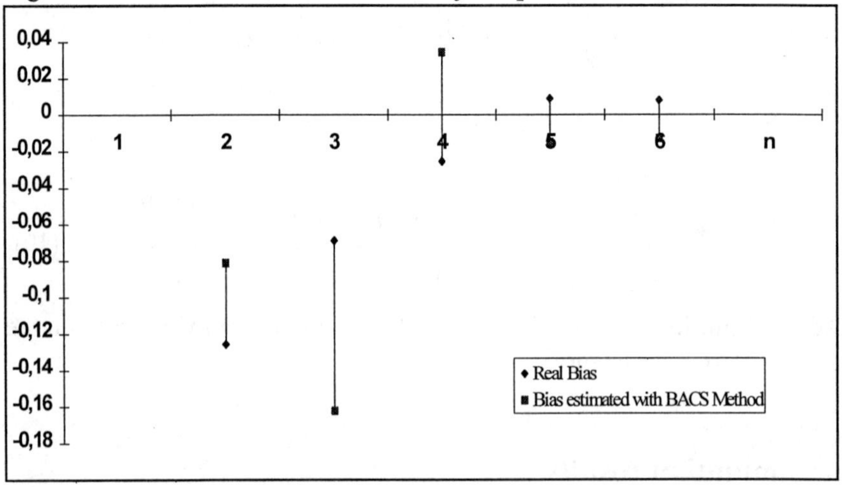

Figure 3. *Real Bias and Estimate Bias of sample mean with c >2 .*

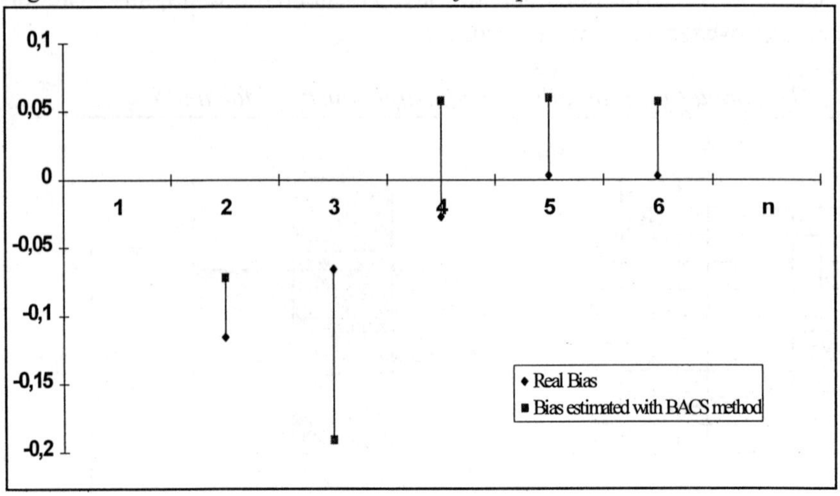

Figure 4. *Real Bias and Estimate Bias of sample variance with* $c > 1$

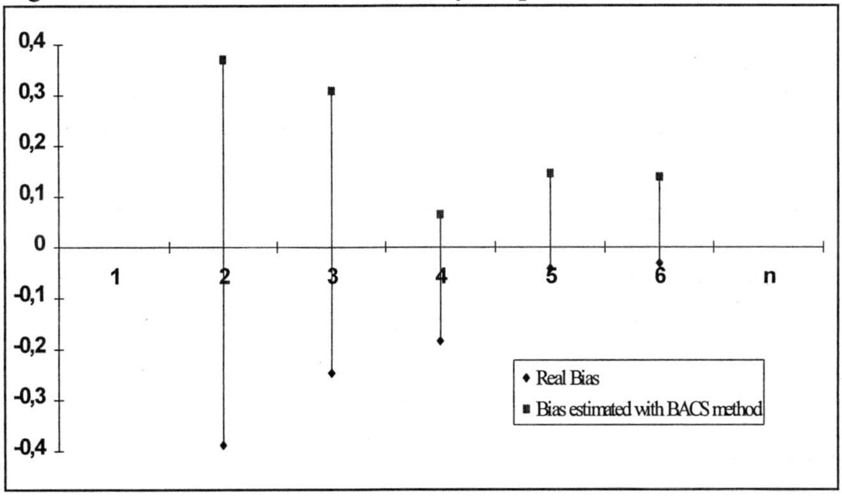

Figure 5. *Real Bias and Estimate Bias of sample variance with* $c > 2$

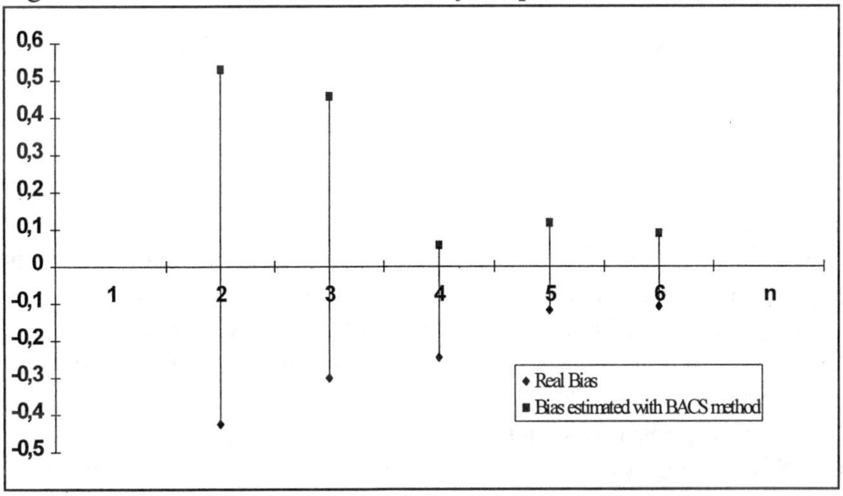

The results shown in Figures 2, 3, 4 and 5 above, verify that the BACS method, starting from small sample size, gives the bias estimate close to the real bias respectively of the population mean and variance.

5. Discussion

From the results obtained in section 3 and 4 we proved that the *BACS* method is able to estimate the bias of the mean and variance estimators computed on the final sample produced by the adaptive procedure.

In particular, in section 3 we proved the asymptotic convergence of the method, while in section 4 we have shown that the method gives a good estimate of bias starting from small sample sizes.
If our T_n estimator applied to initial sampling is biased, the *BACS* method gives information about the total bias of T_n estimator computed on the final sample. Moreover the *BACS* method enables us, through little modifications, to estimate accuracy measures of the T_n estimator computed on the final sample that is different from the bias, for example the standard error or the variance of T_n.
The BACS method coincides with the traditional bootstrap method when the units that belong to the drawn sample do not produce some additional units through an adaptive procedure.

References

Cicchitelli G., Herzel A., Montanari G.E. (1992). *Il campionamento statistico*. Il Mulino, Bologna.

Davison, A.C., Hinkley, D.V. (1996). *Bootstrap Methods and Their Application.* Cambridge University Press, Cambridge.

Efron, B.and Tibshirani, R.J. (1993). *An introduction to the bootstrap,* Chapman

Efron, B. (1979). Bootstrap methods: another look at the jackknife, *Ann.of Statist*. Vol.7, P.1-26.

Hansen, M.M. and Hurwitz, W.N. (1943). On the Theory of Sampling From Finite Populations, *Annals of Mathematical Statistic*, 14, 333-362.

Pace, L. and Salvan, A. (1996) *Teoria della statistica.*CEDAM Padova

Rao, J.N.K. and Wu, C.F.J. (1988). Resampling inference With Complex Survey Data, *JASA*, 83, 231-241.

Shao, J. and Tu, D. (1995). *The Jackknife and Bootstrap.* Springer

Thompson S.K. (1990). Adaptive Cluster Sampling, JASA, 85, p 1050-1059.

Thompson S.K. (1991a). Adaptive Cluster Sampling: Designs with Primary and Secondary units, *Biometrics* 47,, p. 1103-1115.

Thompson S.K. (1992). *Sampling,*. Wiley, New York.

Thompson S.K. Seber G. A. F.(1996). *Adaptive Sampling,*. Wiley, New York

Tukey J.W. (1958). Bias and confidence in not-quite large samples, *Ann. Math. Statist.,* 29, p. 614.

Wolter K.M.(1985). *Introduction to Variance Estimation.*Springer, New York.

Forecasting a Classification

Domenica Fioredistella Iezzi, Maurizio Vichi
Dipartimento di Metodi Quantitativi e Teoria Economica,
Università di Chieti, viale Pindaro 42, 65127 Pescara.
E-mail:vichi@dmqte.unich.it

Abstract: This paper focuses on the problem of forecasting a classification given a panel data set formed by a (multiple) time series of partitions of a same set of units. As far as we know, in classification and time series analysis this is a completely new problem. A methodology based on a vector autoregressive model is here proposed to directly forecast a partition given a (multiple) time series of partitions with the same fixed number of classes for each time. Two real panel data have been analysed with this new procedure. Open problems are discussed in a final section.

Key words: Cluster analysis, partitioning, panel data.

1. Introduction

Let $\mathbf{X}\equiv\{x_{ijt} : i\in\mathbf{I}, j\in\mathbf{J}, t\in\mathbf{T}\}$, be the three-way data array, where x_{ijh} is a real value of the j-th variable observed on the i-th unit (object), according to the h-th situation, and $\mathbf{I}\equiv\{1,...,n\}$, $\mathbf{J}\equiv\{1,...,k\}$ and $\mathbf{T}\equiv\{1,...,T\}$ are the sets of indices of modes i (units), j (variables), and t (occasions), respectively. The most widely collected three-way data set is given when units and variables remain the same and situations are different time points. Thus, units are followed examining the changes of variables over a set of different time points. These collected data are called in field economic a *panel data set*. Primary survey aspects such as measurement errors, not complete coverage of the population of interest, non responses, bias recalls for panel data are not discussed here; see Bailar (1989) for these aspects.

Here we suppose we have observed or computed a particular panel data set formed by T partitions of a same set of objects I examined at T temporal occasions.

A partition at time t may be specified by a $(n \times c)$ (units by classes) matrix $\mathbf{P}_t = [p_{ilt}]$, where p_{ilt} may be: 1) $p_{ilt} \in \{0, 1\}$, and $p_{ilt} = 1$ ($p_{ilt} = 0$) if object i belongs (does not belong) to class l of the partition at time t; 2) $p_{ilt} \in [0, 1]$ denotes the value of the *membership function* of object i to class l of the fuzzy partition at time t; 3) $p_{ilt} = d_{ilt}$, where d_{ilt} is the dissimilarity between object i and the centroid of class l of the partition at time t. In this paper we suppose we have observed or computed the dissimilarities d_{ilt}.

Several types of statistical analyses may be performed on the set of classifications linked by time. Firstly, the classificatory information of the entire period of time, in which the classifications have been observed, may be summarized by a *median consensus* i.e., the best least-squares classification approximating the given set (see: Barthélemy and Monjardet 1981, for *n*-trees; Cucumel 1990, Vichi 1993, for dendrograms). However, a unique consensus does not give information on the evolution of classes of the partitions over time. Further, a single consensus may not be sufficient to synthesise the data when the period of time is long and therefore many changes may occur for many classifications. A solution to this problem is given partitioning the set of classifications into homogeneous subsets regarding classification at contiguous periods of time, and simultaneously to find a consensus classification for each period (Gordon and Vichi 1997).
A completely new problem is to forecast a partition given the set of partitions linked by time. In this paper we are interested to define a new methodology to give an efficient answer to this problem.
An outline of the material in this paper is as follows. Section 2 shows a procedure to obtain the set of partitions from a three-way data set and gives the methodology to forecast a partition. Section 3 analyzes two real panel data and shows the results of the forecasted partitions, computing confidence intervals for the forecasting. A final discussion is given in Section 4.

2. Forecasting a partition of the units

Let c denote the number of classes dividing the same n objects into disjoint classes $C_{1t}, C_{2t}, \ldots, C_{ct}$ for each time point t. Therefore, the number of classes is implicitly assumed constant over time. The choice of c is discussed in section 3.
The partition for each time t is defined by solving the following well-known mathematical programming problem:

$$\begin{cases} \min \sum_{i=1}^{n} \sum_{l=1}^{c} d_{ilt} w_{ilt} \\ \text{subject to} \\ \sum_{l=1}^{c} w_{ilt} = 1, \quad 1 \le i \le n \\ w_{ilt} \in \{0, 1\}, \quad 1 \le i \le n, \quad 1 \le l \le c \end{cases} \qquad \text{[P1]}$$

where, d_{ilt} is the distance between object i and the centroid of class C_{lt} at time t. The distance d_{ilt} is usually the squared Euclidean norm on $\Re^k$, i.e., $d_{ilt} = ||\mathbf{x}_{i.t} - \mathbf{z}_{lt}||^2$, where $\mathbf{x}_{i.t} = (x_{i1t}, \ldots, x_{ikt})'$ and $\mathbf{z}_{lt}=(1/|C_{it}|\sum_{i\in C_{lt}} x_{i1t}, \ldots, 1/|C_{it}|\sum_{i\in C_{lt}} x_{ikt})'$.

A good heuristic solution to problem [P1], which is known to be NP-hard, is given applying the c-means clustering algorithm (Ball and Hall 1967, MacQueen 1967). The initial partition into c clusters:

(i) can be randomly chosen for each time $t \in \mathbf{T}$;
(ii) randomly chosen for t=1 and represents the partition achieved at time t-1, for $t \in \mathbf{T}$ ($t \neq 1$).

In the case (*i*), since classes change composition passing from time t-1 to time t, and partitions at two times are obtained independently, it is necessary to specify for each class at time t-1 which is the corresponding class at time t. However, this correspondence in not necessary using (*ii*) since it is automatically established by the clustering algorithm used.
Note that using (*ii*) it is implicitly supposed that the partition at time t is depending to the partition at time t-1 ([1]) and we believe this is a necessary condition to forecast a partition. For this reason in this paper we will use procedure (*ii*).
If the distances between each object and the c centroids change monotonely, from time t-1 to t, the final partition at time t is equal to the final partition at time t-1. This is because under the monotone transformation of distances each object has minimum distance from the same closest centroid at time t-1. Therefore, a "*stationary partition*" is obtained at time t if d_{ilt} changes momotonely from time t-1 to t. However, often this is not the case and therefore we need to know if and how object i changes, over time, class of the partition. For this purpose we use a vector autoregressive model.
Let $\mathbf{D}_1=[\mathbf{d}_{11},...,\mathbf{d}_{1T}]$, ..., $\mathbf{D}_n=[\mathbf{d}_{n1},...,\mathbf{d}_{nT}]$, denote the n c-dimensional multiple time series, with $\mathbf{d}_{it}=(d_{i1t}, ..., d_{ict})'$ defined by the classification process described above. It is assumed that the c-dimensional multiple time series are generated by stationary stable vector autoregressive processes of order p_i, VAR(p_i):

$$\mathbf{d}_{it} = \mathbf{v}_i + \mathbf{A}_{i1}\,\mathbf{d}_{it-1} + ... + \mathbf{A}_{ip_i}\,\mathbf{d}_{it-p_i} + \mathbf{u}_{it} \quad i=1,...n;\ t=1,...,T, \tag{1}$$

where $\mathbf{v}_i=(v_{i1}, ..., v_{ic})'$ is the c-dimensional vector of intercept terms, the $\mathbf{A}_{im}$ m=1,...,p_i are square coefficient matrices of order c and $\mathbf{u}_{it}$ is a c-dimensional innovation process (white noise), i.e., E($\mathbf{u}_{it}$)=0, E($\mathbf{u}_{it}\,\mathbf{u}_{it}'$)=$\Sigma_u$, E($\mathbf{u}_{it}\,\mathbf{u}_{is}'$)=$\mathbf{0}$, for s≠t, where Σ_u is assumed to be a non-singular covariance matrix. Criteria for selecting the VAR order will be considered in section 3.
The compact form of VAR(p_i) is:

$$\mathbf{D}_i = \mathbf{B}_i\mathbf{Z}_i + \mathbf{U}_i \tag{2}$$

[1] However, in general the solution of the clustering algorithm is not depending on the initial partition.

where $\mathbf{Z}_i = (\mathbf{Z}_{i0}, \ldots, \mathbf{Z}_{iT-1})$ with $\mathbf{Z}_{it}=(1, \mathbf{d}_{11}, \ldots, \mathbf{d}_{1T})$,
$\mathbf{B}_i=(\mathbf{v}_i, \mathbf{A}_{i1}, \ldots, \mathbf{A}_{ip_i})$, $\mathbf{U}_i=(\mathbf{u}_{i1}, \ldots, \mathbf{u}_{iT})$, and $\mathbf{I}_c$ is the identity matrix of order c.
The vec form of the VAR model is:

$$vec(\mathbf{D}_i) = (\mathbf{Z}_i \otimes \mathbf{I}_c)vec(\mathbf{B}_i) + \mathrm{vec}(\mathbf{U}_i) \tag{3}$$

where, $vec(.)$ is the usual stacking operator, $\mathbf{L}\otimes\mathbf{M}$ is the Kronecker product of two matrices $\mathbf{L}$ and $\mathbf{M}$.
The multivariate least squares estimator can be written as:

$$\hat{\mathbf{B}}_i = (\hat{\mathbf{v}}_i, \hat{\mathbf{A}}_{i1}, \hat{\mathbf{A}}_{ip_i}) = \mathbf{D}_i\mathbf{Z}_i'(\mathbf{Z}_i\mathbf{Z}_i')^{-1}. \tag{4}$$

After estimating the parameters of the VAR models we can forecast the values d_{ilT+h} for a given forecast horizon $h>0$ and a forecast origin T.
The optimal h-step forecast of the process (Lütkepohl 1991) is:

$$\hat{\mathbf{d}}_{iT+h} = \hat{\mathbf{v}}_i + \hat{\mathbf{A}}_{i1}\,\hat{\mathbf{d}}_{iT+h-1} + \ldots + \hat{\mathbf{A}}_{ip_i}\,\hat{\mathbf{d}}_{iT+h-p_i} \qquad i=1,\ldots n;\ t=1,\ldots,T, \tag{5}$$

The estimated global solution of [P1] when $\hat{\mathbf{d}}_{iT+h}$ are given is on hand. It is easily obtained assigning object i to class C_{l^*} if l^* is such that $d_{il^*T+h} = \min(d_{ilT+h} : l=1,\ldots,c)$. Notice that the estimated optimal partition is found without applying any clustering algorithm but simply solving an assignment problem in linear computational time complexity $O(nc)$.

3. An application to sea water pollution data

Before partitioning the units of a panel data set a three-way pre-processing is necessary to carry out some basic transformations and return data appropriate for clustering objects and forecasting partitions. Two data pre-processing need: from a cross-sectional point of view, a column fiber standardization is required (Rizzi & Vichi, 1995) to identify clusters not influenced by units of measurement and different variability of the observed variables. On the other hand from a time series point of view the n c-dimensional multiple time series are subject to seasonal effects and trends, which have to be estimated and corrected with the usual methods.
In order to show the proposed model of forecasting, a panel data regarding the sea water pollution on 17 pollution control transects, located at 500 meters from the coast orthogonal to the opening of the rivers, along the Abruzzo coast was considered. The water pollution was measured according to 10 variables, observed for a period of four years on a monthly basis (49 time points).

The values for each station were first standardized so as to have zero mean and unit variance.

Each cross-sectional data matrix was classified separately using the *c*-means clustering algorithm (MacQueen 1967). The criterion proposed by Calinski and Harabasz (1974), and proved to be one of the best by Milligan and Cooper (1985), was used to determine the number of classes in each data set. We found that most of the classifications defined good partitions between 5 and 7 classes, and to allow for comparison the classification in 5 classes (mode) was considered. The initial partition at time t was given by the final partition at time t-1. The LS estimates were computed on the 17 pollution control transects. The order of the VAR model for each pollution control transect, found by using the final predictor error, as summarized by table 1, is equal to one. The stationarity and normality conditions of the VAR models were verified. Table 2 shows forecasts for months 50 and 51. The 95% confidence intervals, $\hat{\mathbf{d}}_{iiT}$ were computed for lower and upper confidence limits as in Table 3. The cluster membership was determined. In the 50th forecasted time, among the 17 pollution control transects, 11 remain with an unchanged cluster membership, and 6 report a change in cluster membership for one of the two limits, while in the 51st forecasted time, 10 remain with an unchanged cluster membership, and 7 report a change in cluster membership for one of the two limits.

Table. 1: *Estimation of the VAR order of the 5 clusters, for 17 pollution control transects. The VAR models of orders p_i=1,2,3,4 are estimated and the corresponding FPE (final predictor error) values are computed. The order minimizing the FPE values is then chosen as estimate for p_i.*

p	1	2	3	4	5	6	7	8	9	10	11	12	13	14	15	16	17
1	0.01	0.01	0.00	0.01	0.01	0.02	0.07	0.02	0.20	0.02	0.02	0.02	0.02	0.01	0.01	0.01	0.02
2	0.03	1.19	2.36	4.14	0.03	0.07	0.04	0.04	0.46	0.04	0.06	0.06	0.04	0.03	0.03	0.03	0.07
3	0.08	3.32	4.31	6.24	0.31	0.12	0.07	0.05	1.06	0.14	0.13	0.24	0.13	0.05	0.06	0.07	0.15
4	0.24	1.32	1.18	2.12	0.01	0.10	0.16	0.18	2.28	0.52	0.41	0.57	0.41	0.11	0.15	0.18	0.53

Table 2 : *Cluster membership for 17 pollution control transects: Forecasting for 2 months; 50th and 51st periods.*

forcast	1	2	3	4	5	6	7	8	9	10	11	12	13	14	15	16	17
50	2	1	1	1	2	2	2	5	2	2	5	2	1	2	1	1	3
51	2	2	3	1	4	2	2	4	2	3	1	3	5	2	2	3	3

Table 3a : 50th period: *Cluster membership for 17 pollution control transects: 95% Confidence Intervals for upper and lower limit forecasts*

transect	1	2	3	4	5	6	7	8	9	10	11	12	13	14	15	16	17
upper	2	3	3	1	2	2	2	4	2	3	5	3	4	2	1	1	3
lower	2	1	1	1	2	2	2	5	2	2	5	2	1	2	1	1	3

Table 3b : 51st period: *Cluster membership for 17 pollution control transects: 95% Confidence Intervals for upper and lower limit forecasts*

transect	1	2	3	4	5	6	7	8	9	10	11	12	13	14	15	16	17
upper	2	3	3	1	4	4	3	4	1	3	1	3	4	2	4	4	3
lower	2	2	3	1	4	2	2	4	2	3	1	3	5	2	2	3	3

The previous data set was characterized by a phenomenon which presents large variability over time. A second example regards the performance of seven most industrialized countries (Canada, France, Germany, Japan, Italy, USA, Great Britain according to changes of some economic indicators GDP, Expenditure, Export, Import, and Industrial production in the period from 1970 to 1996. As it can be expected for industrialized countries all variables have a strong trend with low variability. The per-processing indicated in the previous section was applied.

Table 4 : *Cluster membership for 7 most industrialized countries: Forecasting for 1997 and 1998 (27th and 28th period).*

forecast	Canada	France	Germany	Japan	Great Britain	Italy	USA
28	1	2	2	2	2	1	1
29	1	2	2	2	2	1	2

Table 5a: 28th and 29th period (results are equal for the two periods). Cluster membership for 7 countries: 95% Confidence Intervals for upper and lower limit forecasts

	Canada	France	Germany	Japan	Great Britain	Italy	USA
lower	1	2	2	2	2	1	1
upper	1	2	2	2	2	1	2

In this example clusters do not change cluster membership for the lower and upper limit forecasts. This confirms that when objects present a smoothed trend over time forecasting of the partition is simpler.

4. Discussion

In this paper the problem of forecasting the partition of the units given a set of partitions is discussed. A direct approach has been proposed by Zani, (1981) starting the analysis from a panel data set and forecasting each of the n k-dimensional multiple time series associated to the n units to be partitioned and computing a clustering algorithm, e.g. c-means clustering algorithm to the estimated values of the multiple time series. Here we propose an indirect approach consisting in forecasting the distance between centers and units d_{ilt} by means of vector autoregressive models for each c-dimensional multiple time series of the distances $d_{ilt.}$ When such values are estimated the forecasted classification global optimal solution of [P1] is obtained by an easy assignment problem. There are several reasons for preferring the indirect approach with respect to the direct one: *i*) in indirect approach only VAR processes have to be estimated to achieve a forecasted classification, meanwhile in the direct approach a classification technique has to be applied; *ii*) in economic and social phenomena where the time series are smoothed VAR models for each c-dimensional multiple time series of the distances d_{ijt}, are generally stationary of order 1; *iii)* in real problems, the number of variables is generally much larger than the number of clusters for synthesis reasons, thus making it simpler to solve forecasting problems in the indirect approach.

References

Bailar, A. B. (1989): *Information needs, surverys and measurement errors*, Panel surveys, D. Kasprzyk, G. J. Duncan, G. Kalton, M. P. Singh, Willey.

Ball, G.H., and Hall, D.J. (1967): A Clustering Technique for Summarizing Multivariate Data, *Behavioral Science*, 12, 153-155.

Barthélemy, J.P., and Monjardet, B. (1981): "The median procedure in cluster analysis and social choice theory," *Mathematical Social Sciences*, *1*, 235-267.

Calinski, T. & Harabasz J. (1974), A Dendride Method for Cluster Analysis, *Communications in Statistics*, 3, 1-27.

Cucumel, G. (1990): Construction d'une hiérarchie consensus à l'aide d'une ultrametrique centrale, in: *Recuel des Textes des Présentations du Colloque sur les Méthodes et Domaines d'Application de la Statistique 1990*, Bureau de la Statistique du Québec, Québec, 235-243.

Gordon, A. D., and Vichi, M. (1997): Partitions of Partitions, to appear on *Journal of Classification.*

Lütkepohl, H. (1991): *Introduction to Multiple Time Series Analysis*, Springer Verlag Berlin.

MacQueen, J. B.(1967): Some methods for classification and analysis of multivariate observations. In *Proceeding of the 5th Berkeley Symposium in Mathematical Statistics and Probability*, University of California Press, Berkeley, USA, 281-297.

Milligan, G. W. & Cooper, M (1985):An examination of procedures for determining the number of clusters in data set, in *Psychometrika*, vol. 50, n. 2, 157-179.

Rizzi, A. & Vichi, M. (1995): The Thee-way data set analysis. *In Some relations between matrices and structures of multidimensional data analysis*, Giardini Editori e Stampatori in Pisa, 93-166.

Vichi, M. (1993): Un algoritmo dei minimi quadrati per interpolare un insieme di classificazioni gerarchiche con una classificazione consenso, *Metron*, 51, 139-163.

Zani, S. (1981): Osservazioni sulle serie storiche multiple e l'analisi dei gruppi. In *Analisi Moderna delle Serie Storiche*, Atti del Convegno nazionale, Napoli, 19-22 maggio, Franco Angeli Editore, 263-274.

Neural Networks as a Fuzzy Semantic Network of Events

Antonio Bellacicco

Department of Organizations and Systems Theory
University of Teramo,Viale Crucioli 122,Teramo,e-mail:abellac@tin.it

Abstract: The paper deals with the interpretation of a neural network in terms of a semantic network able to describe a cluster of events.Some general remarks on neural networks and semantic networks are proposed and the interpretation of semantic networks as cluster of events is considered as a new way of understanding cluster analysis in a data base,provided we deal with fuzzy events.

Key words: Cluster,Event,Fuzzy,Interpretation,Neural,Network,Semantic.

1. Introduction

The paper tries to relate four main methodological tools belonging to different disciplines like formal logic, artificial intelligence and statistics.The methodological tools are, respectively, fuzzy set theory, cluster analysis, neural networks and semantic networks.Fuzzy set theory deals with the generalization of the notion of belongness of a given event $\in$ to a set $\in$ of events,cluster analysis deals with the partition Π of a set of events $\in$ in order to minimize/maximize the variability of each subset of Π ,neural networks deal with optimization problems and semantic networks deal with the representation of a set of events related by the usual logic connectives, $\wedge$(*and*),$\vee$(*or*),$\neg$(*not*) and by the material implication,$\rightarrow$ (if..then).
The relation among the mentioned tools can be clear when we consider the general target of identifying a subset of events characterized by a relationship of mutual dependence so that the occurrence of a given event is conditioned by the occurrence of the other events following the scheme of a network,that is a graph. Moreover each event is identified by vague attributes and its occurrence can be described in terms of qualitative ordering,as for instance:*extremely low, low, high, extremely high* and so on as well as by a numerical scale provided the upper and the lowest level are defined. Neural networks play a role as tools for identifying a subset whose main feature is the presence of mutual dependence among its members. Finally,cluster analysis occurs because the aim of the neural

network is to build up the best partition(covering) of the whole set of events identifying the set of mutual dependence relationships.

As a first example we can consider the set of variables occurring in a data base.We have to build up a partition of the set of variables so that each subset of variables is characterized by a set of linear regression equations.Each equation identifies a statistical dependence and we distinguish between the set of independent variables and the dependent variable,in case we assume a multiple regression equation.In other problems we build up subsets of units whose main feature is their geometrical shape in the multidmensional variable space.As another example,we can consider the spatial accessibility among a sebset of towns.Another example can be considered from propositional calculus where we expect each subset of propositions is charachererized by a set of material implications so that each proposition implies all the remaining propositions.

Fuzzy logic,clustering algorithms,semantic networks and neural networs, generally speaking ,are not related and the main aim of the paper is to show that it is possible to define a common background able to relate the previous tools in a unique point of view bringing to a common target. Some consequence can be derived both from the theoretical point of view and from the operational point of view.In this paper we will explore both the consequences and we will propose a new class of algorithms able to solve new problems in all the implied mentioned fields,like fuzzy set theory,neural networks,semantic networks and cluster analysis.

A simple example able to show the logical connections among the previous concepts can be considered in order to understand easily the subsequent consequences.

Let us consider a data base on books stocked in a big library.We like to build up an optimal partition of the books so that the books belonging to each class are characterized by a set of attributes related each other by fuzzy implications.In other terms,each attribute is presented in a fuzzy way and for each couple of attributes we can define a fuzzy implication able to specify the highest level of fuzziness. Fuzziness means that each attribute can occur at a given degree and the implication means the implication can attain the maximum degree provided both the attributes overcome a given treshold.If the books are highly devoted to the statistical methodology and are a low technical degree,then the implication means that we expect a class without advanced books in statistical methodology.

We need to build up classes so that any attribute implies all the other attributes defining the class Finally,neural network is a tool so that we can use imbodies the described logical structure ad is able to find the optimal partition.

In the coming paragraphs we will show how to relate the four tools and we will propose a new algorithm both in neural network and in cluster analysis.

2. Clustering Algorithms

The first generalization regards the implication rules.*Fuzzy implications* can be considered as set of fuzzy relations whose membership functions give a different meaning to the underlying concepts. *Clusters* of events can be considered a set of events linked by implication rules represented by the arcs of a graph. As far as the events can be described by a proposition ,we can consider a cluster of events as a set of propositions related by a set of material implications and by the logic connectives,*and,or,not.*,so that a function F can be optimized. The problem is to give a weight to the implications,to choose the function F and to show the isomorfism among the clustering algorithms,the semantic networks and the neural networks.

Let us consider a set of events E and a graph G(E,W) where E is the set of vertices representing the events and W a subset of E *x* E,representing the arcs joining the vertices.Let us consider for each vertex a weight representing its belongness to the set E .We can consider the fuzzy interpretation of the vertices as fuzzy propositions and the fuzzy interpretation of the arcs as fuzzy implications.We have two types of fuzzy implications: fuzzy material implication and fuzzy propositional implications,where the operator φ is a fuzzy function :

$$\varphi(p \rightarrow q) = \min(1, \varphi P + \varphi q) \tag{1}$$

$$\varphi(p \rightarrow q) = \max(0, \varphi p + \varphi q - 1) \tag{2}$$

It is easy to see that the fuzzy evaluation of the material implication and of the propositional calculus is based on the fuzzy evaluation of the propositions p and q. Kasabov (1996), Kosko (1992),

The main problem to face is the fuzzy evaluation of the vertices of the graph representing the propositions and to define a partition of the graph so that it is minimum the fuzzy evaluation of the arcs relating the subsets corresponding to semantic networks.

Clustering algorithms can be considered as a sequence of operations belonging both to logic and to algebra, able to produce either a partition or a covering of the set of the vertices of the graph representing a set of relations between couples of vertices, so that a function F can be optimized under some constraints.The operations,usually called *modifiers*, can be listed as follows :

i. $(\varphi P)^{k}$, for $k = u/v$, $u = 1,2,3,..,n$ and $v = 1,2,...,n$; (3)

ii. $\varphi P = 1$ if $\varphi P \geq z$ and $\varphi P = 0$ otherwise, $0 \leq z < 1$ (4)

The implication rules (1) and (2) and the modifiers (3) and (4) arc algebraic operations which have some consequences on the arcs like their weighting and their cutting.More esplicitily,*modifier* (3) is able to enlarge the value of fuzziness attached to each attribute and therefore we can modify the value of fuzziness of the implication rule in the cluster to which the attribute is assigned.*Modifier* (4) is a treshold and is able to shift the fuzzy interpretation of a table connecting attributes to units to a boolean interpretation.
It is obvious that *modifier* (4) reduce all the previos considerations on the interpretation on the operation of fuzzy implication to the interpretation of classical boolean implication.

We have to solve three problems :

- the definition of cluster;
- the choice of the function to be optimized;
- the introduction of logical and algebraic constraints.

In Bellacicco and Tulli (1996) a general definition of cluster is given in terms of balanced graphs which are characterized by the assignment of a constant weight to all the arcs.The previous definition is able to identify many types of graphs,including cliques and circuits which seem the most reasonable shapes of graphs representing clusters and own some optimality criterion.
We generalize our previous definition of cluster by the introduction of fuzzy weights which are neither a distance nor a dissimilarity index.
As a general gramework we will consider graph theotetical language in order to simplify and unify all the concepts.
First at all, a cluster can be represented by an oriented graph G(S,X), where S is a set of vertices representing the units,X the arcs,that is the ordered couple of vertices by a suitable weight which can be an integer number,both positive and negative ones,a rational number and a real number,Marshall (1971), Bellacicco and Labella (1979).Following the previous definition,the arch connecting two vertices can represent a logic implication and the vertices of the graph can represent the attributes.
An *interpretation* of the cluster is the mapping of each cluster on the set of books and more generally on the set of the units U:

$$f: G(S,X) \rightarrow U$$

The mapping f can be an homomorphism if f satisfies the obvious request so that the same frame on the attributes can be obtained in the units.Actually,the same attribute can identify many units and a cluster of attributes identifies a cluster of units and viceversa ,Bellacicco and Labella (1979).We show in the mentioned book that in the boolean case the mapping is one-to-one and therefore the mapping f can be an isomorphism.

We will generalize the mapping f to the fuzzy evaluation and therefore to the fuzzy implication.The fuzzy evaluation of an implication by (1),bounded sum, and by (2),bounded product,can be used as a weight of the corresponding arc in the graph.We can introduce the triangular membership function and we can give a fuzzy evaluation of the fuzzy implication .As it is well known,the triangular function is the following one:

$$\Phi\varphi = 1 - \frac{\varphi - \alpha}{\beta} \qquad (5)$$

where α is the minimum evaluation of φ and β is the range of the evaluation φ. It is easy to see that (5) reduces to the evaluation of the complement of the fuzzy implication when $\beta = 1$,which is the length of the range of a fuzzy evaluation.Modifier (4) can reduce in any case a fuzzy evaluation to a fuzzy implication and we expect that the modifier can operate without changing the interpretation.Actually,we can obtain a boolean interpretation for every value of fuzziness and therefore the mapping f is preserved.

A clustering algorithm can be reduced to the sequential use of the modifier (3) for $u = n$ and $v = 1$,and a tresholding operation (4) ,where the treshold **z** can be choosen in case we like to delete the lowest percentile of arcs.

The function to be minimized is :

$$f = \Sigma_i \Sigma_r \ (1 - \varphi_{i/r}) \qquad (6)$$

where i/r means : proposition p_i in the cluster C_r.

It is easy to see that each cluster is characterized in terms of minimum fuzziness.In other terms we like to maximize the best distinction between couples of clusters.We can now analyse the semantic network interpretation of the previous clustering approach and its neural network representation.

3. Some Isomorphisms

In the previous paragraph we gave a general outline of a clustering algorithm in terms of logical operations on a graph G whose vertices are fuzzy propositions representing events and whose arcs are fuzzy implications weighted by their fuzziness evaluated by (1) or by (2).The choice of the weight depends from the type of optimization and in case we prefer to minimize the fuzziness and we adopt a triangular function ,we can adopt equation (1) which push the evaluation near the value 1,that is the absence of fuzziness.We must spend some words about the evaluation of the fuzziness of the elements of the table TN,M and therefore of the vertices of the graph obtained by the introduction of a non symmetric weighting of the arcs.We can follow the subsequent transformations:

Step1.normalize each element x_{ij} by the ratio $z_{ij} = x_{ij}/max(j)$, where by max(j) we mean the maximum value in the column j ;
Step 2.take the average of the elements of each row S_i of the transformed data z_{ij} ,denoted z_{io} ;
Step 3.adopt the modifier (3), for $k = 1/64$, on the values $z_{ij} >?$;
Step 4.adopt the modifier (3), for $k = 5$,on the values $z_{ij} \leq ?$;
Step 5.use two tresholds in order to obtain a crisp set of binary data,c_{ij} ;
Step 6.Consider each couple of S in $T_{n,m}$ and introduce an ordering relationship based on the heighest frequency of true material implications of the type:
$c_{ij} \rightarrow c_{rj}$ and $c_{rj} \rightarrow c_{ij}$,in the set of the m variables.

Step 7. Evaluate the fuzzy implication by (1) considering the data z_{ij} .

We can now state that a clustering algorithm based on the previous steps on an oriented graph is actually an algorithm for detecting a set of semantic networks.Some final remarks regards the isorphism between a neural network NN and a semantic network SN.We can state some relations between neural networks and semantic networks provided a learning rule can be considered as a dynamical updating of the weights of the implications in order to accomplish the task.Semantic networks are a static device for describing the a set of events.We do not consider in this paper the role of the learning rule and we limit the analysis to the static behaviour of the two logical structures introducing the following statements :

Statement 1.*A semantic network M has the same computational power of a neural networN with three layers;*

Statement 2. *For a set of well formed formulas and some true propositions,it exists a neural network N with three layers with a suitable set of weights on the arcs;*

Statement 3. *It exist a universal neural network for every set of propositions as inputs and for for any set of well formed formulas.*

We cannot give in this paper a demonstration of the three previous statements without introducing a specific formalization of the arguments which is beyond the aim of the paper. Neural networks are input-output systems able to produce an input-output transformation, which must be either close to a template or sharing some main features of the input.The last case is characterizing the neural networks with three layers where the hidden layer is the set of inferential rules which can both the *modus ponens* and the *modus tollens*,Freeman and Skapura (1992).Those rules are transformations of the input and the input is the set of the propositions describing the events represented by the units of the $T_{n,m}$ table.The usual non linear transformation of the data is interpreted in terms of fuzzy sets theory as the mentioned inferential rules.*Modus ponens* is defined as follows: if (S_i is p) then (S_j is $q=1-p$), and S_i is w) then S_j is $v = 1 - w$. In terms of fuzzy sets the previous inference rule implies a transformation of the fuzziness able to confirm the consequence both for the fuzzy evaluation and for its complement evaluation.We can see easily that the implication is modified from rule (1) and from rule (2) ,as we mentioned before.*Modus tollens* is just the reverse of the previous statement and in some way it is not a new inferential rule.Non linearity occurs because we can shift from a fuzzy evaluation to its complement and the previous rule can be considered like a step function in a unitary square which can be approximated by the logistic function;both the two functions are usually adopted in neural networks.

References

Bellacicco A.and Labella A (1979), *Le Strutture Matematiche dei Dati,* Feltrinelli,Milano.

Bellacicco A.& Tulli V.(1996), Cluster identification in a signed graph by eigenvalue analysis, in *Matrices and Graphs,* Camiz S & Stefani S.(Eds), World Scientific,233-242.

Freeman J.and Skapura D.M.,(1992), *Neural Networks* ,New York, Addison Wesley.

Kasabov N.K.,(1996),*Foundations of Neural Networks,Fuzzy Systems ,and Knowledge Engineering,*
MIT Press,Cambridge.

Kosko B.,(1992),*Neural Networks and Fuzzy Systems. A Dynamical Approach to Machine Intelligence,* Englewood Cliffs,N.J.,Prentice Hall.

Marshall C.W.(1971), *Applied Graph Theory,* Wiley-Interscience,New York.

Hierarchical Fuzzy Clustering: An Example of Spatio-Temporal Analysis

Loredana Cerbara
Istituto di Ricerche sulla Popolazione - CNR
Viale Beethoven, 56 - 00144 Roma
e-mail:Cerbara@irp.rm.cnr.it

Abstract: This work describes the hierarchical classification procedure called 'fuzzy average linkage' which provides a fuzzy partition of a group of units. The basic principle is that the average similarity of units linked to the same group must be greater than or equal to a certain pre-set similarity level. This method is applied to mortality rates by cause of death for men and women in the 1970s, 1980s and 1990s.

Key words: Hierarchical Cluster Analysis, Similarity, Fuzzy average linkage.

1. Method

The method proposed here, called, 'fuzzy average linkage', is a fuzzy hierarchical classification method producing *fuzzy partitions* of a set of units to be classified. In order to clarify, what we mean by fuzzy partition, we shall use a scheme (Ricolfi 1992) in Table 1 which enables us to distinguish between the clustering methods based on the results they give, i.e., based on the *membership function*. This is, for each unit *i-th*, a multivariate function with G values μ_{ig} ($g=1,...,G$) where G is the number of classes of the partition or clumping, and μ_{ig} is called *grade of membership* of object i to group g. If all the grades of membership range between 0 and 1, we have a fuzzy classification; when these values are strictly 0 or 1, the classification degenerates into a hard or crisp one.

Table 1: *Classification of clustering methods based on propriety of membership function.*

	Range of definition	
Sum of grade of membership	Hard methods: {0,1}	Fuzzy methods: [0,1]
Partition: $\Sigma\mu_{ig}=1$	**Hard partition**	**Fuzzy partition**
Clumping: $\Sigma\mu_{ig}>1$	**Hard clumping**	**Fuzzy clumping**

When the sum of grade of membership of *i-th* object to all clusters is exactly 1, the classification is a partition, however when this sum is greater than 1, the classification is a clumping.
The similarity index between objects i and j, used in this paper is the Gower measure (Gower 1971):

$$S(i,j) = \sum_{k=1}^{K} v_{ij}(k) \cdot p(k) \tag{1}$$

where K is the overall number of variables, $p(k)$ is the non-negative weight associated with the *k-th* variable, $V_{ij}(k)$ is one minus the relative distance (i.e. the distance compared to the maximum distance calculated for that item) between pairs of units. The fuzzy methodology here proposed uses centroids of the variables and therefore only quantitative and ordinal variables can be used. The distance used here is the generalised Hamming distance (Ponsard 1985).

2. The classification procedure

Let **S** be a similarity matrix to which the following classification procedure is applied.
Step 1: A search is made for the pairs of units with the maximum value of the similarity index (for some matrixes **S**, there may be more than one pair). These pairs form the initial groups. The centroids for each group are calculated as well as the values μ_{ij} for each unit, which will be inversely proportional to the distance of the unit from the group centroid (the further away the unit *i-th* is from the g-centroid the less the value of μ_{ig} will be). It has to be observed that if the unit belongs to one group only, its membership function is equal to 1 only for this group, and is equal to 0 otherwise, to satisfy the necessary condition for having a fuzzy partition.
Step 2: Excluding the values already taken into consideration, a further search in **S** is made for pairs of units with the maximum value of similarity. Let α be this maximum. The pairs identified may be composed of units which have already been considered in the previous step. For the sake of simplicity, we can assume that we have identified pair (i',j') and that unit i' has already been inserted in a group. In this case, j' is inserted in the group to which i' belongs only if the average similarity between j' and all the units composing the group is greater than α. Otherwise, units i' and j' will form a new group. Once groups have been formed, the centroids and membership function are calculated as specified in step 1.
Step 3: Step 2 is repeated until all the units form a single group.

3. An application to adult mortality

The fuzzy average linkage technique has been applied to the study of mortality rate by cause of death in some European countries (Table 2). Standardised female and male mortality rates by cause of death (Table 3) have been used for the 30-64 age group for the 1970s, 1980s and 1990s.

Table 2. *European Countries*

Austria	Denmark	Greece	Luxemburg	Sweden
Belgium	Finland	Hungary	Netherlands	Switzerland
Bulgaria	France	Ireland	Norway	United Kingdom
Czechoslovakia	West Germany	Italy	Spain	Jugoslavia

Table 3. *Causes of Death*

Cancer:	Diseases of the digestive system:
Stomach cancer	Hepatic cirrhosis
Colon and rectum cancer	Other diseases of the digestive system
Tracheal, bronchial and lung cancer	
Other cancers	Ill-defined causes and senility
Diabetes	External causes:
	Accidents
Diseases of the circulatory system:	Suicide
Ischemic heart diseases	Other external causes
Cardiovascular diseases	
Other diseases of the circulatory system	All other causes
Diseases of the respiratory system:	
Pneumonia	
Bronchitis	

Most traditional cluster analysis methods combine similar geographical areas into a single group, so that the group indicates a specific mortality profile and each area may belong to a single group even if it is somewhat similar to the geographical areas of other groups. The traditional classification implicitly assumes that there are no similarities in the mortality models of bordering geographical areas, while it seems reasonable to assume that neighbouring population groups are subject to causes of death related to similar types of individual behaviour or characteristics of the environment where they live. A "fuzzy" classification method enables us not only to identify groups of European countries having mortality profiles which are similar for most of the causes of death considered, but also to specify the degree of similarity linking geographical areas belonging to the same group. Since this method enables a country to be assigned to more than one group, we obtain a panorama of the geography of mortality in which the similar groups are not separated by rigid borders, but which can also overlap.

Since the application can not be fully described in this paper, we will provide a summary of the main results. For example, we can show the result obtained for data referring to women in 1990.
We have to chose the level of similarity at which the groups obtained should be analysed, and this choice represents one of the most debated problems concerning hierarchical methods of clustering. Different rules can be followed, but in general, in order to ensure sufficient internal homogeneity in the group, rather high levels should be chosen. The aggregative procedure for fuzzy average linkage produces a number of groups which, being fuzzy, can be different for each level. Therefore, we suggest choosing one of the highest levels since it is more probable that a lesser number of groups will be formed and less fuzziness shall be found. This does not mean that we cannot use a less subjective technique of choosing such as the ones proposed for hard clustering.
We have therefore considered the similarity level α=0.82, where α indicates the maximum threshold allowed for average similarity between two units belonging to the same group, the 4 groups shown in Table 4 were formed.
In order to provide the best interpretation for this result, we can calculate the averages - weighted with the membership function - of the variables for countries belonging to the same group (Table 5). We obtain the profile of a unit representing that group. For example, group 1, consisting of Ireland and United Kingdom only, has rather high rates of mortality due to cancer and circulatory diseases. In the other three groups, the differences are less evident since they are fuzzier, showing that while some differences remained in the 1990s, Northern and Southern Europe have reached comparable mortality rates for almost all causes of death.
To conclude, we can briefly mention the result of the entire application.
Countries showing a nil linkage value are the ones isolated. As we can see, the Eastern European countries and Denmark are isolated. This happens because their profiles are not similar enough with respect to the other European countries considered (i.e. their similarity does not exceed 0.82).
The classifications obtained for the various years of observation range from a situation in which there are groups with little overlapping, thus indicating a certain degree of dissimilarity, to a situation in which there are few groups and a certain overlapping with high degrees of similarity. Some European countries remain exceptions with respect to the others. This applies to Ireland and the United Kingdom, almost always related, but especially Ireland. This also applies to some Eastern European countries, since this region is behind with respect to the changes occurring in the 20-year period (1970-1990). In South-Central Europe there is a gradual move towards "northern" mortality profiles, which explains the greater degree of standardisation among the European countries in 1990. This is due to the fall in mortality rate from diseases due to poor sanitary and health conditions, together with the rise in mortality rate due

to other types of diseases linked to the lifestyles of the more highly developed countries (Caselli 1993).

Table 4: *Woman fuzzy classification in 1990 at level α=0,82*

European countries	Groups			
	g1	g2	g3	g4
Austria	0	0,52	0	0,48
Belgium	0	0,33	0,38	0,3
Bulgaria	0	0	0	0
Czechoslovakia	0	0	0	0
Denmark	0	0	0	0
Finland	0	0	1	0
France	0	0,51	0,49	0
West Germany	0	0,44	0,56	0
Greece	0	0	0	1
Hungary	0	0	0	0
Ireland	1	0	0	0
Italy	0	0,48	0	0,52
Luxembourg	0	0	1	0
Netherlands	0	0,53	0	0,47
Norway	0	0,56	0	0,44
Spain	0	0,47	0	0,53
Sweden	0	0,33	0,31	0,36
Switzerland	0	0,34	0,27	0,39
United Kingdom	1	0	0	0
Jugoslavia	0	0	0	0

Table 5: *Average of variables in groups.*

Variables	Groups			
	g1	g2	g3	g4
Stomach cancer	4.77	5.31	4.94	5.44
Colon and rectum cancer	15.23	12.43	11.76	10.88
Tracheal, bronchial and lung cancer	25.58	11.69	8.00	11.52
Other cancers	121.09	100.95	105.01	95.87
Diabetes	4.14	4.69	3.44	4.37
Ischemic heart diseases	52.72	21.56	23.23	21.58
Cardiovascular diseases	20.33	14.12	18.17	15.71
Other diseases of the circulatory system	17.80	18.96	20.03	20.21
Pneumonia	4.56	1.90	2.38	1.74
Bronchitis	5.89	3.58	3.82	2.92
Hepatic cirrhosis	4.69	10.61	14.16	7.85
Other diseases of the digestive system	6.68	5.03	6.16	4.44
Ill-defined causes and senility	0.74	5.82	6.96	5.57
Accidents	1.39	1.55	1.82	1.50
Suicide	6.13	11.29	15.18	8.84
Other external causes	11.17	12.36	15.91	12.19
All other causes	34.56	21.42	24.22	19.74

References

Caselli, G. (1993). L'evolution à long terme de la mortalité en Europe, *Procedings of the European Conference*, Vol. 2, INED, Paris.

Gower, J.C. (1971). *A general coefficient of similarity and some its properties*, Biometrics, 27, 857-871.

Ponsard, C. (1985). Fuzzy data analysis in a spatial contest, in P. Nijkamp et al., *Measuring the unmensurable*, Martinus Nijoff Publishers, Dordrecht.

Ricolfi, L. (1992). Helga - *Nuovi pricipi di analisi dei gruppi*, FrancoAngeli, Milano.

A New Algorithm for Semi-Fuzzy Clustering

Giampaolo Iacovacci
INA, Via Sallustiana 52, 00187 Rome, e-mail: yaco@writeme.com

Abstract: This paper presents a new algorithm for semi-fuzzy clustering that allows objects to belong not necessarily to all the clusters, but also to only one of them. The advantage of this new method is that fuzziness is not introduced for all objects but only for those that cannot be classified as belonging to a single cluster. The performance of the new algorithm compared to the fuzzy c-means algorithm is showed by an application on a data set.

Key words: Fuzzy *c*-means algorithm, semi-fuzzy classification, soft clustering.

1. Introduction

In order to solve a clustering problem, it is necessary to distinguish between two types of data clustering. Traditional "hard" clustering, in which objects are allowed to belong to only one cluster, and "fuzzy" clustering in which each object belongs to all available clusters with different grades of membership.
During the last few years many fuzzy clustering methods have been developed because the nature of most practical problems is “fuzzy”, that is so complex and diversified that it can hardly be interpreted using hard clustering.
The problem with many fuzzy clustering algorithms is that they often give excessively fuzzy classifications which are difficult to interpret. This is mainly due to the fact that each object must belong to all the clusters.
In order to solve this problem, some “semi-fuzzy” clustering methods have recently been proposed which allow objects to belong to several clusters with various grades of membership, but not necessarily to all clusters (Selim and Ismail 1984 and Kamel and Selim 1991).
In the following paragraph a new semi-fuzzy clustering algorithm is proposed which, by modifying the fuzzy *c*-means method (Bezdek 1981), appears to provide better results without altering its main properties.
Section 3 gives an example in which the performances of the two methods are compared.

2. The semi-fuzzy *c*-means method

One of the most well-known fuzzy clustering algorithms is the fuzzy *c*-means (FCM) method. It is appreciated for some properties (Bezdek and Hathaway

1988) and is particularly suitable for applications on large data set (Iacovacci, 1995).

One of the chief defects of the FCM method lies in the fact that, given the number c of clusters in which n objects must be classified, it is necessary to determine each object's grade of membership of each of the c clusters so that it is greater than zero. It is thus not permissible for the relationship between a object and one or more clusters to be nil, except in the (very rare) case in which a object coincides with the center of a cluster, as in this case it is assigned to that cluster only.

The chief consequence of this characteristic of the FCM method is that, at times, it produces classifications that are excessively fuzzy, i.e. whose overall fuzziness is far greater than that existing in reality.

To reduce this disadvantage, a semi-fuzzy c-means (SFCM) method is proposed that allows not only the objects coinciding with the center of a cluster, but also other objects having given characteristics to be assigned entirely to a single cluster.

This method proposes that object i be assigned totally to cluster k if, taking d_{ik} to indicate the distance between them, $d_{ik} < (1/\alpha)\ d_{rk}$ where α is a parameter (>1) determined a priori, and d_{rk} indicates the distance between the center of the k-th cluster and the center of the r-th cluster nearest to it.

This formulation meets the intuitive need to classify as totally belonging to a cluster any object whose distance from the center of the said cluster is not only reasonably small, but which is at the same time sufficiently far away from all the other clusters.

Accordingly, the SFCM method determines for each cluster a zone of gravitational attraction (determined as a function of the other clusters) and assigns all the objects located inside this zone to the said cluster.

Let n be the total number of objects, c be the number of clusters and μ_{ik} be the grades of membership of object i to cluster k. The final classification is obtained using the same iterative process followed by the FCM method, with a further constraint. It calculates the optimal centers of the c clusters and the optimal values of the grades of membership μ_{ik} that minimize the following function:

$$J_m(\mathbf{U},\mathbf{V}) = \sum_{k=1}^{c}\sum_{i=1}^{n} \mu_{ik}^m \left\| \mathbf{x}_i - \mathbf{v}_k \right\|^2 \qquad (1)$$

subject to:

$$\mu_{ik} \geq 0 \qquad i=1,\ldots,n \quad k=1,\ldots,c \qquad (2)$$

$$\sum_{k=1}^{c} \mu_{ik} = 1 \qquad i=1,\ldots,n \qquad (3)$$

$$\mu_{ik}=1 \quad \text{if } \left\|\mathbf{x}_i - \mathbf{v}_k\right\| < 1/\alpha \left\|\mathbf{v}_r - \mathbf{v}_k\right\| \qquad i=1,\ldots,n \quad r,k=1,\ldots,c \tag{4}$$

where $\| \cdot \|$ is an appropriate norm on $\mathbf{R}^p$; $\mathbf{x}_i \in \mathbf{R}^p$ is the vector of the i-th object; $\mathbf{v}_k \in \mathbf{R}^p$ is the vector of the center of the k-th cluster and $\mathbf{v}_r \in \mathbf{R}^p$ is the vector of the center of the r-th cluster; m is a scalar, $m > 1$; α is a scalar, $\alpha > 1$; $\mathbf{U} = \{\mu_{ik}\}$ is the matrix (n x c) of the grades of membership; $\mathbf{V} = [\mathbf{v}_1,..,\mathbf{v}_c]$ is the matrix (p x c) of the c centers of the clusters.

The parameter m that appears in (1) is particularly important, since with m=1 we obtain a hard classification (and the FCM algorithm coincides with the traditional c-means algorithm), whereas with $m > 1$ a classification is obtained whose fuzziness increases as the value of m rises.

The algorithm that describes the SFCM method is as follows:

Step 1: choose the value of m, α, and a small positive scalar δ. Select an arbitrary membership matrix $\mathbf{U} = \{\mu_{ik}\}$.

Step 2: calculate cluster centers using the following formula:

$$\mathbf{v}_k = \sum_{i=1}^{n} \left(\mu_{ik}\right)^m \mathbf{x}_i \Big/ \sum_{i=1}^{n} \left(\mu_{ik}\right)^m \qquad k=1,\ldots,c$$

Step 3: compute the distance matrix $\mathbf{D}_1=\{d_{ik}\}$, where $d_{ik}=\left\|\mathbf{x}_i - \mathbf{v}_k\right\|$ ($i=1,\ldots,n$; $k=1,\ldots,c$) and the distance matrix $\mathbf{D}_2=\{d_{rk}\}$, where $d_{rk}=\left\|\mathbf{v}_r - \mathbf{v}_k\right\|$ ($r=1,\ldots,c$; $k=1,\ldots,c$).

Step 4: update the membership matrix **U** computing the matrix $\tilde{\mathbf{U}}$
for each $i=1,\ldots,n$ and $k=1,\ldots,c$ **do**
if $d_{ir}=0$ **then** $\mu_{ir}=1$ and $\mu_{ik}=0$ for all $k \neq r$
else **if** $d_{ir} < (1/\alpha)\, d_{rk}$ for some r, with $k \neq r$, **then** $\mu_{ir}=1$ and $\mu_{ik}=0$ for all $k \neq r$
else compute

$$\mu_{ik} = 1 \Big/ \sum_{j=1}^{c} \left(d_{ik}/d_{jk}\right)^{2/(m-1)}$$

Step 5: if $\left|\tilde{\mathbf{U}} - \mathbf{U}\right| < \delta$ **then** stop; **else goto** step 2.

The main advantage of the SFCM algorithm is that it produces a mix between a "hard" classification and a "fuzzy" classification, thus introducing the fuzziness not for all objects, but only in those in which the fuzziness is actually found.

The SFCM method and the FCM method are compared in the following example.

3. Experimental results

To compare the performance of FCM algorithm with that of SFCM algorithm, both algorithms have been applied to the data set composed of 22 objects shown in Figure 1 below.

Figure 1: *Example data set.*

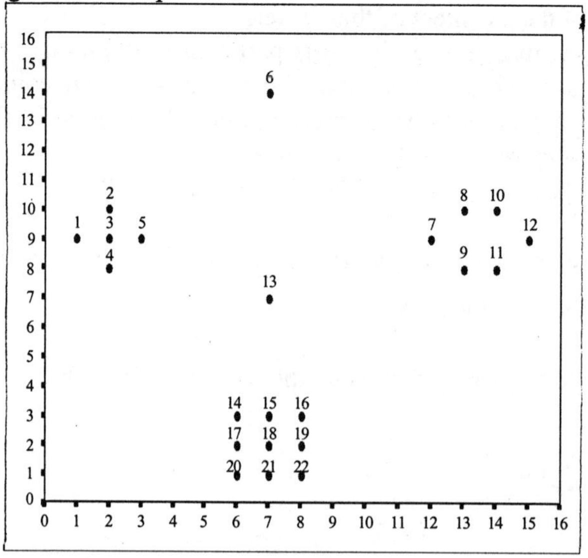

It can be seen from Figure 1 that objects of data set are distributed in three separate and distinct clusters, except for objects 6 and 13, which are located in an intermediate position in relation to these clusters (note the position of object 13 in particular). The groups are: cluster 1: object 1-5; cluster 2: object 7-12; cluster 3: object 14-22.
These three clusters are clearly identified by applying to the data the FCM algorithm with c=3 and m=1. Choosing m=1, we obtain the hard classification reported in Table 1.

Table 1: *Result of FCM algorithm* (δ=0.001, c=3, m=1).

	Grades of membership				grades of membership		
Object	μ_{i1}	μ_{i2}	μ_{i3}	Object	μ_{i1}	μ_{i2}	μ_{i3}
1	1.00	0.00	0.00	12	0.00	1.00	0.00
2	1.00	0.00	0.00	13	1.00	0.00	0.00
3	1.00	0.00	0.00	14	0.00	0.00	1.00
4	1.00	0.00	0.00	15	0.00	0.00	1.00
5	1.00	0.00	0.00	16	0.00	0.00	1.00
6	1.00	0.00	0.00	17	0.00	0.00	1.00
7	0.00	1.00	0.00	18	0.00	0.00	1.00
8	0.00	1.00	0.00	19	0.00	0.00	1.00
9	0.00	1.00	0.00	20	0.00	0.00	1.00
10	0.00	1.00	0.00	21	0.00	0.00	1.00
11	0.00	1.00	0.00	22	0.00	0.00	1.00

We just need to look at the classification in Table 1 to see that hard clustering is unsuitable to describe the fuzzy situation of objects 6 and 13 which are both assigned to cluster 1.
Using the FCM method with various values of $m>1$, it is with $m = 2.7$ that we obtain the classification that seems best able to represent the situation, as objects 6 and 13 are assigned to the three clusters to different extents depending on their greater or lesser nearness to them, and the remaining objects have a high grade of membership of the clusters, as they are located very near to their centers. The classification is reported in Table 2.

Table 2: *Result of FCM algorithm* (δ=0.001, c=3, m=2.7).

	grades of membership				grades of membership		
Object	μ_{i1}	μ_{i2}	μ_{i3}	Object	μ_{i1}	μ_{i2}	μ_{i3}
1	0.86	0.06	0.08	12	0.07	0.83	0.10
2	0.89	0.05	0.06	13	0.36	0.26	0.38
3	0.97	0.01	0.02	14	0.11	0.08	0.81
4	0.85	0.06	0.09	15	0.07	0.06	0.87
5	0.90	0.04	0.06	16	0.09	0.09	0.82
6	0.43	0.35	0.22	17	0.08	0.06	0.86
7	0.08	0.82	0.10	18	0.00	0.00	1.00
8	0.05	0.89	0.06	19	0.06	0.07	0.87
9	0.06	0.86	0.08	20	0.10	0.08	0.82
10	0.06	0.88	0.06	21	0.07	0.06	0.87
11	0.06	0.86	0.08	22	0.08	0.09	0.83

It is necessary to point out that the result of the FCM method too is not completely satisfactory. As a matter of fact, in order to improve the classification of fuzzy objects 6 and 13, the FCM algorithm introduces fuzziness for all other objects with the exception of object 18 which coincides with the center of the cluster. None of the other objects which nevertheless clearly belong to a single cluster obtain the maximum grade of membership value. To obtain a good classification of these objects, it is necessary to take m with a value near 1, but in this case objects 6 and 13 tend to belong only to cluster 1. This type of disadvantage is not present in the SFCM method which, using the same value $m = 2.7$ and $\alpha = 3$, gives the result shown in Table 3.

Table 3: *Result of SFCM algorithm* (δ=0.001, c=3, m=2.7, α=3).

	Grades of membership				grades of membership		
Object	μ_{i1}	μ_{i2}	μ_{i3}	Object	μ_{i1}	μ_{i2}	μ_{i3}
1	1.00	0.00	0.00	12	0.00	1.00	0.00
2	1.00	0.00	0.00	13	0.36	0.26	0.38
3	1.00	0.00	0.00	14	0.00	0.00	1.00
4	1.00	0.00	0.00	15	0.00	0.00	1.00
5	1.00	0.00	0.00	16	0.00	0.00	1.00
6	0.43	0.35	0.22	17	0.00	0.00	1.00
7	0.00	1.00	0.00	18	0.00	0.00	1.00
8	0.00	1.00	0.00	19	0.00	0.00	1.00
9	0.00	1.00	0.00	20	0.00	0.00	1.00
10	0.00	1.00	0.00	21	0.00	0.00	1.00
11	0.00	1.00	0.00	22	0.00	0.00	1.00

This classification seems to be the best to represent the real classification of the objects, as it does not necessarily attribute fuzziness to all of them, but only to objects 6 and 13, i.e. to those objects whose positions are really uncertain in relation to the various clusters.
It is important to point out that both the FCM and SFCM algorithms are apparently independent of the choice of the initial matrix **U**. In both applications, for a fixed value of m and α, the initial matrix **U** has been changed many times achieving the same result.
Lastly, for testing the performances of the SFCM algorithm on a real case, this method has been used to grade Italy's communes (municipalities) according to their degree of urban/rural characteristics. The resulting classification, which has not been included here for reasons of space, appears easier to read and more immediately comprehensible when compared with the analogous classification obtained by the FCM method. Moreover, it corresponds more to reality than the official classification of ISTAT (ISTAT, 1986) obtained using the traditional *c*-means method.
Many initial matrix **U** were randomly generated to test the stability of the solution obtained. The result was always the same final classification.

References

Bezdek J. C. (1981). *Pattern recognition with fuzzy objective function algorithms*, Plenum, New York.

Bezdek J. C. and Hathaway R. J. (1988). Recent convergence results for the fuzzy *c*-means clustering algorithms, in: *Journal of Classification*, Vol. 5, 237-247.

Iacovacci G. (1995). Sull'utilizzo del metodo delle *c*-medie sfocato per la classificazione dei comuni italiani a seconda del grado di urbanità e ruralità, in: *Statistica Applicata*, Vol. 7, n°1, 33-48.

Istat (1986). Classificazione dei comuni secondo le caratteristiche urbane e rurali, *Note e relazioni*, n° 2, Istat, Roma.

Selim S. Z. and Ismail M. A. (1984). Soft clustering of multidimensional data: a semi-fuzzy approach, in: *Pattern Recognition*, Vol. 17, n° 5, 559-568.

Selim S. Z. and Kamel M. S. (1991). A thresholded fuzzy *c*-means algorithm for semi-fuzzy clustering, in: *Pattern Recognition*, Vol. 24, n° 9, 825-833.

Fuzzy Classification and Hyperstructures: An Application to Evaluation of Urban Projects

Antonio Maturo, Barbara Ferri
Dipartimento di Scienze, Storia dell'Architettura e Restauro,
viale Pindaro, 42 - Pescara, Italy -

Abstract: In this paper we have considered the possibility of applying the theories of fuzzy sets and algebraic hyperstructures to feasibility evaluation of urban new qualification projects. We have studied a mathematical model to help detect whether to verify the validity of some choices during each projectual stage, or to single out the best project among a series of possible alternatives by estimating the measure in which the projects attain the economical, psycological, cultural, technological objectives.

Keywords: Fuzzy Methods. Hyperstructures. Evaluation Urban Projects.

1. Evaluation need for urban projects

A prevalent kind of urban planning consists in an "*integrated intervention strategy*" in order to revive the city on the whole, by promoting economic activities through recovery and re-qualification of the existent resources in the urban territory. Preservation no longer involves the single monument, but the whole building tissue; this fact requires the need to achieve various and heterogeneous objectives. So we deem it necessary to accept right procedures for the evaluation of projects, in order to obtain information about the "*feasibility*" of the foreseen works.

The last National Congress of Architects (Florence, March 1997) showed a great interest for a planning directed to quality of urban space and environment. This quality has to be pursued especially through a correct procedure of evaluation and comparison between some projects.

Suitable instruments of feasibility evaluation are indispensable especially at the stages of preliminary planning or programming, when necessary to establish the required amount of resources. These evaluation procedures of projects have to aim at:

1. justifying the destination of the available financial resources to preservation in comparison with other investments;
2. making a selection of projects by estimating the measure in which a program of "integrated maintenance" improves the level of welfare;
3. raising this measure by modifying and integrating the tested project.

At present, the *fuzzy sets* theory is considered useful to minimize the uncertainty about the influence of a decision on the validity of projects.

2. Features of the model

The question about urban re-qualification has to be framed into a general view in which there aren't only economic needs. Others important demands are: psychological, defense of historical-artistic values and of aesthetically visual qualities of areas, improvement of the standard of urbanization, removal of social degradation of areas. It is necessary to attain a "global" or "integrated" evaluation of projects, that is an evaluation inclusive of various essential demands simultaneously.
Therefore we have deemed it necessary the use of "*Multicriteria*" techniques to render the evaluation "all-inclusive". This is seen as the work to express *the measure of attainment of the various objectives by peculiarity of the project.* Our model, in fact, may be represented by a matrix that, in our case, permits to compare the various impacts (*feasibility* categories) with a series of objectives (*criteria*). These ones reflect the needs or the requirements (the aims) of the community with regard to each category.
We deem it necessary to consider the following feasibility categories for an urban project: feasibility *on environment* (evaluation of the project on the effects on natural and built ambient), *aesthetic-cultural* (evaluation of the project on the effects on the historical, artistic and archaeological interests), *economical* (evaluation of the project to verify the increase of general welfare in the considered area), *financial* (evaluation of the project to control the measure of costs-return ratio), *technical* (evaluation of the project on the effects on constraint of construction and regulation), *social* (evaluation of the project to verify the attainment of objects concerning the directly and indirectly involved men), *procedural* (evaluation of project from the contractual point of view and on carrying out).
The feasibility categories or objects may be opportunely weighed so as to consider the importance of the various needs. The weight of each "feasibility" represents the respective importance within the limits of a predefined typology of project with pre-established priorities; whereas, the weight of each "object" represents the degree of importance of the different aims. The assignment of weights has to be made by listening to all of the involved men's opinion: promoters, planners, politicians,.....The "*effectiveness*" of the project is valued with regards to each object included in each category; in other words, we establish the measure in which the tested project attain each object to every single category.

3. A fuzzy set mathematical model to compare projects

Schimpeler and Grecco (1968) consider a particular multicriteria method to estimate urban projects. We summarize this method using our symbology.

(a) We consider a set Ω, called *objective set* and a function W: $\Omega \to [0,1]$, called *weight function*. $\forall \omega \in \Omega$, $W(\omega)$ is the *weight* of the criterion ω with regards to the model to estimate the benefits of the projects. The normalization condition $\sum_{\omega \in \Omega} W(\omega) = 1$ is assumed.

(b) Let $\mathcal{P}$ be the family of the examined projects. For any $P \in \mathcal{P}$, we consider a function E_P: $\Omega \to [0,1]$, called *efficaciousness function* of P. $\forall \omega \in \Omega$, $E_P(\omega)$ is the *grade* in which P satisfies the objective ω. The function $T_P = W\,E_P$ is called *characteristic function* of P.

(c) The function $U{:}P \in \mathcal{P} \to \sum_{\omega \in \Omega} T_P(\omega) \in \mathbb{R}$, called *evaluation or utility function*, measures the global utility of the projects of $\mathcal{P}$.

In this paper we consider some modifications to the previous model. As set Ω of objectives we assume the Cartesian product $F \times C$, where F, called *set of feasibilities*, is the set of the feasibility categories for an urban project and C, called *set of criteria*, is the set of the different points of view to measure the grade in which each project attains any feasibility. The weight function is a matrix independent on the projects considered and, for any $P \in \mathcal{P}$, the efficaciousness function is also a matrix dependent on P. Besides

- we consider the function $\varphi{:}x \in F \to \sum_{c \in C} W(x,c) \in [0,1]$, called *feasibility weight function* and the function $\gamma{:}c \in C \to \sum_{x \in F} W(x,c) \in [0,1]$, called *criterion weight function*. We have the conditions $\sum_{x \in F} \varphi(x) = 1$ and $\sum_{c \in C} \gamma(c) = 1$;
- we fix a *threshold function* S: $F \times C \to [0,1]$ such that $\forall (x,y) \in F \times C$, $S(x,c) \leq W(x,c)$.

Therefore we explain the meanings of all concepts of our model in the fuzzy set theory. Besides we show that many interesting features of the model can be studied with the theory of hyperstructures. For these aims, we give some definitions.

Definition 3.1. Let Ω be a non empty set. A fuzzy set on Ω is a function f: $\Omega \to [0,1]$. For any $x \in \Omega$, f(x) is the *membership grade* of x to f.

In the sequel we denote by C a finite non-empty family of fuzzy sets on a set F. For any $x \in F$, the number $gr_C(x) = \sum_{c \in C} c(x)$ is called *membership grade* of x to C.

Definition 3.2. Let C and C* be two families of fuzzy sets on F and let ψ:C→C* a bijection. Put $c^*=\psi(c), \forall c \in C$. The fuzzy set $T_{c^*}: x \in F \to c(x)c^*(x)$ and the family of fuzzy sets $T_{C^*}=\{T_{c^*}\}_{c \in C}$ are called, respectively, *transmitted component* of c and of C. The pair (C^*,ψ) is called *filter* of C.

By previous definitions we have the following interpretations:

- the weight function W is a family of fuzzy sets on F;
- the pair (E_P, ψ), with E_P efficaciousness function of P and ψ function that to any column of W associates the column of E_P with the same index, is a filter of W and the transmitted component of W is the characteristic function $T_P=WE_P$.

4. An example about the application of the model

The described model has been used to evaluate the social, economical, environmental, aesthetic-cultural feasibility about a project of widening on the main road n. 17, in the stretch connecting Poggio Picenze to Navelli, two towns in province of L'Aquila.

The aim of the study has consisted in defining the measure in which an investment in that transport infrastructure was really feasible and advantageous from every points of view, given a system of constraints and objectives to achieve.

This analysis gains importance because it is necessary to know the effects of accessibility on territorial growth about the optimization of housing, employment, easy access to public services.

For a project in the sector of road system, the proposable feasibility categories, or *objectives*, are:

O_1: benefit for road users with regard to the investment;
O_2: positive influence on urban environment;
O_3: positive influence on natural environment;
O_4: territorial innovation;
O_5: correspondence to territorial problems.

Given the objectives, the specific criterions are:

C_1: benefit for the usual users, in terms of improvement in the quality of transport, that is a higher speed on the roads, less traffic congestion, reduction in travelling time, reduction in risk, reduction in discomfort;
C_2: benefit for users induced to move by the reduction of transport charges in consequence of the improvement for the road;
C_3: influence on urban landscape;
C_4: increase of accessibility to the central zones;
C_5: increase of enjoyment from the central zones;
C_6: fluctuation in pollution;
C_7: extent of vegetation prejudiced by the realization of project;
C_8: extent of damage to landscape in terms of gain or loss of visual amenity;
C_9: break produced in natural environment endowed with unity;

C_{10}: new residential and productive units located in consequence of the urban program raised by the project on transport system

C_{11}: multiple and balanced characteristic of the new planned settlements, on the basis of the multiplicity and organicity of the settled functions;

C_{12}: property of the road network after carrying out the project (property of the network to foster or to oppose a polycentric layout, with satisfactorily scattered settlement on territory);

C_{13}: lack of transport infrastructure;

C_{14}: unsatisfactory economic development;

C_{15}: compatibility of the project to the territorial planning instruments.

The related case of study is an exemplification of the mathematical model; indeed we assume only two alternatives: the situation without intervention, called "null hypothesis", and the proposed project.

Through a close examination of this case, by consulting texts about evaluation methods of public investments in transport, by listening technical opinions, by making inquires about streams of traffic and the variation of travelling time in the various examined sections of road, by estimating the fluctuation of transport charges in the situations "with" and "without" intervention, and by processing opportunely the other available data we have obtained the following *weight matrix* W and the following *efficaciousness matrices* E_p and E_0, respectively for the proposed project and for the null hypothesis; so we have

$$100W = \begin{array}{c} \\ O_1 \\ O_2 \\ O_3 \\ O_4 \\ O_5 \\ \\ \end{array} \begin{array}{ccccccccccccccc|c} C_{01} & C_{02} & C_{03} & C_{04} & C_{05} & C_{06} & C_{07} & C_{08} & C_{09} & C_{10} & C_{11} & C_{12} & C_{13} & C_{14} & C_{15} & \\ 16 & 14 & - & - & - & - & - & - & - & - & - & - & - & - & - & 30 \\ - & - & 6 & 8 & 4 & 2 & - & - & - & - & - & - & - & - & - & 20 \\ - & - & - & - & - & 1 & 3 & 3 & 3 & - & - & - & - & - & - & 10 \\ - & - & - & - & - & - & - & - & - & 9 & 6 & 15 & - & - & - & 30 \\ - & - & - & - & - & - & - & - & - & - & - & - & 3 & 5 & 2 & 10 \\ 16 & 14 & 6 & 8 & 4 & 3 & 3 & 3 & 3 & 9 & 6 & 15 & 3 & 5 & 2 & 100 \end{array}$$

$$10E_P = \begin{array}{c} \\ O_1 \\ O_2 \\ O_3 \\ O_4 \\ O_5 \\ \\ \end{array} \begin{array}{ccccccccccccccc|c} C_{01} & C_{02} & C_{03} & C_{04} & C_{05} & C_{06} & C_{07} & C_{08} & C_{09} & C_{10} & C_{11} & C_{12} & C_{13} & C_{14} & C_{15} & \\ 8 & 7 & - & - & - & - & - & - & - & - & - & - & - & - & - & 15 \\ - & - & 8 & 10 & 7 & 6 & - & - & - & - & - & - & - & - & - & 31 \\ - & - & - & - & - & 6 & 6 & 6 & 7 & - & - & - & - & - & - & 25 \\ - & - & - & - & - & - & - & - & - & 8 & 7 & 8 & - & - & - & 23 \\ - & - & - & - & - & - & - & - & - & - & - & - & 8 & 7 & 8 & 23 \\ 16 & 14 & 6 & 8 & 4 & 3 & 3 & 3 & 3 & 9 & 6 & 15 & 3 & 5 & 2 & 117 \end{array}$$

$$10E_0 = \begin{array}{c} \\ O_1 \\ O_2 \\ O_3 \\ O_4 \\ O_5 \\ \\ \end{array} \begin{array}{ccccccccccccccc|c} C_{01} & C_{02} & C_{03} & C_{04} & C_{05} & C_{06} & C_{07} & C_{08} & C_{09} & C_{10} & C_{11} & C_{12} & C_{13} & C_{14} & C_{15} & \\ 4 & - & - & - & - & - & - & - & - & - & - & - & - & - & - & 4 \\ - & - & 5 & - & 5 & - & - & - & - & - & - & - & - & - & - & 10 \\ - & - & - & - & - & - & - & 6 & 5 & - & - & - & - & - & - & 11 \\ - & - & - & - & - & - & - & - & - & - & 4 & - & - & - & - & 4 \\ - & - & - & - & - & - & - & - & - & - & - & - & 5 & 5 & - & 10 \\ 4 & - & 5 & - & 5 & - & - & 6 & 5 & - & 4 & - & 5 & 5 & - & 39 \end{array}$$

Then if T_P and T_0 are, respectively, the *characteristic matrices* of the two alternatives, we have

$1000T_P$ =

	C_{01}	C_{02}	C_{03}	C_{04}	C_{05}	C_{06}	C_{07}	C_{08}	C_{09}	C_{10}	C_{11}	C_{12}	C_{13}	C_{14}	C_{15}	
O_1	128	98	–	–	–	–	–	–	–	–	–	–	–	–	–	226
O_2	–	–	48	80	28	12	–	–	–	–	–	–	–	–	–	168
O_3	–	–	–	–	–	6	18	18	21	–	–	–	–	–	–	63
O_4	–	–	–	–	–	–	–	–	–	72	42	120	–	–	–	234
O_5	–	–	–	–	–	–	–	–	–	–	–	–	24	35	16	75
	128	98	48	80	28	18	18	18	21	72	42	120	24	35	16	766

$1000T_0$ =

	C_{01}	C_{02}	C_{03}	C_{04}	C_{05}	C_{06}	C_{07}	C_{08}	C_{09}	C_{10}	C_{11}	C_{12}	C_{13}	C_{14}	C_{15}	
O_1	64	–	–	–	–	–	–	–	–	–	–	–	–	–	–	64
O_2	–	–	30	–	20	–	–	–	–	–	–	–	–	–	–	50
O_3	–	–	–	–	–	–	–	18	15	–	–	–	–	–	–	33
O_4	–	–	–	–	–	–	–	–	–	–	24	–	–	–	–	24
O_5	–	–	–	–	–	–	–	–	–	–	–	–	15	25	–	40
	64	–	30	–	20	–	–	18	15	–	24	–	15	25	–	211

The component $U_{i.}$, i=1,...,5 of the marginal column of T_P or T_0, is the *partial utility* of the feasibility O_i. Therefore the component $U_{.j}$, j=1,...15 of the marginal row is the *partial utility* of the criterion C_j. The sum of the $U_{i.}$, equal to the sum the $U_{.j}$, is the *global utility* of the project.
We can observe that, in this example, the proposed project is very preferable to the null hypothesis, even if in different measure for each feasibility or criterion.

5. On some hyperstructures associated to the model

We assume the definitions and the terminology given in (Corsini, 1995), (Vougiouklis, 1994), (Migliorato, 1994), (Maturo, 1997).

Definition 5.1 A **hypergroupoid**. (S, α), is a non empty set S with a function $\alpha: S\times S \rightarrow P^*(S)=P^*(S)-\{\varnothing\}$, called hyperoperation. The image of the pair (x, y) is noted $x\alpha y$ and is called **hyperproduct** by x and y.

For any pair (H,K) of subsets of S different from ∅, we denote by $H\alpha K$ the union of all the sets $x\alpha y$ with $x\in H$ and $y\in K$. The hyperproducts $a\alpha K$ and $H\alpha a$, $a\in S$, are considered equal, respectively, to $\{a\}\alpha K$ and $H\alpha\{a\}$.
$\forall n\in N$ and $\forall(x_1,x_2,..x_n)\in S^n$, the set $\Im_\alpha(x_1,x_2,..x_n)$ of all the hyperproducts generated by $(x_1,x_2,..x_n)$ is given, by induction, as follows: $\Im_\alpha(x_1)=\{x_1\}$ and, for n>1, $\Im_\alpha(x_1,x_2,..x_n)$ is the set of all the hyperproducts $K=F\alpha G$, with $F=\{x_1,x_2,..x_h\}$, $G=\{x_{h+1},x_2,..x_n\}$, $h\in\{1,2,..,n-1\}$. Any element of $\Im_\alpha(x_1,x_2,..x_n)$ is called **block** of S generated by $(x_1,x_2,..x_n)$.

Definition 5.2 A hypergroupoid (S, α) is said to be
(H1) **semihypergroup** if, $\forall x,y,z\in S$, $x\alpha(y\alpha z)=(x\alpha y)\alpha z$;
(H2) **quasihypergroup** if, $\forall x\in S$, $x\alpha S=S\alpha x=S$;
(H3) **commutative** if, $\forall x,y\in S$, $x\alpha y=y\alpha x$;

(H4) **hypergroup** if it is a semihypergroup and a quasihypergroup.

Definition 5.3 Let (S, α) be a hypergroupoid. The pair (S, B), with B set of all the blocks of S, is called **geometric space** associated to (S, α). A **polygonal** of length 1 of B is a block and a **polygonal** of length m>1 is a m-tuple $(K_1, K_2,..,K_m)$ of blocks such that $K_i \cap K_{i+1} \neq \emptyset$, $\forall i \in \{1,2,..,m-1\}$.

Definition 5.4 A hypergroupoid (S, α) is said to be
(W1) **weak commutative** if, $\forall x,y \in S$, $x\alpha y \cap y\alpha x \neq \emptyset$;
(W2) **weak associative** if, $\forall x,y,z \in S$, $x\alpha(y\alpha z) \cap (x\alpha y)\alpha z \neq \emptyset$;
(W3) **feebly associative** if, $\forall n \in N$ and $\forall (x_1,x_2,..x_n) \in S^n$, the intersection of all the blocks belonging to $\Im_\alpha(x_1,x_2,..x_n)$ is different from $\emptyset$.

Definition 5.5 Let (S, α) be a hypergroupoid and let $\emptyset \neq T \subseteq S$. (T, α) is called a **subhypergroupoid** of S if, $\forall x,y \in T$, $x\alpha y \subseteq T$. A subhypergroupoid T of S is called **subhypergroup** if (T, α) is a hypergroup.

Let $\emptyset \neq F^* \subseteq F$, $\emptyset \neq C^* \subseteq C$, $K = F^* \times C^*$. For any $P,Q \in$ P, we put
$P \angle_K Q \Leftrightarrow \forall (x,c) \in K, T_P(x,c) \leq T_Q(x,c)$, $\quad P \approx_K Q \Leftrightarrow P \angle_K Q$ and $Q \angle_K P$.
The $\angle_K$ is a preorder relation and the $\approx_K$ is an equivalence relation.

In the sequel we assume that there exist three ideal projects P_W, P_S and P_O such that T_W, T_S and T_O are respectively, the weight function W, the threshold function S and the null function O. So, for any $P,Q \in$ P, we have some pairs (A,B) of projects such that $A \angle_K P$, $A \angle_K Q$, $P \angle_K B$, $Q \angle_K B$.

Then we can define, for any $P,Q \in$ P, the following hyperproducts:
$P\delta_K Q = \{A \in$ P: $A \angle_K P$, $A \angle_K Q$ and $A \angle_K B \angle_K P$, $A \angle_K B \angle_K Q$ if and only if $A \approx_K B\}$;
$P\sigma_K Q = \{A \in$ P: $P \angle_K A$, $Q \angle_K A$ and $P \angle_K B \angle_K A$, $Q \angle_K B \angle_K A$ if and only if $A \approx_K B\}$.

Let $I_n = \{1,...,n\}$. We can prove the following

Theorem 5.1 For any n-tuple $(P_1,...,P_n)$ of projects, let $\delta_K(P_1,...,P_n) = \{A \in$ P: $A \angle_K P_i$, $\forall i \in I_n$, and $(A \angle_K B \angle_K P_i, \forall i \in I_n) \Leftrightarrow A \approx_K B\}$ and let $\sigma_K(P_1,...,P_n) = \{A \in$ P: $P_i \angle_K A$, $\forall i \in I_n$ and $(P_i \angle_K B \angle_K A, \forall i \in I_n) \Leftrightarrow A \approx_K B\}$. The intersection of all the blocks of $\Im_{\delta K}(P_1,...,P_n)$ contains $\delta_K(P_1, P_2,...,P_n)$ and the intersection of all the blocks of $\Im_{\sigma K}(P_1,...,P_n)$ contains $\sigma_K(P_1, ...,P_n)$.

So we have the following

Theorem 5.2 The hypergroupoids (P,δ_K) and (P,σ_K) are commutative and feebly associative.

We can prove, by examples, that in general (P,δ_K) and (P,σ_K) are not associative.

Let P^+ the set of all the projects P such that $P_S \angle_K P$. We have the

Theorem 5.3 P^+ is closed under both the hyperoperations δ_K and σ_K and so also (P^+,δ_K) and (P^+,σ_K) are commutative and feebly associative hypergroupoids.

The hyperoperations δ_K and σ_K have some interesting interpretations. $P\delta_K Q$ is the set of the better projects than all ones of lower quality with respect to P and Q, relative to the set K. They can be considered, for example, because of some economic necessities. $P\sigma_K Q$ is the set of the minimal projects that have all the good properties of both P and Q, with respect to K.
Theorem 5.1 implies that the intersection of all the hyperproducts δ_K or σ_K of a fixed number of projects is not empty. The projects belonging to such intersections are the ones of greatest interest. We can prove that such intersections may be different, respectively, from $\delta_K(P_1,...,P_n)$ and $\sigma_K(P_1,...,P_n)$.

References

Corsini, P. (1995). *Prolegomena of hypergroup theory*, Aviani Ed., Udine.

Doria, S. & Maturo, A. (1996). Hyperstructures and geometric spaces associated to a family of events, *Rivista di Matematica pura ed applicata*, Udine, n. 19, to appear.

Ferri, B. & Maturo, A. (1996). Un modello matematico per la valutazione di fattibilità dei Programmi Integrati d'intervento urbano, *Ratio Mathematica*, 10, 67-84.

Fusco Girard, L. (1987). *Risorse architettoniche e culturali: valutazioni e strategie di conservazione*, Franco Angeli, Milano.

Klir, G. J. & Yuan, B. (1995). *Fuzzy sets and fuzzy logic*, Prentice Hall.

Maturo, A. (1997). On some hyperstructures of conditional events, *Proceedings of Conference on Algebraic Hyperstructures and Applications '96*, Prague, September 1-9, 1996, pp.115-132.

Miccoli, S. (1995). La valutazione di fattibilità nei programmi complessi d'intervento urbano, *Genio Rurale*, 3.

Migliorato, R. (1994). Some topics on the feebly associative hypergroupoids, *Proceedings of the Conference on Algebraic Hyperstructures and Applications '96*, pp.67-80, Jasi, Rumania.

Schimpeler, C. & W. Grecco, W. (1968). The community: an approach based on community structure and values, *Highway Research Record* 238.

Vougiouklis, T. (1994). Hyperstructures and their representations, Hadronic Press.

Zadeth, A. (1965). Fuzzy Sets, *Inform. Contr.*, 8, 1965.

Variable Selection In Fuzzy Clustering

Maria Adele Milioli

Istituto di Statistica, Facoltà di Economia, Univ. of Parma, Italy

Abstract: The aim of the present paper is to discuss methods for selecting a subset of initially observed variables in the context of fuzzy clustering. The suggested procedure is based on the optimization of an objective function which is differently specified according to the purpose of the selection. Measure of cluster validity, a generalization of Rand index and distance between dissimilarity matrices are then proposed as proper functions to optimize.

Keywords: variable selection, fuzzy clustering, measures of fuzziness, cluster validity.

1. Introduction

In cluster analysis, the importance of variable selection is well recognized. The problem is to select a subset of initially observed variables that would account for cluster pattern present in the data. Solutions to this problem are well known for classical cluster analysis (Fowlkes *et al.*, 1988). The aim of the present paper is the generalization to the context of fuzzy clustering. In particular, two cases will be considered: a) the purpose of the selection is to identify and eliminate the variables that may only constitute a "noise" that masks clear cluster structure delineated by other variables; b) the need of selecting an appropriate subset is strictly a problem of reduction of dimensionality: the purpose is to eliminate redundant variables, i.e. to find a reduced subset of variables which reproduce the cluster structure generated by the complete set of the original ones "as well as possible".

2. Fuzzy clustering

Fuzzy partitions can be defined as follows (Bezdek 1981):

Definition 1 - *A G-fuzzy partition of a set of* n *units is a* (nxG) *matrix* $(1 < G < n)$

$$U = (u_{ig}) \tag{1}$$

where u_{ig} *is the value of the membership function of the* $i-th$ *units to the* $g-th$ *cluster* $(i = 1, 2, ..., n; g = 1, 2, ..., G)$ *with the following constraints:*

$$0 \leq u_{ig} \leq 1 \text{ and } \sum_g u_{ig} = 1$$

If u_{ig} *takes only the values 0 or 1, expression (1) represents a crisp partition of the* n *units.*

Let $\mathbf{X} = \{x_{is}\}$ $(i = 1, 2, ..., n; s = 1, 2, ..., p)$ denote the data matrix containing the values taken by p variables measured on each of n units. The variables of $\mathbf{X}$ could be either numeric or fuzzy.
Starting from the data matrix $\mathbf{X}$ a distance matrix $\mathbf{D} = \{d_{ij}\}(i, j = 1, 2, ...n)$ can be derived following different approaches (Bandemer and Näther, 1992; Leung, 1988; Zani, 1988; Zimmermann,1985). Like in classical cluster analysis, several fuzzy clustering algorithms can be applied to matrix D in order to obtain fuzzy partitions of the n units.

3. Variable selection

Let $k < p$ be the number of variables to be selected (k may be fixed or determinated iteratively letting $k = p - 1$, $k = p - 2, ...$). The data matrix $\mathbf{X}$ can be written as a partioned matrix:

$$\mathbf{X} = [\mathbf{X}_1, \mathbf{X}_2] \tag{2}$$

where $\mathbf{X}_1$ is the matrix of selected variables and $\mathbf{X}_2$ collects the remaining $(p-k)$ eliminated variables.
The problem of variable selection can be formalized as follows (Milioli, 1993):

$$\phi(\mathbf{X} : \mathbf{X}_1, \mathbf{X}_2) = \max \qquad (\mathbf{X}_1, \mathbf{X}_2 \in \Omega) \tag{3}$$

where ϕ is a general objective function and Ω is the set of all partitions of the variables of matrix $\mathbf{X}$ defined in (2).
Taking into account the distinction made in section 1, we consider several objective functions which specify expression (3).
If the aim of variable selection is the elimination of "noise" variables, the function ϕ can be represented by a measure of cluster validity (Bezdek, 1981; Libert and Roubens, 1983). Measure of fuzzy clustering validity are linked to the concept of "separation" and "fuzziness", where more separation implies less fuzziness. Taking into account the opinion of most authors which presume least fuzzy partitions to be most valid, the optimization of function ϕ should lead to select those variables that produce the most separated or the least fuzzy partitions. The most known measure of cluster validity is the *partition coefficient*:

$$F(\mathbf{U}, G) = \frac{tr(\mathbf{U}\mathbf{U}')}{n} = \frac{\sum\sum(u_{ig})^2}{n} \tag{4}$$

which takes value in $[1/G, 1]$. It is $1/G$ only at the equimembership partition $[1/G]$; it is one on all hard G-partitions of $\mathbf{X}$. The normalized expression of the partition coefficient is given by

$$F^*(\mathbf{U},G) = \frac{F(\mathbf{U}, G)G - 1}{G - 1} \tag{5}$$

The objective function can then be specified as follows:

$$F^*(\mathbf{U}_1,G) = \max \qquad (\mathbf{U}_1 \in \Gamma) \tag{6}$$

where Γ is the set of all possible G-fuzzy partitions ($\mathbf{U}_1$) generated by the respective matrices $\mathbf{X}_1$ defined in (2).

Remark 1 - *The measures of cluster validity are affected by the number of clusters, even if a normalized index is used (Bezdek,1981). In this context, we assume the comparison is made among fuzzy partitions with the same number of clusters.*

When the input variables are numeric, the function ϕ can be based on a generalization of a statistic, such as Wilks' Λ statistic or Pillai's trace statistic (Fowlkes *et al.*, 1988), which takes into account the grade of membership of the units to the clusters. In the decomposition of the total sums of crossproducts ($\mathbf{T}$), the within group ($\mathbf{W}$) and between group ($\mathbf{B}$) fuzzy sums of crossproducts matrices are then defined as follows:

$$\mathbf{W} = \{w_{rs}\} \text{ where } w_{rs} = \sum_{g=1}^{G}\sum_{i=1}^{n}(x_{is} - \bar{x}_{sg})(x_{ir} - \bar{x}_{rg})u_{ig} \tag{7}$$

$$\mathbf{B} = \{b_{rs}\} \text{ where } b_{rs} = \sum_{g=1}^{G}\sum_{i=1}^{n}(\bar{x}_{sg} - \bar{x}_{s})(\bar{x}_{rg} - \bar{x}_{r})u_{ig} \tag{8}$$

With reference to Λ statistic, we can define a further objective function:

$$\Lambda(\mathbf{W}_1, \mathbf{T}_1) = |\mathbf{W}_1| \,/\, |\mathbf{T}_1| = \min \qquad (\mathbf{T}_1 \in \Theta) \tag{9}$$

where Θ is the set of all possible total sums of crossproducts matrices associated to the set Γ defined in (6).
If the aim of variable selection is merely the elimination of redundant variables, we give the following suggestions.
A generalization of the Rand index (Milioli, 1994) can be used as an objective function in order to identify a subset of the original variables which allows us to obtain a fuzzy partition of the n units ($\mathbf{U}_1$) which reproduce "as well as possible" the one generated by the complete set of the initial variables ($\mathbf{U}$).
In order to define the index, let's consider a set of n units $a_i \in A$, and let $\mathbf{P}$ and $\mathbf{Q}$ be two different fuzzy partitions of A, with c and r clusters, respectively. The proposed index is based on the following:

Definition 2 *Given two fuzzy partitions* $\mathbf{P}$ *and* $\mathbf{Q}$ *of* A*, the pairs of units in each partition are treated in a similar way if the following conditions are satisfied:*
i) the pairs of units are classified together in both partitions in the same number of clusters;
ii) moreover, the portion of the pairs of units classified together in at least one cluster (or in different clusters) is the same in both partitions.

Let S be the set of all the pairs of units which are classified together in the same number of clusters in $\mathbf{P}$ and in $\mathbf{Q}$, and let s =card(S). The portion of units classified together in the pair $h = (i, j)$ of S in $\mathbf{P}$ and $\mathbf{Q}$ is defined as follows:

$$\varepsilon_h(\mathbf{P}) = \sum_{g=1}^{c} u_{ig}(\mathbf{P}) \cdot u_{jg}(\mathbf{P}) \text{ and } \varepsilon_h(\mathbf{Q}) = \sum_{g=1}^{r} u_{ig}(\mathbf{Q}) \cdot u_{jg}(\mathbf{Q}) \qquad (10)$$

The quantities

$$\delta_h(\mathbf{P}) = 1 - \varepsilon_h(\mathbf{P}) \text{ and } \delta_h(\mathbf{Q}) = 1 - \varepsilon_h(\mathbf{Q}) \qquad (11)$$

are the portions of units of the $h - th$ pair classified in different clusters. The portions of the $h - th$ pair classified either in the same clusters and in different clusters, common to both partitions, are

$$E_h = \min[\varepsilon_h(\mathbf{P}), \varepsilon_h(\mathbf{Q})] \text{ and } \Delta_h = \min[\delta_h(\mathbf{P}), \delta_h(\mathbf{Q})] \qquad (12)$$

The total portion of the s pairs of units treated in a similar way in $\mathbf{P}$ and in $\mathbf{Q}$ is given by

$$C = \sum_{h=1}^{s} (E_h + \Delta_h) \qquad (13)$$

The similarity index between the fuzzy partitions $\mathbf{P}$ and $\mathbf{Q}$ is defined as the ratio between the portion (C) of pairs of units treated in a similar way in both partitions and the total number of pairs of units:

$$s(\mathbf{P}, \mathbf{Q}) = \frac{C}{\binom{n}{2}} \qquad (14)$$

It is zero if all the pairs of units are classified together always in a different number of clusters; it is one if the $\binom{n}{2}$ pairs are treated in a similar way (according to definition 2) in both partitions.
In the context of variable selection the comparison is made between $\mathbf{U}_1$ and $\mathbf{U}$. The maximization

$$s(\mathbf{U}_1, \mathbf{U}) = max \qquad (\mathbf{U}_1 \in \Gamma) \qquad (15)$$

leads to select the subset of k variables which identify the fuzzy partition most similar to the one generated by the full set of the original variables.
To face the problem without a preceding cluster analysis, the selection can be based on the comparison between the distance matrix relating to the p original variables $(\mathbf{D})$ and the one determined with reference to a subset of k variables $(\mathbf{D}_1)$. Relating to this topic, Bandemer and Näther (1992, pp. 153-157) introduce the definition of "neighbourhood" of variables which can suggest a reduction of the number of the original variables by deleting or combining variables allocated in the same narrow neighbourhood.

4. A stepwise algorithm for variable selection

The maximization of expression (3) could involve a great amount of calculations, mostly if the number of variables to select is unknown, as often happens in explorative analysis. To overcome this problem we suggest a stepwise procedure (backward elimination) whose iterative scheme is developed by the following steps:
1) calculate the values of $\phi(\mathbf{X} : \mathbf{X}_{[s]}X_s)$ $(s = 1, 2, ..., p)$, where $\mathbf{X}_{[s]}$ is the data matrix without variable X_s ;
2) delete the variable, say t, for which ϕ is a maximum or a minimum, according to the selected objective function;
3) recalculate the values of $\phi(\mathbf{X}^{[t]}\colon \mathbf{X}^{[t]}_{[s]}X_s)$, where $X_s(s = 1, 2, ...)t(..., p)$ is one of the remaining variables;
4) iterate step 3 until a stopping condition is satisfied, e.g. when the desired number of variables has been reached or when the variation of the function in two successive iterations is less then a fixed threshold.

5. Application

As an example, the problem of variable selection has been analyzed with reference to a set of 10 economic indicators related to the italian regions in 1993 (source: ISTAT):
X_1= employment rate, in %
X_2= female employment rate, in %
X_3= incidence of employees in agriculture on the total employees, in %
X_4= unemployment rate, in %
X_5= female unemployment rate, in %
X_6= gross domestic product per inhabitant, in thousands lire
X_7= final household consumptions per inhabitant, in thousands lire
X_8= incidence of foodstuffs consumption on the total consumption, in %
X_9= private expenditure for public performances per inhabitant, in thousands lire
X_{10}= deposits per inhabitant, in millions lire
The first five indicators reflect specifically the labour force of the italian regions with attention to the women's partecipation in socio-economic activities, while the last five indicators are related to income and wealth. A fuzzy partition of the italian regions according to the previous indicators has been obtained using the fuzzy G-means algorithm FANNY proposed by Kaufmann and Rousseeuw (1990) which has suggested a solution in two fuzzy clusters (see table 1) presenting a value of the partition coefficient equal to 0.41.
Table 1 shows that the first 12 regions, which belong to northern and central Italy, characterize cluster 1, because they have high membership to this group and very low membership to the other one. An inverse situation is shown for the regions of southern Italy, which characterize cluster 2. The region of Abruzzo,

which has almost equimembership to the groups, presents features similar to the regions of both clusters.

Table 1: *Fuzzy partition of the italian regions based on 10 indicators.*

REGIONS	u_{i1}	u_{i2}	REGIONS	u_{i1}	u_{i2}
Piemonte	0.888	0.112	Marche	0.825	0.175
Valle D'Aosta	0.778	0.222	Lazio	0.797	0.203
Lombardia	0.863	0.137	Abruzzo	0.518	0.482
Trentino A.A.	0.757	0.243	Molise	0.275	0.725
Veneto	0.887	0.113	Campania	0.145	0.855
Friuli V.G.	0.848	0.152	Puglia	0.137	0.863
Liguria	0.743	0.257	Basilicata	0.145	0.855
Emilia Romagna	0.805	0.195	Calabria	0.132	0.868
Toscana	0.868	0.132	Sicilia	0.138	0.862
Umbria	0.722	0.278	Sardegna	0.151	0.849

The stepwise algorithm for variable selection illustrated in section 4, choosing firstly the partition coefficient as an objective function to maximize, has been applied in order to reduce the degree of fuzziness of the initial partition, deleting indicators which may constitute a "noise".

Table 2 indicates, for each step of the procedure, the variable whose deletion causes the highest value of the partition coefficient and the related value of the index. For example, at the first step, the deletion of variable X_2 generates the highest increment of the value of the partition coefficient (equal to 0.43). In the second step, after the elimination of X_2 , the deletion of variable X_1 leads the partition coefficient to a maximum value equal to 0.45. If we decide to stop the procedure at step 6, we can see that indicators X_4, X_5, X_6 and X_7 generate a 2-fuzzy partition of the italian regions where the two clusters are undoubtedly more separated than the two groups in the initial classification ($F^*(\mathbf{U}, 2) = 0.54$). Moreover, according to the stepwise procedure, the unemployment rate (X_4) and the gross domestic product per inhabitant (X_6) are the two indicators which generate the partition of the italian regions with the lowest degree of fuzziness ($F^*(\mathbf{U}, 2) = 0.57$).

The generalized Rand index defined in (14) has then been used with the purpose to eliminate redundant indicators. The results of the procedure are summarized in table 3. At the first step, the algorithm shows that the deletion of X_3 leads to a 2-fuzzy partition of the italian regions almost equal to the one obtained from the entire set of indicators ($s(\mathbf{U}_1, \mathbf{U}) = 0.992$). The incidence of employees in agriculture is really redundant with respect to the other 9 indicators. If we stop the procedure at step 6, the solution generated by X_2, X_4, X_6 and X_8 is still very similar to the one shown in table 1 ($s(\mathbf{U}_1, \mathbf{U}) = 0.972$). It is worthwhile to observe that even only two indicators, X_2 (female employment rate) and X_6 (gross domestic product per inhabitant) can reproduce very well the original classification $s(\mathbf{U}_1, \mathbf{U}) = 0.951$).

Table 2: *stepwise variable selection: partition coefficient.*

STEP	DELETED VARIABLE	$F^*(\mathbf{U}_1, 2)_{\max}$
1	X_2	0.43
2	X_1	0.45
3	X_{10}	0.48
4	X_8	0.50
5	X_3	0.52
6	X_9	0.54
7	X_7	0.56
8	X_5	0.57

Table 3: *stepwise variable selection: generalized Rand index.*

STEP	DELETED VARIABLE	$s(\mathbf{U}_1, \mathbf{U})_{\max}$
1	X_3	0.992
2	X_1	0.989
3	X_7	0.989
4	X_5	0.989
5	X_9	0.981
6	X_{10}	0.972
7	X_4	0.968
8	X_2	0.951

In table 4 some results of both criteria are compared. The maximization of the partition coefficient leads to select subsets of indicators that really reduce the fuzziness degree of the obtained partitions, but it disregards their similarity with the initial solution. On the contrary, the maximization of the generalized Rand index leads to select subsets of indicators that reproduce the clusters structure satisfactorily, but it doesn't improve the validity of the solution in the same way. Finally, it should be noted that, for the fuzzy classification of the italian regions, the gross domestic product per inhabitant seems to be a crucial indicator because it reproduces most of the information given by the other economic indicators and, at the same time, it generates well separated groups.

Table 4: *Comparison between procedures of table 2 and table 3.*

subset of selected indicators	$F^*(\mathbf{U}, 2)$	$s(\mathbf{U}_1, \mathbf{U})$
X_4, X_5, X_6, X_7	0.54	0.9356
X_4, X_6 (from table 2)	0.57	0.9191
X_2, X_4, X_6, X_8	0.43	0.9716
X_2, X_6 (from table 3)	0.43	0.9511

References

Bandemer, H. and Näther, W. (1992), *Fuzzy Data Analysis*, Kluwer, Dordrecht.

Bezdek, J. C. (1981), *Pattern Recognition with Fuzzy Objective Function Algorithms*, Plenum Press, New York.

Fowlkes, E. B., Gnanadesikan, R. and Kettenring, J. R. (1988), Variable selection in clustering, *Journal of Classification*, 5, 205-228.

Kaufmann, L. and Rousseeuw, P.G. (1990), *Finding Groups in Data*, Wiley, New York.

Leung, Y. (1988), *Spatial Analysis and Planning under Imprecision*, North Holland, Amsterdam.

Libert, G. and Roubens, M. (1983), New Experimental Results in Cluster Validity of Fuzzy Clustering Algorithms, *New Trend in Data Analysis and Application*, Janssen J., Marcotorchino J. F. and Proth J. M. (eds.), North Holland, 205-218.

Milioli, M. A. (1993), *Variabili ridondanti e osservazioni influenti nell'analisi dei dati multidimensionali*, Collana di Studi e Ricerche della Facoltà di Economia e Commercio, Università degli Studi di Parma, Giuffrè , Milano.

Milioli, M. A. (1994), Confronto fra partizioni sfocate nell'analisi di dati territoriali, *Atti della XXXVII Riunione Scientifica della Società Italiana di Statistica*, 43-50.

Zani, S. (1988), Un metodo di classificazione sfocata, in: G. Diana, C. Provasi and R. Vedaldi (a cura di) *Metodi statistici per la tecnologia e l'analisi dei dati multidimensionali*, Università degli Studi di Padova, 281-288.

Zimmermann, H. J. (1985), *Fuzzy Set Theory and Its Applications*, Kluwer, Dordrecht.

PART II

Other Approaches for Classification

- **Discrimination and Classification**
- **Regression Tree and Neural Networks**

Discriminant Analysis Using Markovian Automodels

Marco Alfò and Paolo Postiglione
Dipartimento di Metodi Quantitativi e Teoria Economica,
Università "G. d'Annunzio" di Chieti

Abstract: Spatially distributed observations occur naturally in a number of empirical situations; their analysis represents a significant source of theoretical challenge due to the multidirectional dependence among nearest observations. The presence of a dependence often causes the standard statistical methods, instead based on independence assumptions, to fail badly. This paper concerns the problem of discrimination and classification of spatial binary data. It presents a suitable discrimination function based on Markovian automodels and suggests a solution to the allocation problem through a Gibbs sampler-based procedure.

Keywords: Binary spatial observations, spatial discrimination and classification, Logistic-Autologistic model, Gibbs sampler.

1. Introduction

Spatially dependent observations arise in a wide variety of applications, including archaeological and agricultural studies, pattern recognition and econometrics, epidemiological spreading and survey analysis (see e.g. Haining, 1990).

Traditional statistical techniques assume independence among the observations, a hypothesis which is obviously violated in all geographical and territorial studies, where "(...) everything is related to everything else, but near things are more related than distant things", as stated in Tobler's first law of geography (Tobler, 1970). Several well-established models have been introduced in the statistical literature to take account of the multidirectional dependence being peculiar to spatial observations. For earlier developments see, for example, the seminal papers by Besag (1974) and Strauss (1977); for a more recent and comprehensive review see Cressie (1991).

In this paper, we concentrate on the definition of a discrimination function for spatial observations, incorporating the notion of contextual information.

Generally speaking, we can recognise in discriminant analysis two different aims: discrimination and classification. The term discrimination concerns the process of deriving classification (or allocation) rules from samples of labelled

units, while the term classification refers to the application of these rules to identify new units group membership (see Krzanowski & Marriott, 1995).
The paper is organised as follows: section 2 is devoted to discuss the limits of classical approach to logistic discriminant analysis when we are dealing with spatial data. Section 3 introduces the definition of a suitable spatial discrimination function based on Logistic-Autologistic model. In particular, we discuss some istances related to parameters estimation using maximum pseudo-likelihood and describe a relevant solution, through Gibbs sampler (Geman & Geman, 1984) based algorithm, to the outlined spatial allocation problem. Section 4 presents the application of the proposed method to a real situation. The aim is to discriminate between pixels from a remote sensed image of Nebrodi Mountains (Sicily) using the information about a number of soil characteristics. The final section is devoted to the presentation of concluding remarks and to outline possible future research lines.

2. The logistic approach

The earliest ideas in logistic discrimination date back to the 60's, but the essential features of the method as it is applied today were developed by Anderson (1982). The method assumes that the posterior probabilities of group membership for the *i-th* unit characterised by covariates vector $\boldsymbol{X}_i = \left(x_{i1}, x_{i2}, \ldots, x_{iq}\right)^t$ are given by (see McLachlan, 1992):

$$\Pr(Y_i = 1|\boldsymbol{X}_i) = \frac{\exp\left(a + \boldsymbol{b}^t \boldsymbol{X}_i\right)}{\left[1 + \exp\left(a + \boldsymbol{b}^t \boldsymbol{X}_i\right)\right]} \tag{1}$$

$$\Pr(Y_i = 0|\boldsymbol{X}_i) = \frac{1}{\left[1 + \exp\left(a + \boldsymbol{b}^t \boldsymbol{X}_i\right)\right]} \tag{2}$$

where a is the intercept and $\boldsymbol{b}=(b_1,\ldots,b_q)^t$ is a q-dimensional vector of parameters. Expressions (1) and (2) can be easily summarised as follows:

$$\Pr(Y_i = y_i|\boldsymbol{X}_i) = \frac{\exp\left[y_i\left(a + \boldsymbol{b}^t \boldsymbol{X}_i\right)\right]}{\left[1 + \exp\left(a + \boldsymbol{b}^t \boldsymbol{X}_i\right)\right]} \tag{3}$$

According to previous expressions, we can write the log-odds as:

$$\log\left[\frac{\Pr(Y_i = 1|\boldsymbol{X}_i)}{\Pr(Y_i = 0|\boldsymbol{X}_i)}\right] = a + \boldsymbol{b}^t \boldsymbol{X}_i \tag{4}$$

The attractiveness of this approach is that the model can be applied empirically, whatever the nature of the involved distributions and even in cases where the distribution of X is not known (cfr. Anderson, 1982).
To apply the method, we can estimate the parameters a e $\boldsymbol{b}$ from the training set (discrimination phase), substitute these estimates in (3) and allocate *i-th* unit into the group having the highest posterior probability (classification phase) or, in term of log-odds, into the group having Y_i=1 if (4) is positive and into the group having Y_i=0 if (4) is negative.
The use of a logistic approach when we are dealing with spatial data is statistically incorrect, in consequence of their mutual dependence. In fact, closeness between units belonging to a certain group modifies our expectation of other units group membership. As a consequence, it seem reasonable to formulate a conditional model that links auxiliary information contained in the covariates vector $\boldsymbol{X_i}$ with the notion of spatial dependence (contextual information).

3. The contextual approach to logistic discriminant analysis

To introduce our model, let us start assuming that we are analysing a study area which can be discretized into M contiguous quadrate cells and that for N of these cells ($N<M$) we are able to assess precisely the group membership (we will refer to these N cells as the training set).
This is only a working simplification, which can be relaxed, when analysing irregular lattices, by imposing a more general connectivity structure between the sample units.
In the case of binary data, each cell i of the training set is black (Y_i=1) if the characteristic under study is present and is white (Y_i=0) otherwise.
Our aim is to label the other (M-N) cells on the basis of the information obtained from the training set (see Figure 1)

Figure 1: *An example of square lattice grid*

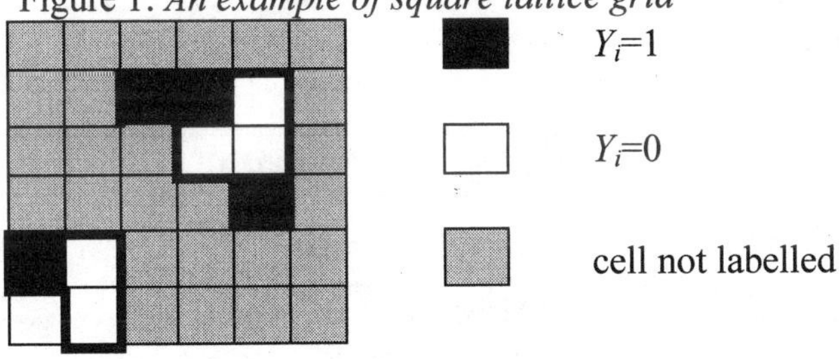

In order to describe a model for spatial discriminant analysis, we can define a Logistic-Autologistic Model (in the following LAM, see also Arbia & Espa, 1996) as:

$$\Pr(Y_i = y_i | Y_j = y_j, j \in N(i); \boldsymbol{X}_i) = \frac{\exp\left[y_i\left(a + \sum_{j \in N(i)} r_{ij} y_j + \boldsymbol{b}^t \boldsymbol{X}_i\right)\right]}{\left[1 + \exp\left(a + \sum_{j \in N(i)} r_{ij} y_j + \boldsymbol{b}^t \boldsymbol{X}_i\right)\right]} \tag{5}$$

where Y_i and Y_j are binary random variables, $N(i)$ is the neighbourhood set of site i, a and $\boldsymbol{b}$ have the same meanings as before and the quantities r_{ij} represent interaction parameters referring to thc cliqucs of sizc two (Haining, 1990).
The LAM is also known in literature as autologistic model for large scale variation and non stationary autologistic model (see Cressie, 1991). But in the present framework, we prefer the LAM notation to show the conjoint nature of the model, obtained composing auxiliary and contextual informations.
If we consider a LAM model based on the first order neighbourhood system (see Figure 2), the spatial interaction parameters r_{ij} can be decomposed as:

$$r_{ij} = r w_{ij} \qquad \text{with} \quad w_{ij} = \begin{cases} 1 & \text{if} \quad w_{ij} \in N_1(i) \\ 0 & \text{if} \quad w_{ij} \notin N_1(i) \end{cases} \tag{6}$$

where r is a constant.

Figure 2: *A first order neighbourhood of site i*

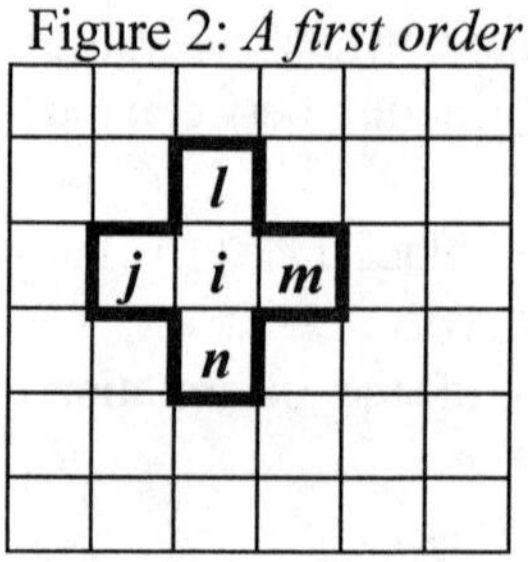

$N_1(i)=\{j,l,m,n\}$

So, the LAM can be written as:

$$\Pr(Y_i = y_i | Y_j = y_j, j \in N(i); \boldsymbol{X}_i) = \frac{\exp\left[y_i\left(a + r \sum_{j \in N_1(i)} y_j + \boldsymbol{b}^t \boldsymbol{X}_i\right)\right]}{\left[1 + \exp\left(a + r \sum_{j \in N_1(i)} y_j + \boldsymbol{b}^t \boldsymbol{X}_i\right)\right]} \tag{7}$$

The previous model is used by physicists in ferromagnetism studies and is known as Ising's Law (Ising, 1925).

With an irregular lattice, w_{ij} can be specified in a number of different forms; for example, as (Cressie, 1991):

$$w_{ij=} = \begin{cases} 1/d_{ij} & \text{if } w_{ij} \in N(i) \\ 0 & \text{if } w_{ij} \notin N(i) \end{cases} \tag{8}$$

where d_{ij} represents the distance between neighbouring sites i and j.
As stated in Arbia and Espa (1996), there are some statistical problems connected with model (5) that are still unsolved. The spatial discrimination function (5) can not be applied directly as function (3); in fact, in the spatial model the binary variable Y_i is present both as a dependent variable (on the left side of (5)) and as an independent one (on the right side of (5)). Hence, in the classification step, we should predict Y_i using the variables $\boldsymbol{X}_i$ (known for all the analysed units) and Y_j $(i \neq j)$, which are not known for units not belonging to the training set. So, when allocating units not belonging to the training set, we should use information about quantities which are not known. To solve our problem of spatial units allocation, we propose the following two-steps procedure:
Step 1 (*Discrimination phase*): We estimate the model parameters via the maximum pseudolikelihood (MPL) procedure (Besag, 1975). Besag coined the term pseudolikelihood to indicate the product of the conditional probabilities (5). We obtain MPL estimates regressing Y_i on the q+1 covariates $(x_{i1}, x_{i2}, \dots, x_{iq}, \sum_{j \in N_1(i)} y_j)$ for the training set. All these variables are known for units belonging to the training set. The maximum pseudolikelihood estimator is consistent (as shown by Geman & Graffigne, 1987) and gives good results in practical applications (Ripley, 1990). The pseudolikelihood estimates can be carried out using standard statistical packages (see Strauss & Ikeda, 1990).
Step 2 (*Classification phase*): We consider the parameter estimates obtained from the training set as representatives of the whole lattice grid. Hence, we allocate the remaining cells through a simulation process that derives the autologistic model, including the covariates effects (known for all M cells). In this way, all cells are classified. This process is based on a simple Gibbs sampler procedure, simulating an autologistic random field with parameters equal to those estimated on the units of the training set.

4. A real example

The proposed method has been applied to a real data set, consisting of a remote sensed image from Landsat TM (Thematic Mapper). The image, covering a surface of 2,016 Km^2, is composed by 2.240 (40×56, vertical×horizontal)

quadrate cells (in the following *pixels*), each representing a ground surface of 30m×30m, which is determined by the satellite spatial resolution.
The aim was to discriminate between farmed surfaces (Y_i=1) and different surfaces (Y_i=0) on the basis of a number of related soil characteristics; the first group is composed by lavic arboreous, citrus fruit, sowing and other arboreous farming (in particular olives and almonds), while the latter represents pastures, woods, thin woods and spontaneous vegetation. The related soil characteristics are: permeation capacity in saturation conditions (measured in millimetres per day), horizon thickness (in centimetres), mean yearly rainfall (in centimetres) and DTM (digital terrain model, in metres).
The dimension of the training set was fixed in 200 pixels and the analysis was repeated adopting 4 different choices for the training set, to check the sensitivity of the proposed approach to different specification of the training set. To choose the training sets, we divided the lattice in four quadrants and sampled a pixel from each quadrant. The training sets were formed by rectangles containing 12×16 pixels (vert.×hor.) and having the choosen pixels as one of their corners.
In the four analysed training sets, the estimated values of interaction parameter r were in the range (0.6-0.7), so outlining a significant spatial dependence; furthermore, the values of -2log(likelihood), even if it can not be considered a fully correct indicator of model fit, for LAM model were always smaller than those obtained by standard logistic approach.
The Gibbs sampler procedure generated realisations of an autologistic random field with parameters a_i and r estimated on the training set. To take into account the local information of each site i, the parameters a_i was defined as:

$$a_i = a + \boldsymbol{b}^t \boldsymbol{X}_i \tag{9}$$

The proposed algorithm was iterated 50 times for each of the 4 training set, so leading to 200 different allocation scenarios. Figure 3 summarises the obtained results in terms of the rightly classified pixels percentage (note that this percentage is less or equal to 71% in the case of logistic discriminant function).

Figure 3: *Distribution of rightly classified pixels percentage*

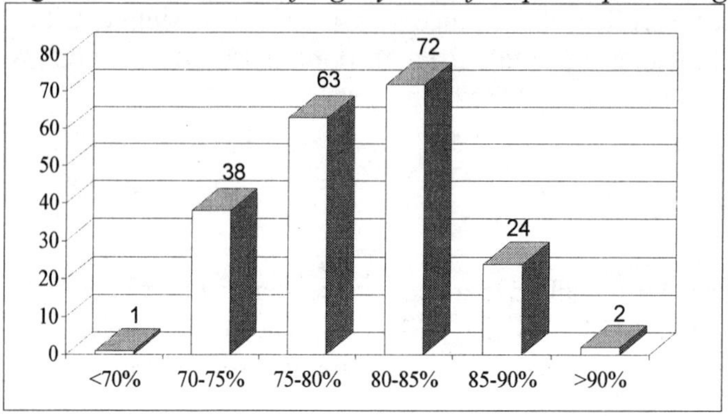

Note that in more than 80% of the analysed cases the solution provided by the LAM model is significantly better than that obtained by a standard logistic approach to discriminant analysis.

5. Concluding remarks

In this paper we have outlined the obvious limitations that arise when applying logistic discriminant analysis to spatially distributed observations. We have contrasted the classical approach with a more congruent one, based on the LAM, which takes into account the mutual and multidirectional dependence of geographical and territorial sites, together with auxiliary information available in the form of covariates measurements on the cells constituting the analysed lattice. In particular, we have suggested a simple algorithm, based on Gibbs sampler, to provide a feasible solution to the allocation problem in the case of a spatially located binary response variable.
The algorithm proposed for the classification phase has been implemented in Matlab code an applied to a real data set, consisting in a portion of a remote sensed image regarding Nebrodi Mountains (Sicily).
In this context, results obtained from LAM have been contrasted with those obtained via a standard logistic discriminant function, and seem to confirm that, when working with spatial data, there is no evident reason for leaving out the contextual information from the model specification. However, this case study should be considered only a particular one, with a major interaction among neighbouring units. Hence, to confirm these conclusions in the general case, more research is needed, in the form of a large-scale simulation study characterised by different values of interaction parameters, and/or by different sizes of the training set and of the whole lattice.
It is worth noting that the proposed algorithm and, in general, the whole method of estimation and classification can be straightforwardly applied to the general situation of spatially located multinomial outcome variables (g groups).

Acknowledgements

The authors would like to thank Prof. G. Arbia and Prof. M. Vichi for helpful comments on earlier version of this paper. Thanks are also due to Prof. A. Bellacicco and Prof. L. Fabbris. Finally we gratefully acknowledge M.L. Paracchini of Space Application Institute Of the Joint Research Centre of the European Commission (Ispra, Varese) for kindly providing the remote sensed image of section 4.

References

Anderson, J.A. (1982). Logistic discrimination, in: *Handbook of Statistics vol.2*, Krishnaiah P.R. & Kanal L.N. (Eds.), North Holland, 169-191.

Arbia, G. & Espa, G.(1996). Forecasting statistical models of archaeological site locations, *Archeologia e Calcolatori*, Roma.

Besag J. (1974). Spatial interaction and the statistical analysis of lattice systems, *Journal of Royal Statistical Society*, B, 36, 192-236.

Besag, J. (1975). Statistical analysis of non-lattice data, *The Statistician*, 24, 179-195.

Cressie, N. (1991) *Statistics for Spatial Data*, Wiley, New York.

Geman, S. & Geman D. (1984) Stochastic Relaxation, Gibbs Distributions, and the Bayesian Restoration of Images", *IEEE Transactions on Pattern Analysis and Machine Intelligence*, Vol. PAMI-6, 721-741.

Geman, S & Graffigne, C (1987). Markov random fields and their application to computer vision, *Proceedings of the International Congress of Mathematicians*, Providence.

Haining, R.P. (1990). *Spatial Data Analysis in the Social and Environmental Sciences*, Cambridge University Press.

Ising, E. (1925) Beitray sur theorie des ferromagnetismus, *Zeitschrift Physic*, 31, 253-258.

Krzanowski, W.J. & Marriott, F.H.C. (1995). *Multivariate Analysis. Part 2-Classification, Covariance Structures and Repeated Measurements*, Edward Arnold.

Mclachlan, G.J. (1992). *Discriminant analysis and pattern recognition*, Wiley, New York.

Ripley B.D. (1990). Gibbsian interaction models in: *Spatial Statistics: Past, Present and Future,* Monograph n.12, Griffith D. (Eds.), Institute of Mathematical Geography.

Strauss, D.J. (1977). Clustering of coloured lattices, *Journal of Applied Probability*, 14, 135-143.

Strauss, D.J & Ikeda, J. (1990). Pseudolikelihood estimation for social networks, *Journal of the American Statistical Association,* 85, 204-212.

Discretization of Continuous-Valued Data in Symbolic Classification Learning

F. Esposito D. Malerba G. Semeraro S. Caggese
Dipartimento di Informatica, Università degli Studi di Bari,
via Orabona 4, 70126 Bari, Italy
e-mail: {esposito | malerba | semeraro | caggese}@lacam.uniba.it

Abstract: Symbolic data analysis aims at extending classical data analysis to data representing classes of individuals instead of single individuals. A major problem in symbolic data analysis is discrimination, that is the generation of data representing classes. Such data can be expressed as classification rules, which are learned from training examples. The paper addresses the problem of learning classification rules from examples described by both numeric and symbolic attributes, so that the discretization of the continuous-valued attributes is performed during the learning process. The proposed technique has been embedded into a classification learning system, named INDUBI/CSL, and tested on several data sets.

Keywords: Symbolic data analysis, discrimination, discretization of continuous-valued attributes.

1. Introduction

Symbolic data analysis aims at extending classical data analysis methods and algorithms to more complex data well adapted to represent classes of individuals instead of single individuals (Diday, 1990). Indeed, in classical data analysis a single individual can be represented as a feature vector, where each feature (or attribute or variable) takes a single value (either discrete or continuous). On the contrary, symbolic objects can be represented as feature vectors with multi-valued variables and interval variables, that is variables that take values in the power set of a given universe of objects. Therefore, a symbolic object is the description of all the properties valid for a class of individuals.

Several studies have been performed on methods for analyzing symbolic data, such as feature selection (Ichino, 1994) and unsupervised classification learning (or clustering) (Brito, 1994). However, one of the main problems remains the generation of symbolic objects from examples. More precisely, given a set of mutually exclusive classes, C_1, C_2, ..., C_r, and a set E of individuals of such classes (i.e., E is a set of training examples), the problem is that of building a set of symbolic objects H_i for each class C_i. Each symbolic object can be interpreted as a classification rule:

$$\textbf{if } y_1=Y_1, y_2=Y_2, \ldots, y_m=Y_m \textbf{ then } class = C_i$$

where y_j are the multi-valued/interval variables used to describe the class C_i, and Y_j are the corresponding set of values. A symbolic object is said to *cover* an individual when the value taken by each variable y_j, i=1, ...,m, in the description of the individual is a member of Y_i. For instance, the following symbolic object:

if color_hat={black,red}, height=[10..20] **then** class = poisonous_mushroom

covers the individual

color_hat=red, height=12, weight=6

but not

color_hat=black, height=12, weight=16

Generated symbolic objects should satisfy two properties:

1. *completeness*, that is all individuals in E of class C_i should be covered by some object in H_i,
2. *consistency*, that is symbolic objects in H_i should not cover individuals in E of class C_j for any $j \neq i$.

Therefore, the generation of symbolic objects from examples is a typical discrimination problem, though traditional techniques of multivariate data analysis, and in particular discriminant analysis, cannot be straightforwardly applied in this case. Indeed, such techniques are not able to handle either multi-valued or interval variables, and above all, are not able to produce classification rules whose body is a logical conjunction of conditions that single variables have to satisfy. Symbolic classification learning methods face this kind of problems and investigate properties of algorithms that generate symbolic classification rules.

In the most part of symbolic classification learning systems, continuous-valued attributes are discretized before starting the learning process. In the paper we address the problem of learning classification rules from both numeric and symbolic data, in such a way that discretization of continuous-valued attributes is performed during the learning process by a specialization operator. This operator has been embedded into a classification learning system, named INDUBI/CSL (Malerba *et al.*, 1997a), and tested on several data sets.

2. The learning strategy

The learning strategy adopted by INDUBI/CSL to generate classification rules is called *separate-and-conquer* (see Figure 1). The *separate* stage of the algorithm is a cyclic process that searches for the current hypothesis H_i, expressed as a set of rules, by checking for its completeness. If the hypothesis is complete, the learning process for the class C_i is over, otherwise those examples in E covered by H_i are marked and the search for a new consistent rule to add to H_i is started. In contrast, the *conquer* stage performs a general-to-specific search to construct a new consistent rule. In each conquer phase a *seed* example e^+ of class C_i is chosen from unmarked examples in E. The seed guides the learning process. Each rule generated in the conquer phase should classify correctly at least e^+

```
procedure separate_and_conquer(Examples, Ci)
    Ei := set of positive Examples of Ci
    E- := set of negative Examples of Ci
    Hi := Ø
    while Ei ≠ Ø do
        Consistent_Rules := Ø
        randomly select the seed e+ from Ei
        Consistent_Rules := Beam_Search_for_consistent_rules(e+, Ei, E-, m)
        Best := FindBest(Consistent_Rules)
        Hi := Hi ∪ {Best}
        Ei := Ei - Covers(Best, Ei)
    endwhile
return Hi
```

Figure 1: *High level separate-and-conquer search strategy.*

and, possibly, other training examples of class C_i. Actually, the seed example plays the role of a prototype: Thus it could be appropriately chosen as the most representative case of C_i in E.
It can be proven that the number of rules generated and tested by the procedure *separate_and_conquer* is polynomial in the number of training examples, in the beam search, and in the number of attributes used to describe a training example. The generation of the consistent rules is made by progressively adding conditions to the body of the rules, that is, by progressively specializing rules until they become consistent. More precisely, INDUBI/CSL starts with the rule having an empty body:

if true **then** class = C_i

that is complete but inconsistent since it classifies all training examples in class C_i. Then the system adds some conditions that make the rule more specific.
The choice of conditions to be added is affected by conditions already selected in previous steps. This is an elegant way of handling interactions between attributes that can become complicated in traditional discrimination methods.
When the condition to be added is of the form "attribute ∈ interval" the problem becomes that of determining the best interval for the continuous-valued attribute involved in the condition. This is a problem of *discretization* of continuous-valued data. In symbolic classification learning, several methods have been proposed for the discretization of continuous-valued data, most of which operate *off-line*, that is discretize before starting the learning process. *On-line* methods for discretizing while learning have been proposed for classification tree learning algorithms (Fayyad & Irani, 1992), which adopt a different search strategy named *divide-and-conquer*.
The discretization procedure implemented in our symbolic learning system operates on-line and has been designed having three distinct goals in mind:

1. On-line discretization of continuous-valued data should be performed by a specialization operator, since our symbolic learning algorithm performs a

general-to-specific search in the conquer stage. The discretization procedure should help to specialize a clause by adding conditions of the type

attribute ∈ [a..b]

where [a..b] denotes a closed interval.

2. The discretization procedure should always guarantee to cover the *seed* example that guides the induction process.

3. The heuristic function used to choose among different discretizations should satisfy some property that reduces the computational complexity of the discretization procedure.

Details of the discretization process are illustrated in the next Section.

3. The discretization procedure

The discretization procedure starts with the construction of a table containing a pair ⟨*Value, Class*⟩ for each example, where *Value* is the value taken by the attribute in the example while *Class* can be either + or – according to the fact that the corresponding example is of class C_i, or not (we are supposing that the rule currently learned concerns class C_i). The table is then sorted in ascending order on the *Value* field.

Now the problem is that of finding the interval that best discriminates positive from negative entries in the table. Any threshold value α laying between two consecutive distinct values defines two disjoint intervals: The left interval $[l_1, l_2]$ and the right interval $[r_1, r_2]$. The lower bound l_1 of the left interval is the smallest value in the table with sign +, while the upper bound l_2 is the largest value in the table that does not exceed the threshold α. On the contrary, the lower bound r_1 of the right interval is the smallest value in the table that exceeds α, while the upper bound r_2 is the largest value with sign +. When one of the two intervals contains no positive entry, then it is set to *undefined.* However, at least one of the two intervals must be defined, since the table contains at least a + in correspondence of the value taken by the attribute for the seed example. Not all definite intervals are to be considered, but only those definite intervals that include the value taken by the attribute for the seed example. Such intervals are said *admissible*, because they guarantee that the corresponding specialized rule still covers e^+.

The best admissible interval is selected according to an information-theoretic heuristic, the *information gain*. By looking at the table as a source of messages labeled + and –, the expected information on the class membership conveyed from a randomly selected message is:

$$inf\,o(n^+, n^-) = -\frac{n^+}{n^+ + n^-}\log_2\frac{n^+}{n^+ + n^-} - \frac{n^-}{n^+ + n^-}\log_2\frac{n^-}{n^+ + n^-}$$

where n^+ and n^- are the number of entries in the table with positive and negative sign, respectively. If we partition the table into two subsets, S_1 and S_2, the former containing entries falling within an admissible interval, while the latter containing the remaining entries, the information provided by S_1 will be close to zero when almost all cases have the same sign, + or –. Although the information prefers partitions that cover a large number of entries with the same sign, we must bias such a preference towards intervals with a high number of *positive* entries, as well. The following *weighted entropy*:

$$E(n_1^+, n_1^-) = \frac{n_1^-}{n_1^+} info(n_1^+, n_1^-)$$

penalizes those admissible intervals with a low percentage of positive entries. The quantity

$$gain(n^+, n^-, n_1^+, n_1^-) = info(n^+, n^-) - E(n_1^+, n_1^-)$$

measures the information gained by replacing the table with S_1. A heuristic criterion to choose the admissible interval is that of maximizing the information gain, that is minimizing $E(n_1^+, n_1^-)$.

This criterion differs from that adopted in other well known learning systems, such as ID3 (Quinlan, 1986) and CN2 (Clark & Niblett, 1989), which do not weight the entropy. This difference is essentially due to the diverse search strategies (separate-and-conquer vs. divide-and-conquer) adopted by the systems.

Note that not all cut points have to be considered. Indeed, only those between two consecutive distinct entries with different sign (*boundary points*) are examined. This choice is due to the following

Theorem. If a cut-point α minimizes the measure , then α is a boundary point.

This result helps to discard several computations of the gain by considering only boundary points, so improving the efficiency of the discretization procedure. Actually, the theorem above is similar to that proved by Fayyad and Irani (1992) for a different measure, namely the "unweighted" class information entropy computed in some decision tree learning systems. The proof of our theorem can be obtained electronically from http://lacam.uniba.it:8000/pagine/proof.html.

4. Comments on experimental results

The proposed discretization procedure has been tested on three data sets taken from the UCI repository of machine learning databases (http://www.ics.uc.edu/~mlearn), namely *Iris*, *Glass* and *Hepatitis*. The performance of INDUBI/CSL is compared to that of *C4.5rules* (Quinlan, 1993), another symbolic classification learning system. The design of the experiment is based on 10-fold cross-validation. For each of the ten trials we collect three statistics, namely, the number of omission and commission errors, and the number of rules generated. Note that while INDUBI/CSL can distinguish between commission and omission errors, C4.5rules always classifies an instance in some class, thus a misclassification error automatically implies both a commission and an omission error. Statistics are summarized in Table 1. The predictive accuracy of INDUBI/CSL is comparable with that of C4.5rules: the detected difference is not statistically significant with respect to the non-parametric Wilcoxon signed-ranks test (Orkin & Drogin, 1990). As to the number of clauses, our system always produces a higher number of rules than C4.5rules does. This can be attributed to the fact that INDUBI/CSL outputs sets of rules that are complete and consistent, while Quinlan's system produces rules that are not necessarily correct with respect to the set of training examples.

Table 1: *Experimental results*

Dataset	Average no. of errors		p-value Wilcoxon signed-rank test	Number of rules	
	INDUBI/CSL	C4.5rules		INDUBI/CSL	C4.5rules
Iris	2.0	1.2	0.1282	9.4	4.0
Glass	14.4	13.8	0.7670	54.1	13.6
Hepatitis	8.2	6.8	0.3329	15.1	5.7

INDUBI/CSL has also been applied to a domain involving structured objects, namely document understanding, where traditional statistical techniques are not straightforwardly applicable because of the difficulty to deal with relations between subparts of an object. According to the ODA/ODIF standard (Horak, 1985), any document is characterized by two different structures representing both its internal organization and its content: The layout (or geometrical) structure and the logical structure. The former associates the content of a document with a hierarchy of *layout* objects, such as text lines, vertical/horizontal lines, graphic/photographic elements, pages, and so on. The latter associates the content of a document with a hierarchy of *logical* objects, such as sender/receiver of a business letter, title/authors of an article, and so on (see Figure 2). The term *document analysis* denotes the extraction of the layout structure from the bitmap of a document, while the term *document understanding* denotes the process of mapping the layout structure of a document into the corresponding logical structure (Tang *et al.*, 1994). The document understanding process is based on the assumption that documents can be understood by means of their layout structures alone.

The mapping of the layout structure into the logical structure can be represented as a set of rules. Traditionally, such rules have been handcoded for particular kinds of documents (Nagy *et al.*, 1992), requiring much human effort. We proposed the application of symbolic classification learning techniques in order to automatically generate the rules from a set of training documents whose layout components have been possibly labeled according to their logical meaning (Esposito et al., 1994). Actually, each training document generates as many training examples as the number of layout components. Classes of training examples correspond to the distinct logical components to be recognized in a document. The unlabelled layout objects play the role of counterexamples for all the classes to be learned.

The description of each training example includes three numeric attributes (*height, width, vertical_position* and *horizontal_position* a of a block), one symbolic attribute (*type* of block) and four binary relations with other components of the layout structure (*part_of, ontop, to_right* and *alignment*) (Malerba *et al.*, 1997b).

In order to compare the performance of our learning system with that of FOIL6.2 (Quinlan & Cameron-Jones, 1993), another symbolic classification learning system that is able to deal with both numeric and symbolic attributes and relations, we have decided to organize an experiment as follows. We have considered a set of 30 business letters containing 364 layout components. The number of attributes and relations used to describe each document is variable, ranging from fifty to more than one hundred. For each layout component, we have associated at most one of the following labels: Logotype, sender, receiver, date, reference number, body of the letter, signature. Those layout components whose content is not significant have been left unlabelled. The experimental design has been based on a 10-fold cross-validation. Once again, for each of the ten trials we have collected three statistics, namely, the number of omission and commission errors, and the number of rules generated. Experimental results are summarized in Table 2.

Briefly, we can observe that the average error rate, as well as the average number of rules, is almost the same for both the systems. The difference is in the type of errors: Foil6.2 discretizes into larger intervals than INDUBI/CSL, thus causing more commission errors. Although commission and omission errors are generally considered equally important, it is worthwhile to observe that in automatic document processing systems omission errors are deemed to be less serious than commission errors, which can lead to unrecoverable errors in storing information unless a heavy human intervention. Moreover, we have also shown that a significant recovery of omission errors can be obtained by relaxing the definition of matching (Esposito et al., 1997).

Table 2: *Experimental results*

Average no. of omission/commission errors		Average no. of errors		Number of rules	
INDUBI/CSL	Foil6.2	INDUBI/CSL	Foil6.2	INDUBI/CSL	Foil6.2
2.6/0.3	1.3/1.5	2.9	2.8	11.5	11.0

References

Brito, P. (1994). Order structure of symbolic assertion objects, *IEEE Transactions on Knowledge and Data Engineering*, 6, 5, 830-835.

Clark, P. & Niblett T. (1989). The CN2 induction algorithm, *Machine Learning*, 3, 261-283.

Diday, E. (1990). Knowledge representation and symbolic data analysis, *Knowledge, Data and Computer-Assisted Decisions*, Schader, M. & Gaul, W. (Eds.), Springer-Verlag, 17-34.

Esposito, F., Malerba, D., & Semeraro, G. (1994). Multistrategy learning for document recognition, *Applied Artificial Intelligence*, 8, 1, 36-84.

Esposito, F., Caggese, S., Malerba, D. & Semeraro, G. (1997). Classification in noisy domains by flexible matching, in *Proceedings of the European Symposium on Intelligent Techniques*, Bari, Italy, 45-49.

Fayyad, U. M. & Irani K. B. (1992). On the handling of continuous-valued attributes in decision tree generation, *Machine Learning*, 8, 87-102.

Horak, W. (1985). Office document architecture and office document interchange formats: Current status of international standardization, *IEEE Computer*, 18, 10, 50-60.

Ichino, M. (1994). Feature selection for symbolic data classification, in *New Approaches in Classification and Data Analysis*, Diday, E. *et al.* (Eds.), Springer-Verlag, 423-429.

Malerba, D., Semeraro G. & Esposito F. (1997a). A multistrategy approach to learning multiple dependent concepts, in: *Machine Learning and Statistics: The Interface*, Taylor, C. & Nakhaeizadeh, R. (Eds.), Wiley, 87-106.

Malerba, D., Esposito, F., Semeraro, G., & Caggese, S. (1997b). Handling Continuous Data in Top-Down Induction of First-Order Rules, in *AI*IA 97: Advances in Artificial Intelligence*, M. Lenzerini (Ed.), Lecture Notes in Artificial Intelligence, 1321, Springer, 24-35.

Nagy, G., Seth, S. C., & Stoddard, S. D. (1992). A prototype document image analysis system for technical journals, IEEE Computer, 25, 7, 10-22.

Orkin, M. & Drogin, R. (1990). *Vital Statistics*, McGraw Hill, New York.

Quinlan, J. R. (1986). Induction of decision trees, *Machine Learning*, 1, 81-106.

Quinlan, J. R. (1993). *C4.5: Programs for induction*, Morgan Kaufmann, San Mateo.

Quinlan, J. R. & Cameron-Jones R. M. (1993), FOIL: A midterm report, in *Machine Learning: ECML-93*, Brazdil, P. B. (Ed.), Lecture Notes in Artificial Intelligence, 667, Springer-Verlag, 3-20.

Tang, Y. Y., Yan, C. D., and Suen, C. Y. (1994). Document processing for automatic knowledge acquisition, *IEEE Transactions on Knowledge and Data Engineering*, 6, 1, 3-21.

Logistic Discrimination by Kullback-Leibler type Distance Measures

Salvatore Ingrassia
Istituto di Statistica, Facoltà di Economia, Università di Catania
Corso Italia 55 - 95129 Catania (Italy)
ingrax@dipmat.unict.it

Abstract: We consider the problem of parameter estimation in logistic discrimination. Our approach exploits the minimization of an error function based on distance measures between posterior probability distributions of the classes. In this context we analyze statistical properties of the Kullback-Leibler directed divergence and the euclidean distances from both theoretical and applied point of view.

Keywords: Logistic discrimination, separability measures, parameter estimation.

1. Logistic discrimination

Classification is a wide area of statistical problems and methods. An interesting survey is given in Mineo (1986) who distinguishes three kind of classification problems: cluster analysis, discrimination and mixture decomposition. Here we focus on discrimination, that is on the process of deriving classification rules from samples of classified objects, while the term *classification* refers to applying the rules to new objects of an unknown class.

Let Ω_1 and Ω_2 be two populations of objects and, for $\omega \in \Omega_1 \cup \Omega_2$, let $\mathbf{x} = \mathbf{x}(\omega)$ be the m-dimensional feature vector of ω (according to some suitable criterion). The classical approach to linear discrimination between Ω_1 and Ω_2 is based on some linear function $g(\mathbf{x}, \mathbf{w}) = \mathbf{a}^t\mathbf{x} + b$ such that ω is classified as coming from Ω_1 when $\mathbf{a}^t\mathbf{x} + b > 0$ and from Ω_2 when $\mathbf{a}^t\mathbf{x} + b < 0$, where $\mathbf{w} = (\mathbf{a}, b) \in \mathbb{R}^{m+1}$ are parameters called *weights* and the notation ^{t} denotes vector transpose; more generally the function g could be linear in some function $\mathbf{h} : \mathbb{R}^m \to \mathbb{R}^p$, that is $g(\mathbf{x}, \mathbf{w}) = \mathbf{a}^t\mathbf{h}(\mathbf{x}) + b$.

Linear discrimination can be also approached by considering the *logistic sigmoid* transformation $\psi(z) = (1 + \mathrm{e}^{-z})^{-1}$ which acts on the function $\mathbf{a}^t\mathbf{x} + b$. In this case we obtain the discriminant function:

$$y(\mathbf{x}, \mathbf{w}) = \psi(\mathbf{a}^t\mathbf{x} + b) . \qquad (1)$$

Linear discriminant functions having the form (1) are known in statistical literature as *logistic discriminant functions*; in neural networks domain they are realized by simple perceptrons, see e.g. Bishop (1995). The form (1) is still regarded as a linear discriminant since the decision boundary which it generates is linear as a consequence of the monotonic nature of $\psi(z)$. The fundamental assumption of this approach is that the logarithm of the ratio of the group conditional densities of Ω_1 and Ω_2, say f_1 and f_2, is linear:

$$\ln(f_1(\mathbf{x})/f_2(\mathbf{x})) \;=\; \mathbf{a}^t\mathbf{x} + b\,, \tag{2}$$

see e.g. Anderson (1982), Chapter 8 in McLachlan (1992) for details. The parameters $(\mathbf{a}, b)$ in equation (1) are generally estimated according to the maximum likelihood method. The main problem of this approach is that the likelihood function can be unbounded, for example if Ω_1 and Ω_2 are linearly separable then the likelihood function has not a unique maximum attained for finite $\mathbf{a} \in I\!R^m$.

2. Parameter estimation by probability distance measures

Another approach to the estimation of the parameters $(\mathbf{a}, b)$ in (1) can be constructed considering that, under the assumption (2), the values of $y(\mathbf{x})$ given by (1) can be interpreted as probabilities. Let $\mu(\Omega_1\,|\,\mathbf{x})$ denote the a posteriori probability of class Ω_1, then it results:

$$\mu(\Omega_1\,|\,\mathbf{x}) \;=\; y(\mathbf{x}) \qquad\qquad \mu(\Omega_2\,|\,\mathbf{x}) \;=\; 1 - y(\mathbf{x})\,. \tag{3}$$

Moreover we can introduce a *target probability distribution* ν on $\Omega_1 \cup \Omega_2$ defined as

$$\nu(\Omega_1\,|\,\mathbf{x}) \;=\; \mathbf{1}_{\Omega_1}(\omega) \qquad\qquad \nu(\Omega_2\,|\,\mathbf{x}) \;=\; 1 - \mathbf{1}_{\Omega_1}(\omega) \tag{4}$$

where $\mathbf{1}_{\Omega_1}(\omega)$ is the indicator function of Ω_1. Let $\omega_1, \ldots, \omega_N$ be N given objects from $\Omega_1 \cup \Omega_2$, in particular $\omega_1, \ldots, \omega_{n_1}$ come from Ω_1 and $\omega_{n_1+1}, \ldots, \omega_N$ come from Ω_2 (separate sampling). For each $\omega_n \in \Omega_1 \cup \Omega_2$, we set $\xi_n = \nu(\Omega_1\,|\,\mathbf{x}_n)$; the value ξ_n represents the *identification value* of ω_n. The set

$$\{(\mathbf{x}_n, \xi_n)\}_{n=1,\ldots,N}$$

is called *design set* or *training set*.

In this case the parameters $\mathbf{w} = (\mathbf{a}, b)$ in (1) are selected by minimizing the following quantity called *error function*:

$$\mathcal{E}_N(\mathbf{w}) \;=\; \sum_{n=1}^{N} \phi(y_n, \xi_n) \tag{5}$$

where $y_n = y(\mathbf{x}_n, \mathbf{w}) = \psi(\mathbf{a}^t\mathbf{x}_n + b)$ and $\phi(y, \xi)$ is a function which satisfies the following conditions:

1. $\phi(y, \xi) \geq 0$ for $(y, \xi) \in [0, 1] \times \{0, 1\}$,

2. $\phi(y, \xi)$ is C^2 with respect to y for each $\xi \in \{0, 1\}$,

3. $\phi(y, \xi) = 0$ if and only if $y = \xi$,

4. $(2\xi - 1)\dfrac{\partial \phi}{\partial y} < 0$ for all $y \in (0, 1)$.

We notice that assumption 4 means that if $\xi = 0$ then ϕ is increasing for $y \in (0, 1)$ and if $\xi = 1$ then ϕ is decreasing for $y \in (0, 1)$. In the spirit of Vapnick (1982), the function ϕ is here called *loss function*. Moreover we point out that, by (3) and (4), the loss function can be considered as a distance measure between probability distributions.

In practical problems, a question of interest concerns the convexity of the error function (5). In fact, when the error function $\mathcal{E}_N(\mathbf{w})$ is convex, the absolute minimum can be attained using simple steepest descent algorithms; on the contrary when it is not convex more complicated minimization strategies should be adopted.

Theorem 1 given below is a necessary and sufficient condition for the error function $\mathcal{E}_N(\mathbf{w})$ given in (5) to be convex; it generalizes analogous results given in Devouge (1992). The result is based on the following condition for a real valued function defined on euclidean spaces to be convex (see e.g. Appendix to Chapter 3 in Cecconi and Stampacchia (1983) for details).

Proposition 1 Let B be an open convex set of $I\!R^g$ and $f : B \to I\!R$ be C^2. Then f is strictly convex if and only if its hessian matrix $\mathcal{H}(f)$ is positive definite, namely if it results $\mathbf{v}^t\mathcal{H}(f)\mathbf{v} > 0$ for each $\mathbf{v} \in I\!R^g$, $\mathbf{v} \neq \mathbf{0}$.

Some preliminary notations concern matrix derivatives, see e.g. Chapter 10 in Lütkepohl (1996). The gradient vector of $\mathcal{E}_N(\mathbf{w})$ with respect to the weights $\mathbf{w} \in I\!R^k$, for some integer k, is the vector of the first order partial derivatives of $\mathcal{E}_N$ given by:

$$\frac{\partial \mathcal{E}_N(\mathbf{w})}{\partial \mathbf{w}^t} = \left(\frac{\partial \mathcal{E}_N(\mathbf{w})}{\partial w_1}, \ldots, \frac{\partial \mathcal{E}_N(\mathbf{w})}{\partial w_k} \right) \tag{6}$$

and the Hessian matrix of the second order partial derivatives of $\mathcal{E}_N(\mathbf{w})$ is the $k \times k$ matrix given by:

$$\frac{\partial^2 \mathcal{E}_N}{\partial \mathbf{w}\, \partial \mathbf{w}^t} = \left(\frac{\partial^2 \mathcal{E}_N}{\partial w_i\, \partial w_j} \right) . \tag{7}$$

Theorem 1 The error function (5) is a convex function of the weights for every learning set $\{(\mathbf{x}_n, \xi_n)\}$ of size $N \geq 1$, if and only if the loss function ϕ satisfies the condition:

$$y(1-y)\frac{\partial^2 \phi}{\partial y^2} + (1-2y)\frac{\partial \phi}{\partial y} \;>\; 0$$

for each $(y, \xi) \in (0, 1) \times \{0, 1\}$.

Proof. Without loss of generality, we can consider logistic discriminant function $y(\mathbf{x}) = \psi(\mathbf{a}^t\mathbf{x})$; in fact if we set $\tilde{\mathbf{x}} = (\mathbf{x}^t, 1)^t$ and $\tilde{\mathbf{a}} = (\mathbf{a}^t, b)^t$, then we have $\mathbf{a}^t\mathbf{x} + b = \tilde{\mathbf{a}}^t\tilde{\mathbf{x}}$. By Proposition 1, the function $\mathcal{E}_N(\mathbf{a})$ is convex if and only if it results $\mathbf{v}^t\mathcal{H}(\mathcal{E}_N)\mathbf{v} > 0$ for each $\mathbf{v} \in \mathbb{R}^m$, $\mathbf{v} \neq \mathbf{0}$. Then we have to compute the hessian matrix of $\mathcal{E}_N$ with respect to $\mathbf{a}$.

Preliminarly let us compute the gradient vector of $\mathcal{E}_N$. Let us denote $z_n = \mathbf{a}^t\mathbf{x}_n$ (and thus $y_n = \psi(z_n)$); then (6) yields:

$$\frac{\partial \mathcal{E}_N(\mathbf{a})}{\partial \mathbf{a}^t} = \sum_{n=1}^{N} \frac{\partial \phi(y_n, \xi_n)}{\partial \mathbf{a}^t} . \tag{8}$$

Applying the chain rule of the gradient of real functions with vector arguments (see e.g. Lütkepohl, 1996) to each term of (8), we obtain:

$$\frac{\partial \mathcal{E}_N(\mathbf{a})}{\partial \mathbf{a}^t} = \sum_{n=1}^{N} \frac{\partial \phi(y_n, \xi_n)}{\partial y_n} \frac{\mathrm{d} y_n}{\mathrm{d} z_n} \frac{\partial z_n}{\partial \mathbf{a}^t} = \sum_{n=1}^{N} \frac{\partial \phi(y_n, \xi_n)}{\partial y_n} \psi'(z_n)\, \mathbf{x}_n^t .$$

Afterwards we can compute the hessian matrix $\mathcal{H}(\mathcal{E}_N)$ of $\mathcal{E}_N$ from (7). For simplicity, in the rest of the proof we set $\phi_n = \phi(y_n, \xi_n)$. Thus:

$$\begin{aligned}
\mathcal{H}(\mathcal{E}_N) &= \frac{\partial^2 \mathcal{E}_N(\mathbf{a})}{\partial \mathbf{a}\, \partial \mathbf{a}^t} \\
&= \sum_{n=1}^{N} \frac{\partial}{\partial \mathbf{a}} \left[\frac{\partial \phi_n}{\partial y_n} \psi'(z_n) \right] \mathbf{x}_n^t \\
&= \sum_{n=1}^{N} \left[\frac{\partial}{\partial \mathbf{a}} \left(\frac{\partial \phi_n}{\partial y_n} \right) \psi'(z_n) + \frac{\partial \phi_n}{\partial y_n} \frac{\partial \psi'(z_n)}{\partial \mathbf{a}} \right] \mathbf{x}_n^t .
\end{aligned}$$

Applying again the chain rule of the gradient of real functions with vector arguments, we get:

$$\begin{aligned}
\mathcal{H}(\mathcal{E}_N) &= \sum_{n=1}^{N} \left[\frac{\partial^2 \phi_n}{\partial y_n^2} \psi'^2(z_n)\, \mathbf{x}_n + \frac{\partial \phi_n}{\partial y_n} \psi''(z_n)\, \mathbf{x}_n \right] \mathbf{x}_n^t \\
&= \sum_{n=1}^{N} \left[\frac{\partial^2 \phi_n}{\partial y_n^2} \psi'(z_n) + \frac{\partial \phi_n}{\partial y_n} \frac{\psi''(z_n)}{\psi'(z_n)} \right] \psi'(z_n)\, \mathbf{x}_n\, \mathbf{x}_n^t
\end{aligned}$$

Moreover, as $\psi(z) = (1 + \mathrm{e}^{-z})^{-1}$, then it results $\psi'(z) = \psi(z)(1 - \psi(z))$ and $\psi''(z) = \psi'(z)(1 - 2\psi(z))$. Then, as we set $y_n = \psi(z_n)$, it follows:

$$\mathcal{H}(\mathcal{E}_N) = \sum_{n=1}^{N} \left[\frac{\partial^2 \phi_n}{\partial y_n^2} y_n(1 - y_n) + \frac{\partial \phi_n}{\partial y_n} (1 - 2y_n) \right] y_n(1 - y_n)\, \mathbf{x}_n \mathbf{x}_n^t .$$

Now, for any $\mathbf{v} \in \mathbb{R}^m$, let us consider the quantity:

$$\mathbf{v}^t \mathcal{H}(\mathcal{E}_N) \mathbf{v} = \mathbf{v}^t \sum_{n=1}^{N} \left[\frac{\partial^2 \phi_n}{\partial y_n^2} y_n(1 - y_n) + \frac{\partial \phi_n}{\partial y_n} (1 - 2y_n) \right] y_n(1 - y_n)\, \mathbf{x}_n \mathbf{x}_n^t \mathbf{v}$$

$$= \sum_{n=1}^{N} \left[\frac{\partial^2 \phi_n}{\partial y_n^2} y_n(1-y_n) + \frac{\partial \phi_n}{\partial y_n}(1-2y_n) \right] y_n(1-y_n)\, \mathbf{v}^t \mathbf{x}_n \mathbf{x}_n^t \mathbf{v}$$

$$= \sum_{n=1}^{N} \left[\frac{\partial^2 \phi_n}{\partial y_n^2} y_n(1-y_n) + \frac{\partial \phi_n}{\partial y_n}(1-2y_n) \right] y_n(1-y_n)\, (\mathbf{v}^t \mathbf{x}_n)^2 .$$

By Proposition 1, $\mathcal{H}(\mathcal{E}_N)$ is convex for every learning set $\{(\mathbf{x}_n, \xi_n)\}_{n=1,\dots,N}$ of size $N \geq 1$ if and only if for each $\mathbf{v} \in \mathbb{R}^m$, $\mathbf{v} \neq \mathbf{0}$, we have $\mathbf{v}^t \mathcal{H}(\mathcal{E}_N)\mathbf{v} > 0$. In this last expression the quantity $y_n(1-y_n)$ is stricly positive, then $\mathbf{v}^t \mathcal{H}(\mathcal{E}_N)\mathbf{v}$ is positive if and only if it results:

$$\frac{\partial^2 \phi}{\partial y^2} y(1-y) + \frac{\partial \phi}{\partial y}(1-2y) \;\; > \;\; 0$$

for each $(y, \xi) \in (0,1) \times \{0,1\}$. This completes the proof. ■

Usually one considers the classical squared distance $\phi_2(y,\xi) = (y-\xi)^2$. In recent years, some authors have proposed distance measures based on the Kullback-Leibler directed divergence, say $\phi_{KL}(y,\xi)$. In the next section we compare the properties of the error function $\mathcal{E}_N(\mathbf{w})$ based respectively on the distances ϕ_2 and ϕ_{KL} from both theoretical and applied point of view.

3. Error functions based on Kullback-Leibler type distances

The Kullback-Leibler directed divergence between two probability distributions $P = \{p_1, \dots, p_n\}$ and $Q = \{q_1, \dots, q_n\}$ on a finite set of n points is defined as:

$$\delta_{KL}(P,Q) \;\; = \;\; -\sum_{k=1}^{n} p_k \ln \frac{q_k}{p_k} . \tag{9}$$

Properties of the distance measures of Kullback-Leibler type can be found in Kannappan and Sahoo (1992). In our case, expression (9) gives:

$$\begin{aligned} \delta_{KL}(\mu,\nu) &= -\nu(\Omega_1 \,|\, \mathbf{x}) \ln \frac{\mu(\Omega_1 \,|\, \mathbf{x})}{\nu(\Omega_1 \,|\, \mathbf{x})} - \nu(\Omega_2 \,|\, \mathbf{x}) \ln \frac{\mu(\Omega_2 \,|\, \mathbf{x})}{\nu(\Omega_2 \,|\, \mathbf{x})} \\ &= -\nu(\Omega_1 \,|\, \mathbf{x}) \ln \mu(\Omega_1 \,|\, \mathbf{x}) + \nu(\Omega_1 \,|\, \mathbf{x}) \ln \nu(\Omega_1 \,|\, \mathbf{x}) + \\ &\quad -\nu(\Omega_2 \,|\, \mathbf{x}) \ln \mu(\Omega_2 \,|\, \mathbf{x}) + \nu(\Omega_2 \,|\, \mathbf{x}) \ln \nu(\Omega_2 \,|\, \mathbf{x}) . \end{aligned}$$

We have that $\nu(\Omega_i \,|\, \mathbf{x}) = 1$ for $\mathbf{x} \in \Omega_i$, with $i = 1,2$; moreover we set $0 \ln 0 = 0$ as $u \ln u \to 0$ for $u \to 0^+$. Then, by (4) and (3), it follows:

$$\begin{aligned} \delta_{KL}(\mu,\nu) &= -\nu(\Omega_1 \,|\, \mathbf{x}) \ln \mu(\Omega_1 \,|\, \mathbf{x}) - \nu(\Omega_2 \,|\, \mathbf{x}) \ln \mu(\Omega_2 \,|\, \mathbf{x}) \\ &= -\xi \ln y - (1-\xi) \ln(1-y) . \end{aligned}$$

Given that $\xi \in \{0,1\}$, we can rewrite this expression as:

$$\delta_{KL}(\mu,\nu) \;\; = \;\; -\ln(1 + 2y\xi - \xi - y) .$$

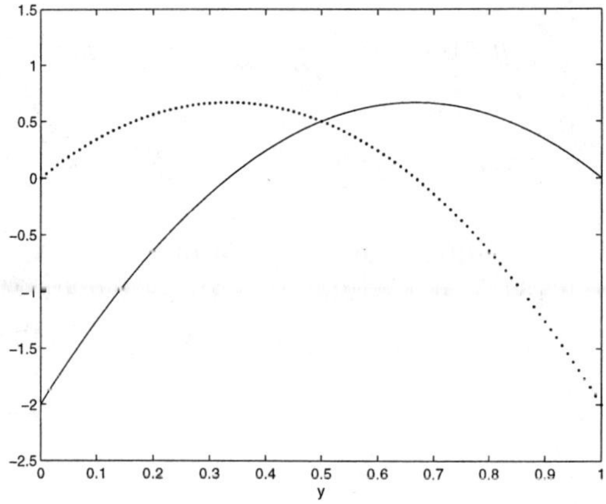

Figure 1: Plot of the functions $4y - 6y^2$ ($\xi = 0$, dotted line) and $8y - 6y^2 - 2$ ($\xi = 1$, solid line)

We can introduce a loss function $\phi_{KL}(y, \xi)$ equal to the Kullback-Leibler directed divergence between ν and μ, that is $\phi_{KL}(y, \xi) = \delta_{KL}(\mu, \nu) = -\ln(1 + 2y\xi - \xi - y)$. By an application of Theorem 1, we find the following result.

Proposition 2 The error function $\mathcal{E}_N(\mathbf{w})$ is in general not convex when the distance $\phi_2(y, \xi)$ is adopted, whereas it is in general convex when the distance $\phi_{KL}(y, \xi)$ is considered.

Proof. The proof follows directly from Theorem 1. When $\phi_2 = (y - \xi)^2$, then $\partial\phi_2/\partial y = 2(y - \xi)$ and $\partial^2\phi_2/\partial y^2 = 2$. In this case it results:

$$y(1-y)\frac{\partial^2\phi_2}{\partial y^2} + (1-2y)\frac{\partial\phi_2}{\partial y} = \begin{cases} -6y^2 + 4y & \text{if } \xi = 0 \\ -6y^2 + 8y - 2 & \text{if } \xi = 1 \end{cases}$$

These two functions are plotted in Figure 1, in particular when $\xi = 0$ it results $-6y^2 + 4y < 0$ for $y > 2/3$ and when $\xi = 1$ it results $-6y^2 + 8y - 2 < 0$ for $y < 1/3$. Hence the condition of Theorem 1 is not satisfied. In the other case $\phi_{KL}(y, \xi) = -\ln(1 + 2y\xi - \xi - y)$, after some algebra we obtain:

$$y(1-y)\frac{\partial^2\phi_{KL}}{\partial y^2} + (1-2y)\frac{\partial\phi_{KL}}{\partial y} = 1 .$$

In this case condition of Theorem 1 is satisfied for all $(y, \xi) \in (0, 1) \times \{0, 1\}$. This completes the proof. ∎

4. Numerical studies

From a theoretical point of view, Proposition 2 implies that the use of distance

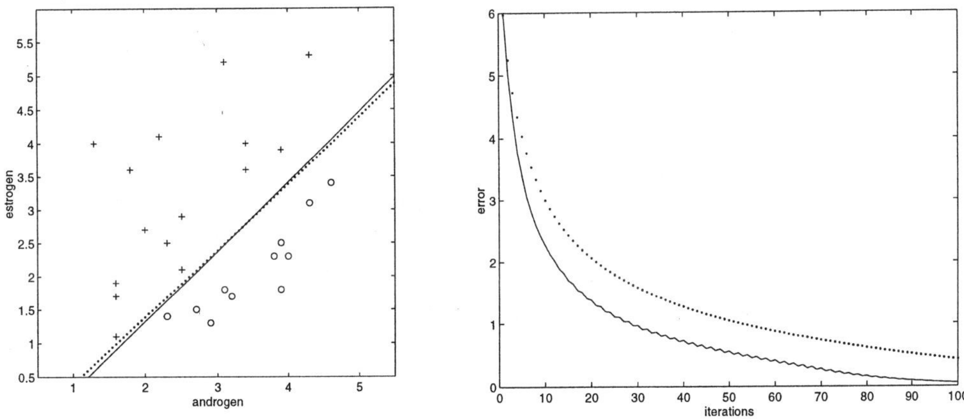

Figure 2: a) Values of urinary androsterone and etiocholanolone in in 11 healthy heterosexual (∘) and 15 healthy homosexual males (+) in mg/24 hours. Decision surfaces obtained by minimizing error function based on ϕ_{KL} (solid line) and on ϕ_2 (dotted line). b) Sum of squared errors vs iteration: ϕ_{KL} (solid line) and on ϕ_2 (dotted line).

measure ϕ_{KL} is preferable with respect to the euclidean distance ϕ_2 in order to estimate the weights $\mathbf{w} = (\mathbf{a}, b)$ in (1). We have investigated also their properties from a practical point of view using some sets of real data.

To begin with, we analyzed the speed of convergence towards the absolute minimum of the error function (5) based on the loss functions ϕ_2 and ϕ_{KL}. The Figure 2a shows the values of urinary androsterone and etiocholanolone in healthy heterosexual and homosexual males in mg/24 hours (data from Margolese (1970) reprinted in Hand (1981)), it is $m = 2$, $n_1 = 1$ and $n_2 = 15$. In both cases the error function (5) was minimized 100 times by a steepest descent algorithm starting from a random point randomly chosen with uniform distribution on $[-1, 1]^{m+1}$.

The numerical experiments showed that, adopting the ϕ_{KL} distance, the convergence to the absolute minimum of the error function is generally obtained in a smaller number of iterations than in the ϕ_2 case. A typical output is plotted in Figure 2, in particular we remark that the Figure 2b gives the plots of the sum of the squared errors vs iteration for both the loss functions. Actually, in order to make a congruent comparison between the two loss functions ϕ_2 and ϕ_{KL}, we compared the related sum of squared errors: this justifies the "flickering" in the error curve concerning ϕ_{KL}.

Another numerical aspect of the minimization of the error function (5) concerns the importance of standardization of the data when they have different order of magnitude. It has been exhibited by using the data concerning three species of flea beetle *Chaetocnema* (data from Lubischew (1962)). The discrimination is based on the maximal width of aedeagus in the forepart and the front angle of the aedeagus. When the original data were taken into account the minimization of expression (5) was critical; on the contrary a complete separation has been obtained when the input data were preliminarly standardized.

Acknowledgements
The author thanks G.Lunetta, A. Mineo for helpful suggestions and comments. Thanks are also duc to S. Zani who suggested to deepen the concepts here proposed in the context of the analysis of contingency tables by parameter estimation method based on the minimum discrimination information (see Read and Cressie (1988)). This provides ideas for a successive paper.

References

Anderson J.A. (1982). Logistic discrimination, in P.R. Krishnaiah and L.N. Kanal (Eds.), *Classification, Pattern Recognition and Reduction of Dimensionality*, Vol. 2 of *Handbook of Statistics*, 169-191, North Holland, Amsterdam.

Bishop C.M. (1995). *Neural Networks for Pattern Recognition*, Clarendon Press, Oxford.

Cecconi J. and Stampacchia G. (1983). *Analisi Matematica 2: Funzioni di più Variabili*, Liguori Editore, Napoli.

Devouge C. (1992). Quelques aspects mathématiques de l'auto-organisation neuronale et des perceptrons multicouches, Thèse de Doctorat, Univ. Paris 11.

Lubischew A.A. (1962). On the use of discriminant functions in taxonomy, *Biometrics*. **18**, 455-477.

Lütkepohl H. (1996), *Handbook of Matrices*, John Wiley & Sons, Chichester.

Kannappan P.L. & Sahoo P.K. (1992). Kullback-Leibler type distance measures between probability distributions, *Journal of Mathematical and Physical Sciences*, **26**, n. 5, 443-454.

McLachlan G.J. (1992). *Discriminant Analysis and Statistical Pattern Recognition*, John Wiley & Sons, New York.

Margolese M.S. (1970). Homosexuality: a new endocrine correlate, *Horm. and Behaviour*, **1**, 151-155.

Mineo A. (1986). Problemi e metodi di classificazione, in *Atti XXXIII Riunione Scientifica della Società Italiana di Statistica*, 1986, vol. 1, 83-108, Cacucci Editore, Bari.

Read T.R.C. & Cressie N.A.C. (1988). *Goodness-of-fit Statistics for Discrete Multivariate Data*, Springer-Verlag, New York.

Vapnick V. (1982). *Estimation of Dependencies Based on Empirical Data*, Springer-Verlag, New-York.

An Empirical Discrimination Algorithm based on Projection Pursuit Density Estimation

Angela Montanari Daniela G. Calò
Dipartimento di Scienze Statistiche, Università di Bologna,
via Belle Arti 41, 40126 Bologna, Italy;
montanar@stat.unibo.it, calo@stat.unibo.it.

Abstract: In this paper a nonparametric method for discriminant analysis is proposed, based on a group separation oriented version of projection pursuit density estimation. Each population is separated in turn from the remaining ones, considered as a whole, by approximating the boundary between them through the composition of some informative directions, chosen according to an appropriate discrimination criterion. A coherent allocation rule is proposed, too. Simulation studies have shown that this method represents a valid solution for problems when the parametric approaches are not flexible enough and sample sizes are too small to use classical nonparametric methods.

Keywords: Projection Pursuit, Projection Pursuit Density Estimation, Nonparametric Discriminant Analysis.

1. Introduction

The principal goal of discriminant analysis is to assign a new object to one of two or more groups $\Pi_1, \Pi_2, ..., \Pi_g$, on the basis of m measured characteristics $\boldsymbol{x}' = (x_1, x_2, \ldots, x_m)$ associated with the object. An allocation rule is generally determined by optimizing an objective function, describing group separation, which is based on character densities in the various groups (McLachlan, 1992). When nothing is a priori known about such densities, nonparametric density estimation is called for (Hand, 1982). However, this approach is not free from serious drawbacks, as classical multivariate nonparametric density estimators, such as kernel and nearest neighbour, are heavily biased by what is generally called "the curse of dimensionality".

To overcome this problem Friedman, Stuetzle and Schroeder (1984) propose the use of projection pursuit methods for density estimation (*ppde*). The basic ideas of projection pursuit can be traced back to 1974 when Friedman and Tukey formulated the method as the search for the linear projections (whose generic vector coefficient will be denoted by $\boldsymbol{a}$) of a multivariate data set which maximize a user defined measure of interestingness which they called projection index (denoted by $I(\boldsymbol{a})$. See Montanari, Guglielmi (1996) and Montanari, Lizzani (1997) for a thorough discussion). Within the density estimation context this method suggests to approximate an m-variate density $p(\boldsymbol{x})$ by a density of the form:

$$p^K(\boldsymbol{x}) = p^0(\boldsymbol{x})\prod_{k=1}^{K} f_k(\boldsymbol{a}_k'\boldsymbol{x}), \qquad (1)$$

where $p^0(\boldsymbol{x})$ is a "null" model (e. g. a normal density with the same mean and covariance as p) and f_k are the so called "augmenting functions" with $\boldsymbol{a}_k \in \mathbf{R}^m$ determining univariate projections.
As Huber (1985) suggests, the sequence $\{f_k, \boldsymbol{a}_k\}$ can be viewed either "synthetically", as a series of modifications to p^0, or "analytically", as a series of modifications to p that "strips away" its structure step by step.
The synthetic approach determines $\{f_k, \boldsymbol{a}_k\}$ such that the sequence $p^k(\boldsymbol{x}) = p^{k-1}(\boldsymbol{x}) f_k(\boldsymbol{a}_k'\boldsymbol{x})$, expressed as a recursion relation derived from (1), converges to p as fast as possible: at each step, it tries to attain the largest improvement of the current model p^{k-1}. Symmetrically, the analytic viewpoint starts with $p^K = p$ and sequentially looks for $\{f_k, \boldsymbol{a}_k\}$ such that $p^{k-1}(\boldsymbol{x}) = p^k(\boldsymbol{x}) f_k^{-1}(\boldsymbol{a}_k'\boldsymbol{x})$ converges to p^0 using the smallest number of steps.
When, according to Friedman, Stuetzle and Schroeder, relative entropy (*RE*) is used to measure the quality of the approximation, in the former approach the optimization problem is solved by

$$f_k(\boldsymbol{a}_k'\boldsymbol{x}) = p(\boldsymbol{a}_k'\boldsymbol{x}) / p^{k-1}(\boldsymbol{a}_k'\boldsymbol{x}) \quad \text{and} \quad \boldsymbol{a}_k = arg \max_{\boldsymbol{a}, \|\boldsymbol{a}\|=1} RE\left(p_{\boldsymbol{a}}, p_{\boldsymbol{a}}^{k-1}\right):$$

in other words, since this optimal form of f_k establishes marginal agreement along $\boldsymbol{a}_k$, the augmenting function is best applied to p^{k-1} along the direction in which the current model and the objective function differ most. Its analytic counterpart is

$$f_k(\boldsymbol{a}_k'\boldsymbol{x}) = p^k(\boldsymbol{a}_k'\boldsymbol{x}) / p^0(\boldsymbol{a}_k'\boldsymbol{x}) \quad \text{and} \quad \boldsymbol{a}_k = arg \max_{\boldsymbol{a}, \|\boldsymbol{a}\|=1} RE\left(p_{\boldsymbol{a}}^k, p_{\boldsymbol{a}}^0\right).$$

Thus, while in the synthetic algorithm the projection index is based on a comparison between the data (coming from p) and the current estimate (updated at each step of the procedure), the analytic approach only involves a comparison between the density estimate (whose non-normal features are cleaned away step by step) and the null model p^0. For this reason, which will

be reconsidered later, in what follows reference will be made to the analytic approach and not to the synthetic one which has had a wider use in statistical literature.
As it has been shown above, in Friedman Stuetzle and Schroeder's approach the projection index (i.e. marginal relative entropy) is oriented to reconstruct the multivariate density in an "optimal way", but this is not necessarily the best choice when the estimated density has to be used in the context of discriminant analysis.

2. Projection pursuit and discriminant analysis

The nonparametric method for discriminant analysis we present in this paper is based on projection pursuit density estimation, but resorts to a new projection index, more closely oriented to group separation.
The idea of projection pursuit density estimation has already been applied to discriminant analysis by Polzehl (1995), who uses the expected overall loss of the allocation rule as a projection index. That solution generalizes a proposal due to Posse (1992), which was limited to the derivation of a two-group linear discriminant function.
We share Polzehl's view that in the classification context there is some interest in choosing the same updating direction for all populations and corresponding densities, but we start from this consideration to derive a proposal whose methodological aspects profoundly differ from Polzehl's.
Instead of estimating g distinct densities for the g groups, our estimation approach regards the density mixture. This choice, which also has the further effect of giving more stable density estimates, is made necessary by our particular projection index.
In order to define it we must first introduce the concept of "false neighbouring" observations along the direction $\boldsymbol{a}_k$. Two units i and j, coming from two different populations, are said to belong to the set of "false neighbours" in the $\boldsymbol{a}_k$ direction, $\mathrm{FN}(\boldsymbol{a}_k)$, if their distance along $\boldsymbol{a}_k$ is less than a threshold ${}_{\mathrm{FN}}t$ that has to be specified by the researcher: that is,

$$i, j \in \mathrm{FN}(\boldsymbol{a}_k) \text{ if } \left|\boldsymbol{a}_k'\boldsymbol{x}_i - \boldsymbol{a}_k'\boldsymbol{x}_j\right| < {}_{\mathrm{FN}}t .$$

As the goal of discriminant analysis is to separate groups as much as possible, those directions along which the number of "false neighbours" is minimized will be preferred. This requirement is however not strong enough for a discriminant analysis purpose as, the number of false neighbours being equal (Fig. 1), those directions along which the height of the density insisting on them is smaller will be preferred.

Figure 1: *Kernel density estimates of two homoscedastic mixtures having the same number of "false neighbours".*

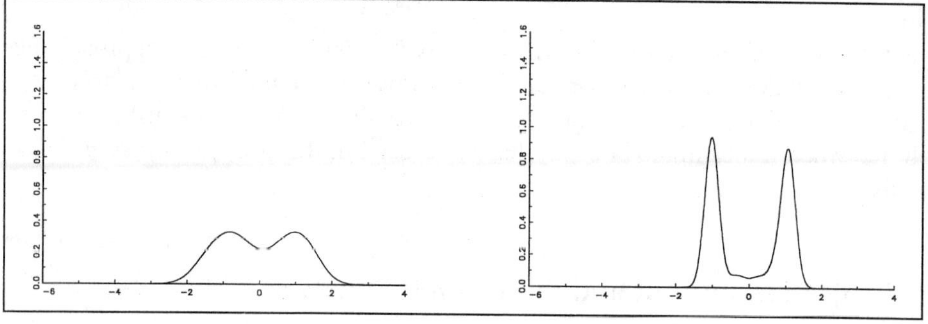

This suggests to compute the distance between "false neighbours" along $\boldsymbol{a}_k$ as

$$d_{ij}(\boldsymbol{a}_k) = \frac{1}{2}\left(\frac{1}{p^k(\boldsymbol{a}_k'\boldsymbol{x}_i)} + \frac{1}{p^k(\boldsymbol{a}_k'\boldsymbol{x}_j)}\right) \quad \text{where } i, j \in \mathrm{FN}(\boldsymbol{a}_k), \tag{2}$$

and to use an average distance as the projection index to be maximized

$$I(\boldsymbol{a}_k) = \underset{i,j \in \mathrm{FN}(\boldsymbol{a}_k)}{average}\left[d_{ij}(\boldsymbol{a}_k)\right]. \tag{3}$$

In order to make meaningful comparisons of different directions, the projected data should be standardized before computing the index.

The projection index (3) closely resembles the measure suggested by Wong and Lane (1983) to solve a cluster analysis problem (see Calò (1997) for an application of *ppde* in this context). However, their measure, aimed at detecting Hartigan's high density clusters (1977), was computed on the multivariate nearest-neighbour density estimate and defined for "any" neighbouring observations, the distinction between "true" and "false" neighbours being meaningless, in that context.

The *ppde*-based discriminant procedure using (3) as the projection index envisages two nested phases: a population separation phase and a population boundary approximation one.

Within each step of the separation stage, the boundary approximation involves the search for those directions along which a given population is best separated from the remaining ones considered as a whole. The number of directions may be different for the different separation steps, and can be chosen by inspecting the projection index trend, stopping when it becomes stable or decreases. At the generic k-th step the highlighted structure is incorporated in p^k through the augmenting functions and cleaned away from the data by "gaussianizing" them (Friedman, 1987).

The separation phase is stopped after a number of steps which is the number of the populations under study (however, when only two populations have to be separated, one separation step is enough). This sequential procedure leads to $\hat{p}$, that is the discrimination oriented density estimate of p, which will be used to derive an allocation rule.

What has been said up to now shows that the projection index we propose involves the data only (and not a current model); this is the reason why, in this context, the *ppde* has to be performed following the analytic approach.

When the results derived so far have to be used to devise an allocation rule the concept of "true neighbours" of a new unit z, that has to be classified, must be introduced.

Supposing that q_l directions have turned out to be necessary in order to optimally approximate the boundary between population Π_l and the remaining ones, the generic unit j of the training set is said to belong to the set of z's true neighbours (TN) if it lies within a ${}_{\mathrm{TN}}t_l$ neighbourhood of z, for $l = 1, 2, \ldots, g$; that is

$$(1/q_l)\sum_{i=1}^{q_l}\left[\left|\boldsymbol{a}_i'\boldsymbol{x}_j - \boldsymbol{a}_i'\boldsymbol{x}_0\right| / \hat{sd}(\boldsymbol{a}_i'\boldsymbol{x})\right] < {}_{\mathrm{TN}}t_l, \qquad l = 1, 2, \ldots, g$$

where $\boldsymbol{x}_0$ denotes the m-dimensional observation on z.

Coherently with Wong and Lane's use of the single linkage method in cluster analysis, the new unit z is then allocated to the population to which the nearest unit belonging to its TN belongs:

$$z \to \Pi_l \quad \text{if} \quad \exists j \in \Pi_l : \; \frac{1}{2}\left(\frac{1}{\hat{p}(\boldsymbol{x}_0)} + \frac{1}{\hat{p}(\boldsymbol{x}_j)}\right) = \min_{j \in \mathrm{TN}},$$

but different allocation rules could be devised, too.

The performances of the method are heavily conditioned by the choice of ${}_{\mathrm{FN}}t$ and ${}_{\mathrm{TN}}t_l$, $l = 1, 2, \ldots, g$. As ${}_{\mathrm{FN}}t$ plays the role of a smoothing parameter, analogous for instance to the window width of kernel density estimation, it can be selected through the classical criteria suggested for the latter density estimation method. On the contrary the definition of ${}_{\mathrm{TN}}t_l$, $l = 1, 2, \ldots, g$, is a little more tricky. Each element can be chosen, through cross validation, as the value minimizing the classification error rate within the training set.

3. Concluding remarks

The proposed methodology (*ppda*), translated into a GAUSS algorithm, has been applied to a simulated data set for the two group case and compared to the classical linear and quadratic allocation rules and to two nonparametric procedures based on kernel density estimation and on Polzehl's solution.
The following situation has been considered:

$$\Pi_1 : \frac{1}{2} N_5\left(\begin{pmatrix}-6\\0\\0\\0\\0\end{pmatrix}, S\right) + \frac{1}{4} N_5\left(\begin{pmatrix}0\\-6\\0\\0\\0\end{pmatrix}, S\right) + \frac{1}{4} N_5\left(\begin{pmatrix}0\\6\\0\\0\\0\end{pmatrix}, S\right)$$

$$\Pi_2 : \frac{1}{2} N_5\left(\begin{pmatrix}0\\0\\0\\0\\0\end{pmatrix}, S\right) + \frac{1}{4} N_5\left(\begin{pmatrix}6\\-6\\0\\0\\0\end{pmatrix}, S\right) + \frac{1}{4} N_5\left(\begin{pmatrix}6\\6\\0\\0\\0\end{pmatrix}, S\right)$$

where $S = \text{diag}(1,1,25,25,25)$, with sample size 60 for each population. The last three variables represent noisy variables and have been purposely included in order to prove the ability of projection pursuit methods to overcome the "curse of dimensionality". The "false neighbourhood" threshold ${}_{\text{FN}}t$ has been determined by the "rule of thumb" suggested for kernel density estimation and the projection index has been computed as the arithmetic mean distance between "false neighbours". Along each direction, the data density has been estimated by univariate normal kernel estimators with bandwidth chosen according to Sheather and Jones (1991) method. Within the unique separation step, two directions have turned out to be necessary in order to approximate the boundary between Π_1 and Π_2; further directions do not give a relevant improvement in the projection index value. Probability of misclassification has been estimated by classifying 1000 additional observations for each population. The simulation results, reported in Table 1, are expressed in terms of mean probability of misclassification (MPMC).

Table 1: *Simulation results on 100 replications*

	MPMC	*std(MPMC)*
linear rule	0.24	0.0006
quadratic rule	0.24	0.0012
nonparametric rule (kernel)[1]	0.12	0.0015
Polzehl's ppda	0.03	0.0011
ppda rule	0.03	0.0017

1. The gaussian kernel window width has been estimated by maximum likelihood cross validation.

The obtained results clearly highlight the good performances of the *ppde*-based discrimination rules. With respect to the other parametric and nonparametric considered methods, our solution is in line with Polzehl's.

We have also compared the performances of our method and of Fisher's discriminant analysis in the case of two 4-variate homoscedastic normal populations with identity covariance matrix and expectation given by $\mu_{1j} = 1 = -\mu_{2j}$, $j = 1,\ldots,4$, from which 10 samples of size $n_1 = n_2 = 100$ have been generated. Table 2 shows the angles in degrees between the true optimal direction and its estimates by Fisher's linear discriminant rule and by our method. Fisher's results are only slightly more accurate than ours, but this is not surprising, since Fisher's discrimination takes into account that both populations are multivariate homoscedastic normals while our approach is fully nonparametric.

Table 2: *Angles (in degrees) between the true optimal direction and its estimates by Fisher's method and our ppde-based discrimination procedure.*

Fisher's method	*Proposed ppda*
5.077	10.053
8.289	6.695
4.484	6.300
8.633	15.403
4.600	8.055
3.580	4.846
5.805	8.814
7.408	9.491
7.845	11.211
5.187	8.783

The simulation results seem therefore to support the conclusion that the proposed classification method represents a valid solution for problems when the parametric approaches are not flexible enough and sample sizes are too small to use classical nonparametric methods.

References

Calò, D. G. (1997). *La stima di densità non parametrica secondo la metodologia projection pursuit: proposta di un suo possibile impiego nella cluster analysis*, Phd thesis, Dipartimento di Scienze Statistiche, Università degli Studi di Bologna, Italia.

Friedman, J.H. (1987). Exploratory Projection Pursuit, *Journal of the American Statistical Association*, 82, 249-266.

Friedman, J. H. & Stuetzle, W. & Schroeder, A. (1984). Projection Pursuit Density Estimation, *Journal of the American Statistical Association*, 79, 599-608.

Friedman, J. H. & Tukey, J. W. (1974). A Projection Pursuit Algorithm for Exploratory Data Analysis, *IEEE Transactions on Computers*, C-23, 881-889.

Hand, D. J. (1982). *Kernel Discriminant Analysis*, Chichester: Research Studies Press, Letchworth, England.

Hartigan, J. A. (1977). Distribution problems in clustering, in *Classification and Clustering* (J. Van Ryzin Ed.).

Huber, P. J. (1985). Projection Pursuit, *The Annals of Statistics*, 13, 435-475.

McLachlan, G. J. (1992). *Discriminant Analysis and Statistical Pattern Recognition*, John Wiley & Sons, New York.

Montanari, A. & Guglielmi, N. (1996). Gli indici di proiezione nella projection pursuit, *Statistica*, 1, 63-86.

Montanari, A. & Lizzani, L. (1998) Projection pursuit and departure from unimodality, *Metron*, in press.

Polzehl, J. (1995). Projection Pursuit Discriminant Analysis, *Computational Statistics and Data Analysis*, 20, 141-157.

Posse, C. (1992). Projection Pursuit Discriminant Analysis for two groups, *Communications in Statistics*, Theory and Methods, 21 (1), 1-19.

Sheather, S. J. & Jones M. C. (1991). A Reliable Data Based Bandwidth Selection Method for Kernel Density Estimation, *Journal of the Royal Statistical Society* B, 53, 683-690.

Wong, M. A. & Lane, T. (1983). A *k*-th Nearest Neighbouring Clustering Procedure, *Journal of the Royal Statistical Society* B, 45, 362-368.

Notes on Methods for Improving Unstable Classifiers

Rossella Miglio - Marilena Pillati
Dipartimento di Scienze Statistiche, Università di Bologna
Via Belle Arti, 41 - 40126 Bologna
email: miglio@stat.unibo.it - pillati@stat.unibo.it

Abstract: Methods for improving the predictive power of unstable classifiers based on combining multiple versions of these have received much attention in the last few years. The aim of this paper is to compare some of the proposed methods with a focus on neural network classifiers. Experimental results are provided to illustrate, in different data sets, the performances of different methods of combining the output of several neural classifiers.

Keywords: supervised classifiers, combining, neural networks.

1. Introduction

The idea of combining predictors instead of selecting the single best has been studied by the statistical community for a long time (see, for example, Granger, 1989). This idea implicitly assumed that one could not identify the "true" model, but different forecasting models were able to capture different aspects of the available information, with gains in accuracy.

More recently, this concept has been extended to the combination of multiple versions of the same predictor. In fact, it is well known that some classification and regression methods "*are unstable in the sense that small perturbations in their training sets or in construction may result in large changes in the constructed predictor*" (Breiman, 1994). Improvement may occur by generating multiple versions of the predictor, by perturbing the training set or the construction method and then pooling the available outputs into a single predictor. Stacked generalization, introduced by Wolpert (1992), provides a method that uses cross-validation data and least squares to determine the coefficients of a linear combination of the different predictors. In the context of regression and classification trees, Breiman (1994) suggests to generate different training sets by making bootstrap replicates of the original learning set. Multiple classifiers are constructed using these different sets and then combined to obtain the so called "bagging" predictor by majority voting, whereas for regression problems the final solution is obtained by averaging all the available predictors.

When classification is performed using a neural network, multiple classifiers can derive from bootstrap samples. However, it is always necessary to train a multitude of models for each sample, because different starting values for the parameters may lead to different predictors, and may result in differences in performances. A multitude of models is thus available using the same training set, just starting from different points in the parameter space.

The main purpose of this investigation is to compare different ways of generating and combining multiple neural network classifiers. Section 2 introduces neural network classifiers. Section 3 discuss the usefulness of combining and describes different methods of training and combining neural classifiers. Section 4 presents the results obtained in three data sets and discuss the performances of the different solutions. Section 5 gives the conclusions and directions for future researches.

2. Neural network classifiers

In the last few years, artificial neural networks have found an important role in classification. We will focus on the so called multilayer perceptron, which is the most popular type of network for supervised classification problems. For more general accounts of neural computation, see for example Hertz *et al.* (1991).

The multilayer perceptron classifier performs a non linear transformation $g(.)$ of the features vectors $\mathbf{x} = (x_1,...,x_s)'$ into K outputs $o = (o_1,...,o_K)'$, that define the class membership of the objects. More specifically, in the so called single hidden layer perceptron the output function is defined as

$$o = g(\mathbf{x},\theta) = \mathbf{F}[\alpha \cdot \Psi(\gamma'\mathbf{x})] \tag{1}$$

where $\theta=(\alpha_1,...,\alpha_K,\gamma_1',...,\gamma_h')'$, $\gamma=(\gamma_1,\gamma_2,...,\gamma_h)$ and $\alpha=(\alpha_1,\alpha_2,...,\alpha_K)'$ are weights matrices of dimensions $(s+1)\times h$ and $K\times(h+1)$ respectively, h is the number of hidden units, Ψ is some given non linear mapping from R^h to R^h and F is a vector function from R^K to R^K. For classification problems, if the activation function of the k-th output unit is defined as follows

$$o_k = g_k(x,\theta) = \exp(\alpha_k \cdot \Psi(\gamma' x))\left[\sum_k \exp(\alpha_k \cdot \Psi(\gamma' x))\right]^{-1} \tag{2}$$

the output of the classifier can be interpreted as an estimate of the *a posteriori* probability of the k-th class.

The role of the learning or training process is to determine the best values for network weights θ on the basis of a set of n observations, the training sample. This is generally done by optimizing some appropriate objective function over all observations. Different optimization procedures can be used and to avoid overfitting, the performance can be better measured either by cross-validation on the training data or by an independent validation set.

3. Combining neural networks

Within the past decade considerable literature regarding the combination of classifiers has been accumulated. Besides the empirical evidence that shows that classification error rate can be reduced by learning and using multiple models, there are also relevant theoretical results (see, for instance, Perrone & Cooper, 1993; Hashem *et al.*, 1994; Breiman, 1994; Tibshirani, 1996).
The improvement of the ensemble classifier performance relates to the bias-variance trade-off. Generalizing the decomposition of the prediction error in the regression context, several authors define the concept of bias and variance of a classifier. Tibshirani (1996), for instance, gives a decomposition of the prediction error of a classifier under general error measures and derives bootstrap estimates of its components. Unstable classifiers, just as trees and neural networks, characteristically have low bias, but high variance. It was shown that the main effect of combining classifiers is to reduce the variance and then the resultant predictor can be more accurate than the original one. As stated by Breiman (1996) for bagging classifiers, accuracy increases if the prediction method is good but unstable. On the other hand, aggregation can make stable but biased procedures worse.
Generating multiple classifiers via bootstrap replicates of the training set requires a lot of computational efforts, particularly with computer intensive methods such as neural networks. In fact empirical evidence shows that a significant increase of a classifier accuracy cannot be obtained by considering only a few replicates.
As an alternative to bagging neural classifier, we consider a straightforward and less time consuming way to generate a multitude of models using the same training set. Starting the training process of a neural classifier with different weight initializations can lead to classifiers with the same architecture but different performances. Instead of selecting the best one we resume the information of this set of classifiers into a single aggregated predictor.
The most popular way of combining multiple classifiers is via majority voting that can be applied to any type of classifiers.
Consider multilayer perceptron classifiers with output function (2). Since each object is assigned to the class whose output unit has the largest activation value, some information may be discarded by considering only the winning class for the multiple classifier. There can be significant differences in the activation values depending on the classifiers used, even if they select the same class. For this reason, other methods such as the average of the correspondent output values have been considered as combining techniques. This method implicitly assumes that all the networks are equally good but we might expect that some classifiers would make better predictions than others would. As in the regression context, we can think of making a weighted combination of the results and thus estimating the *a posteriori* probability of the k-th class by $\hat{p}(k|x)=\sum_j \alpha_j o_{kj}$, where o_{kj} is the k-th output of the j-th network. For

regression problems, the optimal set of weights can be determined by minimizing a sum-of-squares error function (see, for example, Hashem *et al.*, 1994). In the classification context, this weight estimation does not guarantee that the estimated probability will stay between 0 and 1. The solution to this problem can be obtained by requiring that $\alpha_j \geq 0 \; \forall j$ and $\sum_j \alpha_j = 1$.
Moreover, considering a different error function, such as the cross-entropy is more appropriate with this class of problems and could lead to a better solution. The minimization of this error function, under the two constraints above, represents a more difficult problem that will be analyzed and developed in our future works. A different solution could be reached by relating the weights to the performance of the classifiers. We can assign to each classifier a weight inversely related to the error rate of that classifier in an independent data set. Two different sets of weights can be derived by excluding or not in the calculation of the error rate the observations that are assigned to the same class by each classifier.

4. Some examples

In this section we analyze the combination performance of different methods of combining by studying three different classification problems.
The first data sets are derived from the synthetic waveform recognition problem presented in Breiman *et al.* (1984). It is a three-class problem with 21 dimensional feature vectors. Three hundreds vectors were generated using equal prior probabilities. In order to compare the performance of different classifiers three hundreds new observations were generated for each class as a test set.
The second example, the threenorm problem, is based on a two-class data set with 10 dimensional measurement vectors. Class 1 is drawn with equal probability from a unit multivariate normal with mean (a,..., a) and from a unit multivariate normal with mean ($-a$,...,$-a$). Class 2 is drawn from a unit multivariate normal with mean $(a,-a,...,a,-a)$, where $a=2/(20)^{1/2}$. Two hundreds observations were generated as a training set and one thousand observations as a test set.
The third example is a real classification problem, where the objective is to correctly identify different glass types. It is a six-class problem. Each of the 214 observations consists of 9 chemical measurements on one of 6 type of glass. The data set are in the UCI repository of machine learning databases (ftp ics.uci.edu/pub/machine-learning-databases). The sample was randomly split into a training set (107 units), a validation set (53 units) and a test set (54 units), on which the performance of the classification rules was measured.
Multilayer perceptron, with a different number of hidden units, are used to classify the patterns of the three data sets. All the networks are trained until the classification error rate on a validation set reaches a minimum.
For each data sets we calculate the test set error we will get on average when

we decide to perform only one run (*singles* column). This will be used as a reference with which the other methods will be compared.

Table 1 reports in the *singles* column the average test set performance (standard deviations in brackets) over 10 runs of a single multilayer perceptron (MLP), which starts from different random initial set of weights. The last three columns provide the combining results for three, five and seven classifiers. That is, we have considered all the possible combination of three, five or seven classifiers over the 10 available runs. We also determine on average the test set error of the network with the lowest error on the validation set among the different numbers of classifiers combined (*minimum* column).

Table 1: *Test set error rate (%) for the waveform problem*

Waveform	*Singles*		*Minimum*	*Averaging*	*W. Aver.*	*Voting*
2 hidden units	18.6	3	18.5 (1.4)	16.3 (0.9)	16.2 (0.9)	16.9 (0.9)
	(1.4)	5	18.6 (1.5)	15.7 (0.5)	15.7 (0.5)	16.2 (0.7)
		7	18.4 (1.3)	15.4 (0.2)	15.4 (0.3)	15.8 (0.5)
3 hidden units	17.3	3	17.1 (0.8)	16.1 (0.6)	16.1 (0.6)	16.3 (0.8)
	(1.3)	5	17.5 (0.5)	15.8 (0.5)	15.9 (0.4)	16.0 (0.5)
		7	17.1 (0.3)	15.8 (0.4)	15.8 (0.4)	15.9 (0.3)
4 hidden units	17.3	3	16.9 (0.2)	16.9 (0.2)	16.7 (0.3)	17.2 (0.5)
	(0.6)	5	17.2 (0.5)	16.8 (0.2)	16.8 (0.1)	17.2 (0.3)
		7	16.9 (0.1)	16.9 (0.2)	16.7 (0.1)	17.1 (0.3)
5 hidden units	19.5	3	19.5 (0.2)	18.7 (0.4)	18.7 (0.4)	18.9 (0.4)
	(0.8)	5	19.0 (0.7)	18.9 (0.5)	18.9 (0.5)	19.0 (0.6)
		7	18.4 (0.7)	18.6 (0.3)	18.6 (0.3)	18.8 (0.5)

The results show that the performances of the simple and weighted average are similar. For the waveform problem the simpler structure with 2 hidden units is the one which reaches the greatest error reduction (-15,6%). One important observation that emerges from table 1 is that combining classifiers with the same (erroneous or correct) classification decisions provides little gain, regardless of the chosen scheme. This is particularly evident for the multilayer perceptron with 4 and 5 hidden units. The aggregate predictors perform better than the *minimum* only for 2 and 3 hidden units. These architectures give also the aggregate predictors with the lowest error rates.

The aggregate classifiers for the threenorm problem do not seem outperform significantly the single ones. To improve their performances we generated different training sets, resampling the original one by both cross-validation and bootstrap. The results in Table 3 show that, by combining a few classifiers, cross-validation gives the best results in terms of error reduction. If we compare this results to the ones of table 2 we can observe that with cross-validation applied to 2 hidden units we have an error rate similar to the one obtained by averaging architecture with 5 hidden units. The bootstrap's worse performance probably depends on the structure of each resampled set, which contains only about 63% of the data on the average. The exclusion of a subset of observations may reduce the individual classifier performance, negating any potential gain. We have to combine a large number of classifiers to obtain a substantial improvement, but from a computational point of view this is not a winner strategy.

Table 2: *Test set error rate (%) for the threenorm problem*

Threenorm	*Singles*		*Minimum*	*Averaging*	*Voting*
2 hidden units	24.5	**3**	23.8 (0.4)	22.7 (0.5)	22.7 (0.4)
	(3.0)	**5**	23.9 (0.6)	22.7 (0.5)	22.7 (0.5)
		7	24.1 (0.5)	22.6 (0.4)	22.8 (0.3)
3 hidden units	23.3	**3**	23.0 (0.5)	22.5 (0.5)	22.4 (0.6)
	(1.2)	**5**	22.9 (0.6)	22.4 (0.5)	22.3 (0.5)
		7	23.1 (0.5)	22.2 (0.4)	22.4 (0.3)
4 hidden units	22.9	**3**	22.3 (0.9)	22.2 (0.6)	22.2 (0.7)
	(0.9)	**5**	22.1 (0.8)	22.1 (0.4)	22.0 (0.5)
		7	22.0 (0.7)	22.0 (0.3)	22.1 (0.4)
5 hidden units	22.9	**3**	22.3 (0.5)	21.9 (0.5)	22.2 (0.5)
	(0.9)	**5**	22.2 (0.4)	21.8 (0.3)	22.0 (0.9)
		7	22.3 (0.4)	21.9 (0.3)	21.9 (0.3)

Table 3: *Test set error rate (%) for alternative methods of perturbing the data*

Threenorm	*Singles*		*Average*	*Voting*
2 hidden units				
Training set	24.5 (3.0)	**3**	22.7 (0.5)	22.7 (0.4)
		5	22.7 (0.5)	22.7 (0.4)
		7	22.6 (0.4)	22.8 (0.3)
3-fold CV	23.7 (1.8)	**3**	22.8 (0.8)	22.2 (0.8)
5-fold CV	23.3 (1.0)	**5**	21.8 (0.5)	22.5 (0.7)
7-fold CV	23.1 (0.8)	**7**	21.7 (0.4)	22.3 (0.5)
Bootstrap	24.2 (1.9)	**3**	23.1 (1.1)	23.5 (0.7)
	24.2 (1.9)	**5**	22.8 (0.6)	23.3 (0.5)
	24.2 (1.9)	**7**	22.8 (0.5)	23.4 (0.4)
	24.4 (1.8)	**30**	22.4 (0.4)	22.3 (0.5)
3 hidden units				
Training set	23.3 (1.2)	**3**	22.5 (0.5)	22.4 (0.6)
		5	22.4 (0.5)	22.4 (0.6)
		7	22.2 (0.4)	22.4 (0.3)
3-fold CV	23.1 (1.3)	**3**	21.8 (0.7)	21.9 (0.8)
5-fold CV	23.3 (0.9)	**5**	22.0 (0.4)	22.7 (0.5)
7-fold CV	23.4 (0.7)	**7**	22.1 (0.3)	22.5 (0.4)
Bootstrap	24.2 (1.6)	**3**	22.7 (0.8)	23.0 (0.9)
	24.2 (1.6)	**5**	22.3 (0.6)	22.7 (0.7)
	24.2 (1.6)	**7**	22.2 (0.4)	22.5 (0.5)
	24.2 (1.3)	**30**	21.9 (0.3)	22.2 (0.4)

The results for the glass data sets (Table 4) confirm that the lowest error rate can be obtained by combining classifiers trained on the same training set. The simplest architectures (2 hidden units) give poor classifiers, whose performance are not improved by aggregating. Also for this problem, a good bootstrap solution requires perhaps a great number of replicates.

Table 4: *Test set error rate (%) for the glass problem*

Glass	*Singles*		*Minimum*	*Average*	*W. Aver.*	*Voting*
2 hidden units						
Training set	32.5 (1.9)	**3**	32.0 (1.3)	31.6 (1.9)	31.5 (1.8)	31.6 (1.4)
		5	31.9 (0.8)	31.4 (1.7)	31.4 (1.6)	31.2 (1.2)
		7	31.9 (0.5)	31.6 (1.6)	31.6 (1.6)	31.2 (1.2)
Bootstrap	44.2 (6.3)	**3**		41.2 (2.5)	41.3 (2.5)	42.1 (2.6)
		5		40.6 (1.7)	40.7 (1.7)	41.5 (1.6)
		7		40.0 (1.7)	40.2 (1.5)	41.1 (1.2)
3 hidden units						
Training set	32.1 (2.5)	**3**	32.3 (2.2)	28.2 (2.2)	28.1 (2.2)	28.9 (2.5)
		5	33.2 (1.5)	27.7 (2.0)	27.5 (2.2)	27.5 (1.1)
		7	33.7 (0.9)	27.8 (2.1)	27.2 (2.1)	27.0 (1.3)
Bootstrap	42.6 (5.9)	**3**		38.7 (4.1)	38.2 (4.1)	41.1 (4.3)
		5		36.8 (3.3)	38.7 (3.6)	36.7 (3.6)
		7		35.7 (2.6)	35.4 (2.8)	37.9 (3.0)
4 hidden units						
Training set	31.3 (3.4)	**3**	29.9 (2.6)	28.8 (1.8)	28.7 (1.7)	29.6 (2.3)
		5	29.4 (1.8)	29.1 (1.7)	29.0 (1.8)	29.7 (1.9)
		7	29.6 (1.0)	29.3 (1.3)	29.0 (1.7)	30.1 (1.7)
Bootstrap	38.8 (6.3)	**3**		34.0 (3.9)	33.9 (4.1)	36.0 (3.9)
		5		33.3 (2.7)	33.1 (2.9)	35.4 (2.9)
		7		32.9 (2.0)	32.6 (2.1)	34.4 (2.4)
5 hidden units						
Training set	31.5 (4.5)	**3**	29.2 (4.0)	29.5 (2.2)	29.5 (2.4)	29.8 (2.6)
		5	29.0 (3.7)	29.3 (2.0)	29.3 (2.0)	28.7 (2.0)
		7	28.9 (3.2)	29.4 (1.3)	29.5 (1.8)	28.4 (2.0)
Bootstrap	48.4 (6.7)	**3**		46.9 (4.3)	46.0 (4.9)	48.4 (3.2)
		5		46.3 (2.9)	45.8 (3.4)	47.1 (2.8)
		7		45.7 (2.7)	45.0 (2.7)	46.7 (2.2)

5. Concluding remarks

Combining multiple versions of unstable classifiers by resampling the training set is a variance reduction method, but a substantial improvement requires a lot of computational efforts, particularly with computer intensive methods such as neural networks.

We have already stressed that for a neural network classifier one always needs to estimate multiple versions of the model. Our experimental results seem to confirm that combining the networks corresponding to different random initial conditions can reduce the error rate, while involving no additional computational costs. The combined classifier often performs better than the best single classifier.

Although the classification performances in some case are not dramatically better, all the combined results have a lower standard deviation and this means that they are less unstable and less dependent on initial conditions.

One important observation that emerges from these experiments is that there does not seem to be a particular combiner that can be labeled "best" under all circumstances, although other methods of combining can be fruitfully explored in the context of neural network classifiers. Nevertheless, the results, based on straightforward methods, are quite encouraging because they provide a relatively easy way to improve a neural network classifier.

References

Breiman, L. (1994). Bagging predictors, Technical Report 421, Department of Statistics, University of California, Berkley.

Breiman, L. (1996). Bias, variance and arcing classifiers, Technical Report 460, Department of Statistics, University of California, Berkley.

Breiman, L. & Friedman, J.H. & Olshen, R.A. & Stone, C.J. (1984). *Classification and Regression Trees*, Monterey, Wadsworth and Brooks/Cole.

Granger, C.W.J., (1989). Combining forecasts - twenty years later, *Journal of Forecasting*, 8, 167-173.

Hashem, S. & Schmeiser, B. & Yih, Y. (1994). Optimal linear combinations of neural networks: An overview, in *Proceeding of the 1994 IEEE International Conference on Neural Networks*, vol. 3, IEEE Press.

Hertz, J. & Krogh, A. & Palmer, R. (1991). *Introduction to the theory of neural computation*, Redwood City, Addison Wesley.

Perrone, M.P. & Cooper, L.N. (1993). When networks disagree: ensemble methods for hybrid neural networks, in: *Neural Networks for Speech and Images Processing,* Mammone, R.J. (Eds), Chapman Hall.

Tibshirani, R. (1996). Bias, variance and prediction error for classification rules, Technical Report, University of Toronto.

Wolpert, D. H. (1992). Stacked generalization, *Neural Networks*, 5, 241-259.

Selection of Cut Points in Generalized Additive Models

Francesco Mola
Dipartimento di Matematica e Statistica
Università di Napoli *Federico II.*
Complesso Monte S. Angelo
via Cinthia, I-80126 Napoli, Italy
e-mail: mola@dms.unina.it

Abstract: This paper offers, in the framework of *generalized additive models (GAM),* a proposal of a *cut point selection* for GAM smoothers that stems out of the CART like regression tree procedures. The proposal allows to find a parsimonious bin smoother *(regressogram)*, a new smoother based on the well known *loess* smoother, and provides, moreover, the user with an additional information inherited from the regression tree methodology. The problem of the choice of *span* parameter is considered too.

Keywords: Generalized Additive Models, Loess, Regression Trees, Regressogram, Spline Smoothers.

1. Introduction

Consider the vector of response measurements $\mathbf{y} = (y_1, \cdots, y_n)'$ obtained at design points $\mathbf{x} = (\mathbf{x'}_1 | \cdots | \mathbf{x'}_n)'$, where we assume that y_i's represent measurements of some variable Y and $\mathbf{x'}_i$'s represent measurements of (vector) variable **X**. We assume, moreover, that Y is a response variable and **X** is a (vector) predictor(s).

In Section 2 the idea of GAM is concisely summarized and in Section 3 typical smoothers are reminded. Section 4 presents the same for the recursive partitioning methods (of the CART type). In Section 5 is summarized our new proposal, introducing a new smoother based on the loess smoother proposed by Cleveland in 1979.

Finally, in Section 6 an application on three data sets have been considered, analyzing both real and simulated data sets. An evaluation of different smoothers is considered too.

2. Generalized additive models

The linear model holds a central place in the toolbox of applied statisticians. Simple in structure, elegant in its least squares theory and easily interpretable by its users, it is an invaluable tool. However, with the recent explosion in speed and size (of memory) of computers, it was possible to complement thc linear model with many new methods that assume less conditions and therefore potentially offer to discover more aspects of the data. To this type of methods belong the so called *generalized additive models*, which form a generalization of the linear regression model. The central idea is to replace the usual linear function of respective covariates with an unspecified smooth function. The estimated model consists of a function for each of the covariates. The additive model consists of a sum of such functions. This model is nonparametric in that we do not impose a parametric form on the function but instead estimate them in an iterative manner through the use of the so called *scatterplot smoothers.* This is useful as a predictive model and can also help the data analyst to discover the appropriate shape for each of the covariate effects.
The role played by smoothers in GAM idea is central. This means that we can focalize our attention on smoothers, since the choice of the smooth function becomes fundamental.

3. Smoothers

Scatterplot smoothers used in GAM is typically defined as a smooth function of $\mathbf{x}$ and $\mathbf{y}$ estimated in nonparametric way. The basic idea is to let the data to show the appropriate functional form themselves. More precisely, these methods try to expose the functional dependence without imposing rigid parametric assumptions between Y and $\mathbf{X}$.

Generalized additive models can be, among others, applied also to other data than those usually described by the standard linear regression model. Typical examples are binomial or binary response data, survival data and data from case/control studies. Many estimators of this type are proposed in the literature, see Hastie and Tibshirani (1990) for details; we shall concentrate on some of them, i.e., on *bin smoother* (known also as *regressogram)*, locally weighted running-line smoother (currently called *loess* in *S*+), *kernel smoothers* and *spline smoothers.*

It is well known that the regressogram mimics a categorical smoother by partitioning the predictor values into a number of disjoint and exhaustive regions and averaging the response in each region. Formally, we must fix cut points

$$-\infty = c_0 < c_1 < \cdots < c_K < c_{K+1} = +\infty$$

and define the indices of the data points in each region by

$$R_k = \{i; c_k < x_i \le c_{k+1} \quad , \qquad k = 0,1,\cdots,K .$$

Then the desired smoother *s(.)* is defined as

$$s(x_0) = \text{ave}_{i \in R_k} y_i \qquad\qquad \text{if } x_0 \in R_k .$$

This estimator is, however, quite rough because it has jumps at each cut point, even if it is interesting from a theoretical point of view.

Regression spline smoothers offer a compromise by representing the fit as a piecewise polynomial. Even if several configurations of splines can be used, usually are cubic polynomials that are considered. Respective regions defining the pieces are separated by a sequence of knots (breakpoints). In addition, it is customary to force the smoothness of the curve at the knots. A serious practical problem is the choice of the knots.

Locally-weighted running-line smoother (loess or lowess, Cleveland, 1979) is based on the formula:

$$s(x_0) = \hat{\alpha}(x_0) + \hat{\beta}(x_0)x_0$$

where $\hat{\alpha}(x_0)$ and $\hat{\beta}(x_0)$ are the estimates in the point x_0 using a neighbourhood of x_0. Usually it is easy and natural to think of least squares estimates, but obviously it is possible to define alternative methods.

A crucial point in the use of loess, and for all smoothers of this type, there is the choice of the neighbourhood. In fact it is important not only to specify if the neighbourhoods are symmetric (i.e., running-mean smoother) or not, but how big the neighbourhood has to be. In literature, this is known as the problem of the choice of the span parameter, i.e., the part of data to be considered in each neighbourhood.

It is known that when the span increases, the smooth function is very smooth, but the estimates are not so accurate. On the contrary, when the span decreases, the estimate is very accurate but the smooth function is not so smooth. A lot of attention has been payed to this problem; however, it seems that a final solution has not been reached.

The *kernel smoothers* are based on the local estimates at x_0 points, considering either a "h" parameter to define the amount of data in which the estimate is computed, or a kernel function that identifies the distribution of data in the neighbourhood of x_0.

A good source of the information about all these procedures can be found in the monograph of Hastie and Tibshirani (1990) and in the referenced literature.

4. Recursive partitioning methods

Recently, more and more important role in the regression analysis play different *tree based regression methods*. Let us mention, among others, AID introduced in literature by Morgan and Sonquist (1963), CART described in Breiman et al. (1984), FIRM suggested by Hawkins (1990), two stage procedure of Mola and Siciliano (1992) or RECPAM described in a series of Ciampi's papers.

Tree-structured methods that use a recursive partitioning algorithm provide a powerful analysis tool for both exploration of the structure of the data and for the prediction of the outcomes of new cases. With tree-structured regression techniques, some of the restrictive classical assumptions about the relation between the response variable and the independent variables can be avoided. Moreover, a tree-structure provides easier interpretation than a regression equation or GAM since the regression tree identifies effects of covariates in subgroups whereas regression examines effects in the whole sample.

From the nonparametric point of view, the results of the CART-like regression tree approach is nothing else than the *bin smoother* (*regressogram*), however, the way how to find cut points it is totally different from the classical regressogram estimator typically used by GAM smoothers, cf. Breiman et al. (1984) or Hastie and Tibshirani (1990).

5. The proposed methodology

One of the main problems of the above mentioned smoothers is the *choice of the cut points* for the regressogram, *places of knots* (for the smoothing splines), the *span* parameter for loess and, in different manner, kernel smoother. Several ideas are discussed, e.g., in Hastie and Tibshirani (1990), however, the possibilities mentioned there are not exhaustive. Moreover, almost all of the standard GAM smoothers are based on the classical non-parametric approach which balance the (dis)proportion when reducing both the variance and the bias parallely.

Therefore, our idea is to replace during the construction of the "GAM smoothers" standard cut points resulting from the respective procedures by the cut points resulting from the corresponding (CART-like) regression tree. In this way the results of the regression tree procedure allows us to determine both cut points for regressograms and the knots for spline smoothers, neighbourhoods for loess and kernel smoothers.

In figures 1 and 2 a schema of how our methodology works is shown. The first step is to perform the regression tree analysis on the observed data; in this way we obtain cut points (in the figure c1, c2, c3, c4, c5 respectively). The cut points allow us to identify the five terminal nodes in figure 1 denoted by α, β, γ, δ, ε. In the figure 2 the scatterplot of the data set is shown and in addition the regions in which to perform the loess procedure identified by the regression tree analysis are marked. It is possible to notice that the regions changes as the cloud of points modify.

Figure 1: Regression tree

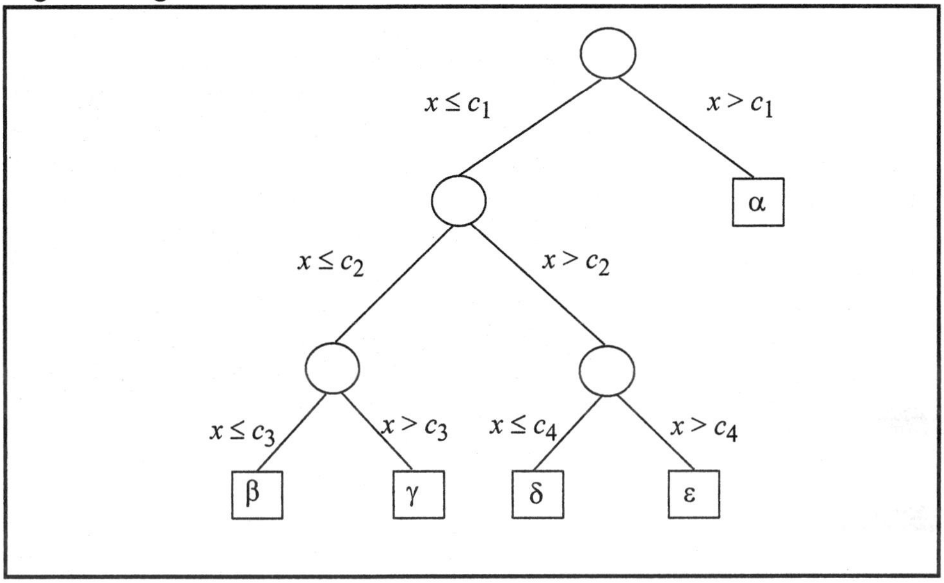

Figure 2: scatterplot and regions

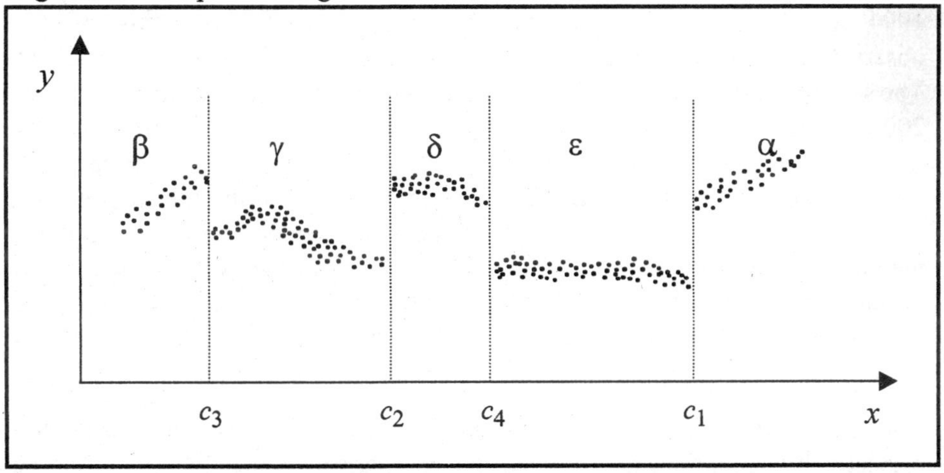

What advantages and disadvantages this kind of methodology brings to us?

On one hand, the solution is complemented by the information resulting from the corresponding regression tree, giving us thus much more information about the problem than the classical GAM solution. Taking into account of the basic idea that the data should "speak" for themselves, this is often an extremely important advantage of our proposal.

With this kind of approach, we do not impose an a priori choice about the cut points identification but we let to determine this choice by the data structure. In fact the regression tree gives us the advantage to detect points in which there are jumps.

Another advantage is that our proposal is much more adaptable to the data and of special practical interest is its ability to cope with the dependence between Y and $\mathbf{X}$. This feature is definitely superior to the classical GAM approach.

On the other hand, both regressogram and classical regression tree (of the CART type) are not very smooth because they have jumps at each of cut points. This disadvantage can be up to some extend overcomed by the use of regression splines or loess smoothers.

Finally, notice that the fast computer is a must when we want to play with this type of estimators. This methodology obviously is faster then the Cleveland one since it uses for the estimation step only the part of data which is required.

6. Application

As an example of application, we have considered three data sets. The first one is a data set distribuited with the S-plus package (Venables and Ripley, 1994 for details) and are due to Silvermann (1985) consisting of 133 observations of acceleration against time for a simulated motorcycle accident. The second and third data sets have been generated by us: the former consists of 200 pairs with four jumps imposed and identifying four regions in which the (x,y) pairs are very regular; the latter consists of 200 pairs and four jumps imposed, but differently by the previous one, imposing in the four region a lot of noise. In particular we compare the accuracy of a smoother obtained applying the proposed methodology (i.e., the loess) with standard smoothers, such as the original loess smoother and the kernel smoother. In order to evaluate the performance of these estimators the following commonly used measure has been considered: $1/n\sum|y_i - y_i^*|$. For sake of brevity figures of the scatterplot smoother have been omitted, but tables 1 to 3 (for the three data sets respectively) summarize the results of the comparison. Notice that the proposed smoother can be directly compared with the loess smoother since the span parameter is the same, whereas the kernel smoother, that depends on the h parameter, cannot be directly compared. Nevethless, a global idea of the performance of this last estimator can be deduced, too.

Comparing the results in the three tables, for motorcycle accident data set, simulated data set 1 and simulated data set 2 (*with no jumps*, *with jumps and regularities*, *with jumps and irregularities* respectively), we can notice that even if the proposed smoother works ever better than the others, in presence of jumps or jumps and irregularities it works very well. This means that the procedure works well every time, but it is particularly appreciable in case of jumps and irregularities. It is also possible to notice that the variability for the proposed smoother is more less than the variability of the other smoothers, denoting the low influence of the choice of the span parameter and the neighbourhoods.

Table 1:Motorcycle data set

Span (%)	Proposed Smoother	Loess Smoother	Kernel Smoother	"h" Parameter
10	14.59	14.94	15.86	1
20	14.84	15.91	18.94	2
30	15.20	16.99	26.81	5
50	15.60	22.51	34.14	10
66	15.94	26.86	37.53	20
99	16.06	33.33	38.52	40

Table 2: Simulated data set 1 (jumps and regularities)

Span (%)	Proposed Smoother	Loess Smoother	Kernel Smoother	"h" Parameter
10	0.15	0.41	0.19	1
20	0.21	0.62	0.28	2
30	0.21	0.81	0.45	5
50	0.23	1.33	0.69	10
66	0.24	1.57	1.24	20
99	0.24	2.01	2.09	40

Table 3: Simulated data set 2 (jumps and irregularities)

Span (%)	Proposed Smoother	Loess Smoother	Kernel Smoother	"h" Parameter
10	0.79	10.81	2.88	1
20	1.16	22.73	5.53	2
30	1.38	33.25	13.82	5
50	2.96	54.91	27.50	10
66	4.98	63.31	49.79	20
99	8.52	81.44	82.47	40

Acknowledgment. Author is very much indebted to J. Antoch and R. Siciliano for reading the manuscript and for valuable comments. This research was partially supported by CNR Research funds number 93.0878.CT10 (Prof. Lauro).

References

Antoch J. and Mola F. (1996). Parsimonious regressograms for generalized additive models, *Sviluppi metodologici e applicativi dell'inferenza computazionale nell'analisi multidimensionale dei dati* (Lauro N.C., ed.), Rocco Curto Editore.

Breiman L., Friedman J. H., Olshen R. A. and Stone C. J. (1984). *Classification and Regression Trees*, Wadsworth, Belmont, CA.

Breiman L. (1989). Discussion of "Linear Smoothers and Additive Models" by Buya et al. *Annals of Statistics* **17**.

Breiman L. (1993). Fitting additive models to regression data:Diagnostics and alternative views, *Computational Statistics & Data Analysis* **15**, 13-46.

Buja A.T., Hastie T. and Tibshirani R. (1989). Linear smoothers and additive models (with discussion), *Annals of Statistics*. **17**, 453-555

Cleveland W. S. (1979). Robust locally weighted regression and smoothing scatterplots, *Journal of the American Statistical Association* **74**, 829-836.

Cleveland W.S. & Devlin S.J. (1988). Locally-Weighted Regression: an Approach to Regression Analysis by Local Fitting, *Journal of the American Statistical Association* **83**, 597-510.

Friedman J.H. and Silverman B.W. (1989). Flexible parsimonious smoothing and additive modeling (with discussion), *Technometrics* **31**, 3-39.

Hastie T. J. and Tibshirani R. J. (1990). *Generalized Additive Models*, Chapmann & Hall, London.

Hawkins D. M. (1990). *FIRM (Formal Inference-based Recursive Modeling).* Technical Report 546, University of Minnesota, School of Statistics.

Mola F. and Siciliano R. (1992), A two-stage predictive splitting algorithm in binary segmentation, *Computational statistics* (Dodge Y. and Whittaker J., eds.), Physica Verlag, Heidelberg, 179-184.

Morgan J.N. and Sonquist J.A. (1963), Problems in the analysis of survey data and, a proposal *Journal of American Statistical Association*, vol.58, 415-434.

Silverman B.W. (1985). Some Aspects of the Spline smoothing Approach to Non-Parametric Regression Curve Fitting, *Journal of the Royal Statistical Society*, B **47**, 1-52.

Venables W.N. & Ripley B.D. (1994). *Modern Applied Statistics with S-Plus*, Springer-Verlag New York.

Vieu P., Pelegrina L. and Sarda P (1996). On multidimensional nonparametric regression, *Computational Statistics* (Prat A., ed.), Physica Verlag, Heidelberg, 149-160.

Latent Budget Trees for Multiple Classification

Roberta Siciliano

Dipartimento di Matematica e Statistica
Università degli Studi di Napoli Federico II
Monte S.Angelo, via Cintia, 80126 Napoli, Italy
r.sic@dmsna.dms.unina.it

Abstract: This paper provides a methodology to grow classification trees when a multiple qualitative response is considered as a criterion variable. The latent budget model is used recursively to find ever finer partitions of cases into a prior fixed number of groups. The Akaike statistic is considered to select the most predictive model at each node of the tree. A fruitful interpretation of the final decision rule is given through the Bayes rule. The proposed approach is also convenient to deal with multiple questions through the use of compound variables in the latent budget model. An application of the proposed approach on a data set taken from a Survey of the Bank of Italy is finally shown.

Keywords: Multiple Response, Latent Variable, Mixing Parameter, Bayes Rule, Akaike Criterion

1. Introduction

In the field of categorical data analysis latent budget model is a reduced-rank probability model to decompose a table with constant row-sum data (i.e., compositional data, time budgets, conditional frequency distributions). The model is a mixture of conditional probabilities known as *latent budgets* and the mixture is defined by the *mixing parameters* (de Leeuw, van der Heijden and Verboon, 1990). The latent budget models can be fruitfully used for describing the dependence of a response variable due to a given predictor in two-way contingency tables (section 2.). For the latent budget model the response as well as the predictor can be also formed by more variables through the definition of suitable compound variables. Particular interactions among the variables can be tested considering restrictions upon the parameters (van der Heijden, Mooijaart and de Leeuw, 1992). An extension of the latent budget model to the analysis of a set of two-way tables is known as simultaneous latent budget analysis (Siciliano and van der Heijden, 1994; Tambrea and Siciliano, 1997). However, as soon as the number of predictors increases the application of the latent budget model, and in general of any parametric model, becomes not feasible due to a large number of parameters to introduce in the specification of the model.

A nonparametric approach is proposed in this paper in order to define a model in the form of a classification tree for prediction of a multiple response variable (section 3.). The tree-growing procedure is described in details providing also an alternative interpretation of the tree through the Bayes rule. An application performed on a data set taken from a Survey of the Bank of Italy on Family Budgets is shown (section 4). The main advantages of the proposed approach are outlined in the final concluding remarks.

2. The latent budget model

For a two-way cross-classification of N observed cases according to the I categories of the predictor X and the J categories of the response variable Y let p_{ij} be the observed proportion in cell (i, j) for $i = 1,...,I$; $j = 1,...,J$ with $\sum_i \sum_j p_{ij} = 1$; the usual dot notation is used for summations, i.e., $\sum_i p_{ij} = p_{.j}$. The interest pertains to the I conditional distributions or *observed budgets* $p_{j|i} = p_{ij}/p_{i.}$ and their departures from the marginal distribution $p_{.j}$. The latent budget model decomposes the *theoretical budgets* $\pi_{j|i} = \pi_{ij}/\pi_{i.}$ as a mixture of K latent budgets: $\pi_{j|i} = \sum_{k=1}^{K} \pi_{k|i}\pi_{j|k}$, where the conditional probabilities $\pi_{j|k}$ for $j = 1,\ldots,J$ form the k-th *latent budget* and the conditional probabilities $\pi_{k|i}$ for $k = 1,\ldots,K$ are the *mixing parameters* for the i-th predictor category. For $K = 1$ the model reduces to the independence model, i.e., $\pi_{j|i} = \pi_{.j}$. In general, it holds $K \leq K^*$ where $K^* = \min(I, J)$ is the maximum rank of the matrix with theoretical budgets (i.e., the saturated model). The latent budget model can be written in matrix formulation as $\mathbf{P} = \mathbf{AB'}$ where the matrix $\mathbf{A}$ includes the parameters $\pi_{k|i}$ whereas the matrix $\mathbf{B}$ includes the parameters $\pi_{j|k}$; all the matrices are of proper order. The latent budget model is not identified: $\mathbf{P} = \mathbf{AB'} = \mathbf{ASS^{-1}B'}$ for any $K \times K$ matrix $\mathbf{S}$ which rows sum to one. Some restrictions are imposed upon parameters to make them identifiable; the number of restrictions depends on the number of free parameters of the matrix $\mathbf{S}$, this number being $K(K-1)$. The degrees of freedom are given by $I(J-1) - [I(K-1) + K(J-1) - T(T-1)] = (I-K)(J-K)$. The estimate of the model can be obtained by maximum likelihood under the product-multinomial sampling scheme as well as by an alternating least-squares method. The rank K is selected to have a good fit of the model to the observed data as tested by the usual likelihood ratio statistic G^2 or alternative goodness of fit measures.

3. The latent budget tree procedure

3.1 Two strategies of analysis

The aim is to define an *exploratory tree* as well as a *tree predictor* for the response variable Y given a set of predictors observed on N cases. Both the response variable and some of the predictors can be defined as compound variables. The idea is to select at each node of the tree a predictor and thus a latent budget model which is used to find a partition of the cases into K groups where K is the number of latent classes. As it concerns the choice of K, two strategies are proposed. In the first strategy, this number can be fixed a-priori at the beginning of the procedure to be either K=2 or K=3 in order to grow *binary trees* or *ternary trees* respectively. In the second strategy, the number K might modify from node to node and it is given by the lowest number corresponding to the most parsimonious model that fits to the data. Such recursive procedure yields to grow a classification tree called *latent budget tree* characterized by a sequence of latent budget models assigned to the nodes of the tree.

3.2 The selection criterion

The procedure selects at each node a predictor and thus a model considering the set of tables which cross-classify the response variable with each predictor at a time. As a selection criterion the Akaike Information Criterion (AIC) is considered in order to compare the fit of models with different number of degrees of freedom. This leads to choose the predictor X with the associated model by maximizing the AIC criterion or equivalently by minimizing the corrected AIC criterion: $\mathrm{AIC}_X^* = G^2(X) - 2\,\mathrm{df}(X)$, where $G^2(X)$ is the likelihood ratio statistic for testing the model X against the saturated model. Obviously, in order to select a model with a fixed number of latent classes (first strategy) it is necessary that both $J \geq K$ and $I \geq K$ are satisfied; in particular for K=3 it is necessary that $J>2$ (*multiclass problem*) and $I>2$; dummy predictors can be combined to form compound variables (*multiple questions*). Instead, according to the second strategy the model with the lowest K that fits to the data is selected applying a selection procedure starting with K=2.

3.3 The partitioning criterion

An important advantage of the latent budget model over other models for categorical data is that the parameter estimates are conditional probabilities adding up to one so that the interpretation simplifies. This property is considered for the definition of a partitioning criterion to apply at each node of the latent budget tree. The sample of N_t cases at node t can be partitioned into K sub-groups on the basis of the estimates of the mixing parameters of the selected model. For the $I \times K$ matrix **A** which rows sum up to one (i.e.,

$\sum_k \pi_{k|i} = 1$) the I predictor categories are summarized into K latent budgets; the i-th predictor category is assigned to the k-th latent budget which presents the highest mixing parameter estimate. The partition of the I categories into K sub-groups induces the partition of the N_t cases at node t into K sub-groups. For K=2 the I categories are divided into two disjoint subgroups which induce the partition of the N_t cases into the left sub-group of cases presenting the categories with $\pi_{i|k=1} \geq 0.5$ and the right sub-group with the remaining cases.

3.4 The stopping rule

Such partitioning procedure continues until the current node cannot be further partitioned either when the number of degrees of freedom of the current best model is equal to zero or when the number of cases is too low according to a fixed number. A node which is not further partitioned is declared *terminal node* and is assigned the *label class* corresponding to the highest proportion of observed cases falling into the node. Such *exploratory tree* describes the conditional interactions of the predictors with the response variable. Furthermore, a *post-pruning procedure* can be adopted to define a *tree predictor* to be used for classification of new cases of unknown class on the basis of the given predictors (Cappelli and Siciliano, 1997).

3.5 The interpretation through the Bayes rule

A further aid to the interpretation of the tree is provided by the latent budget parameter estimates. For the matrix **B'** which rows sum up to one (i.e., $\sum_j \pi_{j|k} = 1$) the K latent budgets are related to the J response categories. Each row of the matrix **B'** can be compared with the independence hypothesis given by the marginal proportions $p_{.j}$ of the current table (prior probability estimates): for the k-th latent budget the response categories which departure more from the independence are those better predicted by the latent budget. Furthermore, the latent budget parameter $\pi_{j|k}$ for each j can be viewed as the posterior probability to fall in class j once that the case is assigned to the subgroup k. Using the bayesian rule we can write $\pi_{j|k}$ as $\pi_{j|k} = \frac{\pi_j \pi_{k|j}}{\sum_j \pi_j \pi_{k|j}} = \frac{\pi_j \pi_{k|j}}{\pi_k}$ which is the posterior probability of a case to belong to class j given that it falls into the descendant node k. Starting from the root node of the tree, where are assigned the prior probability estimates given by the group proportions of the criterion variable, the posterior probability estimates are recursively updated and related through a chain of conditional probability estimates until the definition of the posterior classification in the terminal nodes.

4. Application

As an example of application of the latent budget tree methodology a data set taken from a Survey of the Bank of Italy on the Family Budgets at the year 1994 is considered. This consists of 850 families where the interest pertains to the study of the kind of payment prefered by the head of the family. In particular, four typologies of users are identified: *1. archaic user* (who pays exclusively by cash), *2. classic user* (who pays by cheque*), 3. evolving user* (who pays by cheque and bancomat), *4. modern user* (who pays by cheque, bancomat and credit card). These groups define the categories of a multiple response variable. The following variables with their categories are considered as predictors: *title of study* (1. none, 2. primary school, 3. middle school, 4. seconday school, 5. degree); *age* (1. lower than 35 years, 2. 35 - 45 years, 3. 46 -55 years , 4. 56 - 65 years , 5. over 66 years); *area of residence* (1. North-West, 2. North-East, 3. Center, 4. South, 5. Isle); *city population size* (1. lower than 20 thousands of citizens, 2. 20 - 40 thousands of citizens, 3. 40 - 50 thousands of citizens, 4. over 50 thousands of citizens); *family income* (1. lower than 30 millions of italian lira, 2. 30 - 45 millions of italian lira, 3. over 45 millions of italian lira).

The latent budget tree obtained performing the proposed procedure is shown in figure 1 where the boxes denotes the terminal nodes with the label classes below. The model selected at each node which fits to the data has always K=2 latent classes providing a binary tree. Tables 1-5 describe the estimates of both the mixing parameters (the left block matrix $\mathbf{A'}$) and the latent budgets (the right block matrix $\mathbf{B'}$). For each latent class (defining the sub-group of cases) the circled estimates of the mixing parameters (with value not lower than 0.5) indicate which predictor categories cases belong to in order to fall in such sub-group. The estimates of the latent budget parameters (the posterior probability estimates) are compared with the independent model (the prior probability estimates) shown in the last row of the right block matrix. Cases belonging to a given latent budget have high probability to fall into the response classes associated with the circled estimates of the latent budget parameters.

For example, the city population size was selected as predictor for partitioning the 850 cases of node 1, with model fit G^2=1.7 (df=4). Table 1 shows that the predictor categories are divided in two sub-groups bringing categories 1, 2, 3 to the left (circled estimates for k=1) and category 4 to the right (circle estimate for k=2). The partition of cases is induced by such a split of the predictor categories: families living in cities with 50 thousands citizens or less go to the left subnode whereas families living in large cities with more than 50 thousands citizens go to the right subnode. The left sub-group has high probability to be assigned to the response classes 1 and 2, whereas the right sub-group has high probability to be assigned to the response classes 3 and 4.

Table 6 reports the posterior probability estimates for each terminal node: the values in bold are associated for each terminal node to the final response class.

Table 1: Latent budget model at node 1: G^2=1.7 (df=4), AIC*=-6.3.

node 1	City population size				Typologies of users				
	i=1	i=2	i=3	i=4	j=1	j=2	j=3	j=4	total
k=1	1.00	0.75	0.82	0.00	0.32	0.28	0.25	0.15	1.00
k=2	0.00	0.25	0.18	1.00	0.20	0.13	0.27	0.40	1.00
total	1.00	1.00	1.00	1.00	0.30	0.24	0.26	0.20	1.00

Table 2: Latent budget model at node 2: G^2=5.7 (df=6), AIC*=-6.3.

node 2	Area of residence					Typologies of users				
	i=1	i=2	i=3	i=4	i=5	j=1	j=2	j=3	j=4	total
k=1	0.00	0.00	0.34	0.67	0.74	0.63	0.26	0.04	0.08	1.00
k=2	1.00	1.00	0.66	0.33	0.26	0.18	0.25	0.35	0.23	1.00
total	1.00	1.00	1.00	1.00	1.00	0.30	0.25	0.26	0.19	1.00

Table 3: Latent budget model at node 5: G^2=0.6 (df=2), AIC*=-3.4.

node5	Area of residence			Typologies of users				
	i=1	i=2	i=4	j=1	j=2	j=3	j=4	total
k=1	0.00	0.17	1.00	0.31	0.30	0.21	0.18	1.00
k=2	1.00	0.83	0.00	0.17	0.24	0.37	0.23	1.00
total	1.00	1.00	1.00	0.21	0.26	0.32	0.22	1.00

Table 4: Latent budget model at node 10: G^2=3.9 (df=6), AIC*=-8.4.

node10	Title of study					Typologies of users				
	i=1	i=2	i=3	i=4	i=5	j=1	j=2	j=3	j=4	total
k=1	0.00	0.09	0.53	0.81	1.00	0.00	0.22	0.32	0.46	1.00
k=2	1.00	0.91	0.47	0.19	0.00	0.52	0.34	0.14	0.00	1.00
total	1.00	1.00	1.00	1.00	1.00	0.31	0.29	0.22	0.18	1.00

Table 5: Latent budget model at node 11: G^2=8.1 (df=6), AIC*=-3.8.

node11	Age					Typologies of users				
	i=1	i=2	i=3	i=4	i=5	j=1	j=2	j=3	j=4	total
k=1	0.94	0.94	1.00	0.47	0.00	0.06	0.21	0.45	0.28	1.00
k=2	0.06	0.06	0.00	0.53	1.00	0.39	0.29	0.18	0.13	1.00
total	1.00	1.00	1.00	1.00	1.00	0.18	0.24	0.35	0.23	1.00

Table 6: Posterior classification based on latent budget tree.

terminal node \ *response class*	j=1	j=2	j=3	j=4
3	0.21	0.13	0.25	**0.41**
4	**0.50**	0.25	0.13	0.12
40	0.16	0.31	0.16	**0.37**
41	0.17	0.22	**0.34**	0.27
42	0.30	**0.49**	0.21	0.00
43	**0.73**	0.11	0.08	0.08
44	0.16	0.25	**0.30**	0.29
45	**0.43**	0.31	0.20	0.06
46	0.04	0.18	0.37	**0.41**
47	0.08	0.22	**0.45**	0.24

Figure 1: The latent budget tree of payment instruments users

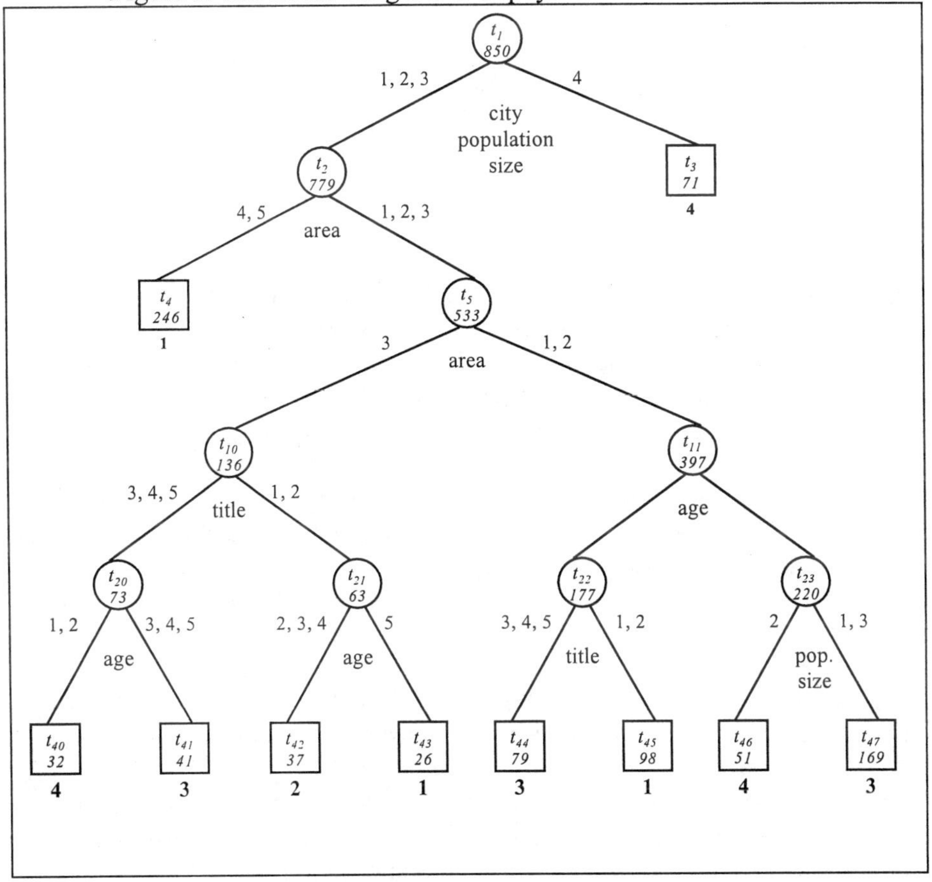

5. Concluding remarks

This paper has provided a methodology to grow a classification tree for a multiple qualitative response variable on the basis of the latent budget model. On one hand, the recursive use of the latent budget model has allowed to extend the latent budget analysis to multidimensional contingency tables; on the other hand, the tree-growing procedure via latent budget analysis has characterized an alternative tree predictor built up by using the bayesian concepts underlining the latent budget parameters. It was beyond the aim of this paper to show the relations of the proposed method with other approaches to classification with prior knowledge of a criterion variable such as supervised neural networks (Mola, Davino, Siciliano and Vistocco, 1997), other tree-structured procedures (Mola and Siciliano, 1997; Siciliano and Mola, 1997). As main advantages resulting from the application on real data sets, the proposed approach has proved to solve the *multi-class problem* of classification trees by improving the classification over all the response classes; it provides a fruitful interpretation of the tree in terms of conditional probabilities and Bayes rule; it deals very easily with *multiple questions* through the use of compound variables as predictors in the model fitted at each node.

Acknowledgements: This research project was supported by MURST 60%.

References

Cappelli, C., Siciliano, R. (1997). On simplification methods for decision trees, *NGUS-97 Conference*, Bilbao, september 11-14, 1997.

de Leeuw, J., van der Heijden, P.G.M., Verboon, P. (1990). A latent time-budget model, *Statistica Neerlandica*, 44, 1, 1-21.

Mola, F., Davino, C., Siciliano, R., Vistocco, D. (1997). Use and Overuse of Neural Networks in Statistics, invited lecture at the *IV International Meeting of Multidimensional Data Analysis*, Bilbao, september 10-12.

Mola, F., Siciliano, R. (1997). A Fast Splitting Procedure for Classification Trees, *Statistics & Computing*, 7, 3, 209-216.

Siciliano, R., Mola, F. (1997). Ternary Classification Trees: a Factorial Approach, *Visualization of categorical data* (Greenacre, M., Blasius, J., eds.), Academic Press, CA.

Siciliano, R., van der Heijden, P.G.M. (1994). Simultaneous latent budget analysis of a set of two-way tables with constant row sum data, *Metron*, 1-2, 155-180.

Tambrea, N., Siciliano, R. (1997). Fixed-value and equality constraints in simultaneous latent budget models, *VIII International Conference on Applied Stochastic Models and Data Analysis,* Anacapri, june 11-15, 1997.

van der Heijden, P.G.M., Mooijaart, A., de Leeuw, J. (1992). Constrained latent budget analysis, *Sociological Methodology* (Clogg, C.C., ed.), 22, 279-320, Cambridge, Basis Blackwell.

PART III

Multivariate and Multidimensional Data Analysis

- **Proximity Analysis and Multidimensional Scaling**
- **Factorial Methods**
- **Spatial Analysis**
- **Multiway Data Analysis**
- **Multivariate Data Analysis**

Methods for Asymmetric Three-way Scaling*

Giuseppe Bove
Dept. of Educational Sciences, University of Rome III,
Via Castro Pretorio 20, 00185–Roma, Italy. e-mail: bove@educ.uniroma3.it
Roberto Rocci
Dept. of Statistics, University of Rome "La Sapienza",
P.le A.Moro 5, 00185–Roma, Italy. e-mail: rocci@pow2.sta.uniroma1.it

Abstract: A review of methods for asymmetric three-way scaling is presented focusing on their graphical capabilities. A general strategy of analysis is outlined with an example of application to import-export data.

Keywords: Three-way Data Matrices, Asymmetric Scaling, Graphical Methods.

1. Introduction

Square data matrices which rows and columns correspond to the same set of "objects" occur frequently in applications: proximities (e.g. similarity ratings), preferences (e.g sociomatrices), flow data (e.g. import-export, brand switching) and contingency tables (e.g. occupational mobility, word associations) are known examples. Not completely random asymmetry is often observed in such matrices and models taking into account this feature are required. To this purpose, multilinear and distance models are suitably modified by increasing the number of parameters in order to represent the asymmetry. A review of possible methods was recently provided in Zielman & Heiser (1996) for the two-way case. In this paper some methods for the simultaneous analysis of k asymmetric square $n \times n$ matrices $\Omega_h = [\omega_{ijh}]$ (three-way case) are discussed focusing on their relationships and graphical capabilities. This will suggest a proposal for a possible strategy of analysis based on the trade-off existing between fit and parsimony of the model. An example of application of the strategy is given by using import-export data.

In what follows the model formulation is specified for metric data, when the model can also deal with the non-metric case this fact will be mentioned. Data are assumed processed in order to make the matrices Ω_h coherent with the different models (similarities for multilinear models or dissimilarities for distance-like models).

---

* This research has been made possible by National Research Council of Italy with grant 96.01350.CT10 for the first author and 97.01192.CT10 for the second author.

2. Methods for intrinsic asymmetry

When relationships are mainly directional (intrinsic asymmetry), each datum is interpreted as a direct estimate of a single relation and less attention is paid to the symmetric component (e.g. journal citation data).
Most of the multilinear models proposed to deal with this asymmetry are particular cases of the following general formulation

$$\omega_{ijh} = x_i' R_h y_j + \varepsilon_{ijh} \tag{1}$$

where x_i and y_j are $q \times 1$ ($q \leq n$) vectors of loadings respectively for row-object i and column-object j in q dimensions (or "aspects"), R_h is a square matrix of order q, representing the underlying asymmetric relationships among the aspects at the occasion h, and ε_{ijh} is the error term. Two sets of latent components (aspects) are given for the n objects, when considered as rows and when considered as columns. The relationships summarised in R_h between the two sets can change across the occasions. The parameters are computed minimising the sum of squared residuals, an alternating least squares algorithm could be obtained by the TUCKER-2 method (Kroonenberg & de Leeuw, 1980). Even if the asymmetric relationships in the data Ω_h are usually simply represented in the R_h matrices, model (1) has some drawbacks: components for rows and columns are different; it does not provide a graphical representation of the objects; it is difficult to deal with non-metric data. In the following we list and comment some particular cases of model (1) known in the literature, in the equations below R is a square matrix and D_h is diagonal.

PARAFAC-CANDECOMP (Harshman 1970, Carroll & Chang, 1970)	$R_h = D_h \geq 0$	(2)
Three-way DEDICOM (Harshman 1978, Kiers 1989)	$x_i = y_i$	(3)
Three-way dual domain ("weak") DEDICOM (Harshman 1978)	$R_h = D_h R D_h$	(4)
Three-way single domain ("strong") DEDICOM (Harshman 1978)	$R_h = D_h R D_h$, $x_i = y_i$	(5)

Models (3)-(5) in different ways simplify the factorial interpretation: introducing (3) we allow only one set of components changing asymmetric relations across the occasions; model (4) allows separate sets of components varying proportionally across the occasions and having the same underlying asymmetric structure. Model (2) is the only one with graphical capabilities, each object is represented by two points in a q-dimensional space in order to represent the two different directions of ω_{ijh} and ω_{jih} by scalar products. On each occasion the q dimensions are weighted according to the diagonal entries of D_h. It follows that

a representation for the occasion-weights can also be obtained.
In the multidimensional scaling (MDS) tradition the distance-like models proposed for intrinsic asymmetry are particular cases of the General Euclidean Model (GEM, Young 1987) that in scalar notation can be expressed as

$$\omega_{ijh}^2 = \left(x_i - y_j\right)' V_i W_h \left(x_i - y_j\right) + \varepsilon_{ijh} \tag{6}$$

where x_i and y_j are q-dimensional vectors of coordinates respectively for row i and column j, V_i is a $q \times q$ positive semi-definite (psd) matrix of weights associated with row object i, W_h is a $q \times q$ psd matrix of weights associated with occasion h and ε_{ijh} is the approximation error. The particular cases of model (6) more relevant for analysing asymmetry are:

ASINDSCAL	$x_i = y_i$, $V_i = \mathrm{diag}(V_i)$, $W_h = \mathrm{diag}(W_h)$	(7)
Weighted Multidim. Unfolding	$V_i = I$, $W_h = \mathrm{diag}(W_h)$	(8)

Both models provide graphical representations for the objects and the occasions and can deal with non metric and missing data. However, while modifications produced by the weights W_h for the q dimensions in model (8) are easily interpretable (as in the INDSCAL model), the interpretation of results for model (7) seems controversial because of the non-Euclidean geometry induced by the introduction of the diagonal matrices V_i. In fact, in spite of the wide availability of software (e.g. the ALSCAL procedure in SPSS), ASINDSCAL has been rarely used.

3. Methods for the analysis of symmetry and skew-symmetry

The decomposition of any square matrix in symmetric and skew-symmetric components has inspired many proposals about two-way asymmetric MDS. Some of these methods were reviewed in Zielman & Heiser (1996). The three-way case has not been so widely explored. We describe two methods for representing separately the symmetric and the skew-symmetric components and one providing their simultaneous analysis.
A method for the first case is provided by Zielman (1991). It is based on the model

$$s_{ijh} = \tfrac{1}{2}\left(\omega_{ijh} + \omega_{jih}\right) = d_{ijh} + \varepsilon_{ijh} = \sqrt{(x_i - x_j)' W_h (x_i - x_j)} + \varepsilon_{ijh} \tag{9a}$$

$$k_{ijh} = \tfrac{1}{2}\left(\omega_{ijh} - \omega_{jih}\right) = u_h'(r_i - r_j) + \varepsilon_{ijh}^* \tag{9b}$$

where:

s_{ijh} and k_{ijh} are the symmetric and the skew-symmetric components of the data,
x_i , r_i are coordinate vectors of object i;
W_h is a psd matrix which can be diagonal (INDSCAL) or not (IDIOSCAL);
u_h indicates the relative importance of each dimension in occasion h.
For this model the symmetric component is represented by the distances d_{ijh} in a weighted Euclidean space. The skew-symmetry is depicted in a separate multidimensional space in which r_i describes the location of object i while u_h represents occasion h: the projections of the objects-point on the occasion-vectors indicate the rank-order of the objects for that particular occasion (as in the vector model of preference data, Carroll 1972). The method allows missing values. A linear model for asymmetry at each occasion is assumed in model (9b), and this could be relaxed allowing a multidimensional representation of skew-symmetry by using the PARAFAC model, however the properties of this approach need further investigation.
Finally, a proposal for simultaneous analysis of the symmetric and the skew-symmetric components is considered in Rocci & Bove (1994). The elementary datum is approximated as

$$\omega_{ijh} = d_{ih} d_{jh} x_i' J x_j + \varepsilon_{ijh}, \quad J = \mathrm{diag}\left(\begin{bmatrix} 1 & 1 \\ -1 & 1 \end{bmatrix}, \begin{bmatrix} 1 & 1 \\ -1 & 1 \end{bmatrix}, \ldots \right) \tag{10}$$

For each pair of dimensions: (1,2), (3,4), ..., a "compromise" representation of objects in a plane (bimension) is obtained. The scalar product among pairs of object-points represents symmetry while twice the area of the triangle they form with the origin describes the skew-symmetry, algebraic sign is associated with the orientation of the plane (positive counter-clockwise, negative clockwise). Object weights allow to display the different occasions by stretching or shrinking the distance of each point from the origin in the compromise configuration. The total amount of symmetry (skew-symmetry) reconstructed by the model is obtained summing algebraically scalar products (triangle areas) in each bimension.

4. A general strategy of analysis

In the previous sections we have shown that multilinear and distance-like methods are hierarchically related. As a matter of fact, they are nested models and optimise the same criterion function, but those lower in the hierarchy optimise this function subject to more severe constraints than those higher in the hierarchy, which results in poorer model fits. On the other hand, the more severely constrained methods correspond to simpler models, yielding easier interpretations.
Distance-like and multilinear methods can be related noting that models (2) and

(8) are equivalent because the first is the scalar product version of the latter. Model (10), under the constraint $d_{ih} = d_h$, can be seen as a constrained version of (5), where R=J and $D_h = d_h I$, or a particular case of (2), where y_j =Jx_j and $D_h = d_h^2 I$. Finally we can summarise the relations among the different methods with the following direct graph

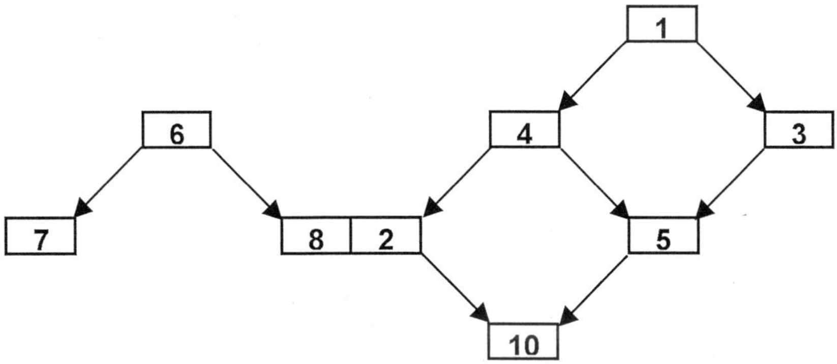

where an arrow goes from a model to a nested one. Some suggestions for possible strategies of analysis can be deduced from the afore mentioned hierarchy. For example in both contexts, multilinear and distance-like, the more parsimonious model could be taken as starting point, possibly with graphical capabilities, moving to a more complex one until an acceptable level of fit is reached.

5. An application to import-export data

Some of the methods previously discussed are now applied to import-export data following the indications of the outlined strategy. The flows in US dollars (O.C.S.E.) in the three years 1981-1985-1989 are analysed for the following 12 European countries: Belgium (BE), Denmark (DE), Finland (FI), France (FR), Germany (GE), Ireland (IR), Italy (IT), Netherlands (NL), Norway (NO), Spain (SP), Sweden (SW), United Kingdom (UK). The entries ω_{ijh} in the data matrices represent the exportation from country i to country j at year h. The iterative proportional fitting algorithm was preliminarily applied to each matrix (column and row sums equal to one) in order to remove the influence of country size and to emphasise the row-column association, that is the factors of interchangeability. Data in each matrix were also centred respect to the overall mean to remove the trivial component. Thus the exportation from one country to the other is evaluated with respect to the overall mean exportation. The diagonal entries of the data matrices were not taken into account because are meaningless.

We first applied method (10), under the constraint $d_{ih} = d_h$, because of its graphical properties and because it requires a minimum number of parameters. At first only one bimension has been fitted to represent symmetry and skew-symmetry simultaneously by using only n points. Unfortunately the model fit was very poor (about 50% of total sum of squares) and we needed two bimensions to improve the fit up to 74%. This implies two planar representations of n points each, with doubled number of parameters. Moreover algebraic sums of areas and scalar products from the two different bimensions are required resulting in difficult graphical interpretation. We could circumvent this inconvenience by adopting an opportune rotation in the multidimensional space (see Rocci and Bove, 1998), but here we will show what was obtained changing model.

Figure 1:*PARAFAC-CANDECOMP*

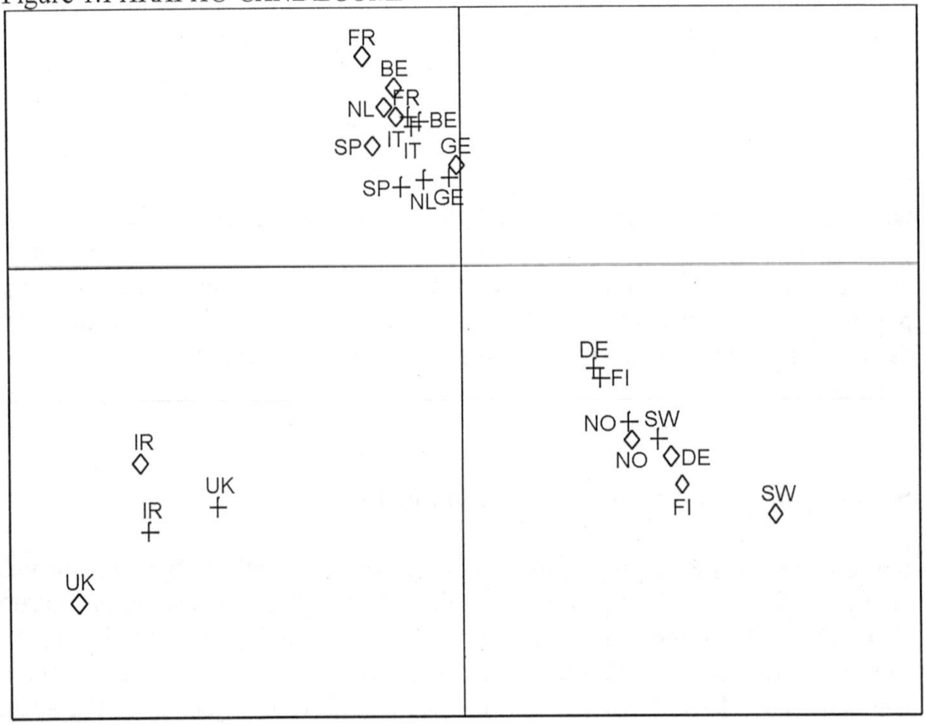

PARAFAC-CANDECOMP model (2) was chosen at this stage for its graphical capabilities. The resulting bidimensional configuration is depicted in Fig.1, it reproduces 76% of the total sum of squares. Three groups of countries are clearly isolated (DE-FI-NO-SW; IR-UK; BE-FR-GE-IT-NL-SP) and we notice that usually the within-group relationships are higher than the overall mean exchange in the data matrices (positive scalar products) while the between-group relationships are less or equal (negative or null scalar products). Each centred oriented flow is represented in the display, for instance the centred flow from UK to IR is approximated by the scalar product between the points UK(+) and

IR(◊) while the opposite flow can be analysed similarly by IR(+) and UK(◊). From their comparison it follows that the skew-symmetry is favourable to IR. Moreover the mean exchange can be also considered by averaging the two scalar products. Of course the graphical analysis of the symmetric and the skew-symmetric components is easier with the methods explicitly based on this decomposition while, on the contrary, for them is more difficult to detect the oriented flows. The configurations obtainable for the three years applying the weights in D_h are very similar to Fig.1 for the general stability of the import-export phenomenon across the time.

Figure 2: *Unfolding*

At this point we decided to apply a distance model of the GEM class (6) in order to simplify the diagram interpretation (usually it is easier to detect a distance rather than a scalar product). After subtracting each entry to the maximum in every occasion, the Unfolding model was chosen being substantially the PARAFAC-CANDECOMP reformulation in the distance models domain. Furthermore it is much easier to be interpreted than ASINDSCAL in spite of their equivalence in term of number of parameters. The fit obtained was 62% of the total sum of squares and the configuration is depicted in Fig.2. The interpretation is now based on point distances rather than on scalar products, capital letters are associated to the rows and small letters to the columns. We have results similar to Fig.1 in terms of between-groups relationships and some minor differences within the groups (e.g. in Fig.2 is less evident the skew symmetry UK-IR).

Finally we outline that when a good approximation is not obtainable by the previous models an approach representing separately the symmetric and the skew-symmetric components can be conveniently applied.

References

Carroll, J.D. (1972). Individual differences and multidimensional scaling, in: *Multidimensional Scaling: Theory and Applications in the Behavioral Sciences*, R.N. Shepard et al. (eds.), Seminar Press, New York.

Carroll, J.D., Chang, J.J. (1970). Analysis of individual differences in multidimensional scaling via an N-way generalization of Eckart-Young decomposition, *Psychometrika*, 35, 283-319.

Harshman, R.A. (1970). Foundations of the PARAFAC procedure: Models and conditions for an 'explanatory' multi-model factor analysis, *UCLA Working papers in Phonetics*, 22, 111-117.

Harshman, R.A. (1978).Models for analysis of asymmetrical relationships among N objects or stimuli, Paper presented at the *First Joint Meeting Psychometric Society and Society of Mathematical Psychology,* Hamilton, Ontario.

Kiers, H.A.L. (1989). An alternating least squares algorithm for fitting the two and three-way DEDICOM model and the IDIOSCAL model, *Psychometrika*, 54, 515-521.

Kroonenberg, P., de Leeuw, J.(1980). Principal Component Analysis of three-mode data by means of alternating least squares algorithms, *Psychometrika*, 45, 69-97.

O.C.S.E. (1981,1985,1989), *Statistical Annual of Foreign Trade*, Rome.

Rocci, R., Bove, G. (1994). A method to analyse asymmetric two-mode three-way data, *Atti della XXXVII Riunione Scientifica Società Italiana di Statistica*, CISU, Roma (in italian), 253-260.

Rocci, R., Bove, G. (1998) Rotation problems in Asymmetric Multidimensional Scaling, submitted for publication.

Young, F.W. (1987). *Multidimensional Scaling: History, Theory, and Applications*, Edited by R.M. Hamer, Hillsdale, NJ: Lawrence Erlbaum.

Zielman, B. (1991).*Three-way Scaling of asymmetric proximities*, Research Report RR-91-01, Department of Data Theory, Leiden University, Leiden.

Zielman, B., Heiser, J.W. (1996). Models for asymmetric proximities, *British Journal Mathematical Statistical Psychology*, 49, 127-146.

Comparison of Euclidean Approximations of non-Euclidean Distances

Sergio Camiz
Dipartimento di Matematica "Guido Castelnuovo",
Università di Roma "La Sapienza", E-mail: camiz@mat.uniroma1.it

Abstract: The different techniques used for Euclidean approximation of distances are discussed. In the special case of points in a Euclidean space, whose distances are biased due to measure errors, accepting negative eigenvalues may help in the interpretation of results that are less biased than those obtained through an additive constant solution. Numerical examples are given.

Keywords: Distances, Eigenanalysis, Euclidean approximations.

1. Introduction: *PCA* and *PCoA*

In exploratory data analysis geometrical representations, individuals are seen as points in an *affine space E*, sustained by the vector space spanned by the observed variables. If no metrics is known, variables may be arbitrarily represented on graphical Euclidean space and individuals are set accordingly. If some measured association exists, reflecting the relationships among either individuals or variables, it may be used as a spatial metrics, according to its properties. If the association matrix is *positive semi-definite (psd)*, either Principal Components (*PCA*) or Principal Coordinates (*PCoA*) Analyses give an exact Euclidean representation, where points distances are interpreted as dissimilarities among corresponding individuals and vectors angles as measures of variables reciprocal agreement (similarity). In this frame, reduced dimensional representations are effective for information synthesis and main factors detection.

When negative eigenvalues occur, i.e. when the available association matrix is not *psd*, Euclidean representation becomes critical, since negative eigenvalues may not be interpreted as points inertia along corresponding eigenvectors. Instead, an *Euclidean approximation* is always possible through different techniques.

A similarity matrix S among variables is a symmetrical bilinear form having maxima on the diagonal; if S is *psd* it induces a *scalar product*, that gets the space E spanned by the variables an *Euclidean space*. Thus (Lang, 1972), orthonormal bases may be formed and eigenanalysis of $S = U'\Lambda U$, with constraint $U'U = I$, once sorted the eigenvalues in descending order, solves the optimization problem

$$\sum_{ij} \left(s_{ij} - \sum_{\alpha=1}^{q} \lambda_\alpha u_{\alpha i} u_{\alpha j} \right)^2 = \text{minimum} \ \forall \ q<p \qquad (1)$$

where p is the order of S, $\lambda_{\alpha} \in \Lambda$, $u_{\alpha i}$, $u_{\alpha j} \in U$. The loss of information is minimised by projection of both vectors and points on the reduced dimensional spaces, spanned by the first q eigenvectors of S. It may be measured, since $\sum_{\alpha} \lambda_{\alpha} = \text{trace}(S)$, and, given the $n \times p$ data table X concerning p variables measured on the n individuals, $\sum_i c_{i\alpha}^2 = \lambda_{\alpha}$ is the points inertia along α-th axis, $c_{i\alpha} \in C = XU$, being the α-th coordinate of i-th individual in the space basis formed by the eigenvectors. Then, the ratio λ_{α} / trace(S) is a relative measure of inertia along the α-th axis. These are the essentials of *Principal Components Analysis.* Given an $n \times n$ distance matrix D among individuals, Torgerson's (1958) formula

$$s_{ij} = -\frac{1}{2}\left(d_{ij}^2 - d_{i.}^2 - d_{.j}^2 + d_{..}\right) \tag{2}$$

for all $i,j = 1,\ldots,n$ defines a similarity between vectors associated to considered points, whose origin is at the points centroid. If the corresponding matrix S is *psd*, the distance is Euclidean and S may be used as a scalar product, having as distance among points D itself. S may be submitted to *Principal Coordinates Analysis* (Torgerson, 1958; Gower, 1966, Mardia *et al.*, 1979): S eigenanalysis under constraint $U'U = \Lambda$, provides an orthogonal basis with the same meaning as *PCA*.

2. Theoretical *PCoA* and distance limits, with solutions

Although partitioning points inertia, *PCoA* **does not** lead to the best least squares approximation of distances in reduced dimensional spaces, since the problem is now

$$\sum_{ij} \left(d_{ij}^2 - \sum_{\alpha\,1}^{q} (u_{\alpha i} - u_{\alpha j})^2 \right)^2 = \text{minimum } \forall\ q<n \tag{3}$$

that may not be solved through eigenanalysis, nor solutions in different dimensional spaces are encapsulated as in the scalar product case (Le Calvé, 1976). Thus, *PCoA* application should be limited to cases in which, far from aiming at minimizing distances bias, one wishes to use available distances to detect angles among vectors, in agreement with given distances. Aiming at maximizing distances approximation in reduced dimensional spaces, numerical techniques (such as *Non-Metric Multidimensional Scaling, NMMDS*, Kruskal, 1964a, b) lead to local optima. With *NMMDS* all existing metric information, such as factors and points inertia, if any, is lost, since it deals with dissimilarities ranks. Both Gower (1966) and Seber (1984) suggest the use of *PCoA* as an initial guess for *NMMDS*, but Kruskal (quoted by Seber, 1984) admits that it does not change the initial configuration very much. Then, one may question the convenience of such an attempt, particularly in exploratory analyses, when a metrics exists.

If the given distance is not Euclidean, the matrix is not *psd*, so that in eigenanalysis negative eigenvalues result: this is usually considered a serious drawback for eigenvalues interpretation in terms of explained inertia, yet, the interest in an *Euclidean approximation* remains, at least for descriptive purposes. One may skip the problem by using *NMMDS*, or override it by adding a suitable constant to all distances (Lingoes, 1971; Cailliez, 1983). In particular, Lingoes (1971) solution, given by $d_{ij}^2 + c$, $c = 2|\lambda_n|$ where λ_n is the negative eigenvalue with maximum module, biases individuals pattern. No theoretical improvement derives from Cailliez (1983) exact solution $d_{ij} + c$, where c is the largest eigenvalue of a particular matrix. Since all distances are biased, Messick and Abelson (1956), Gower (1966), Saito (1978), Mardia (1978), and Critchley (1980) argue that Torgerson solution, ignoring negative eigenvalues, is to be preferred, provided that they are small. In addition, Mardia (1978) shows that Torgerson solution limited to positive eigenvalues has some optimal properties. Additive constant technique seems then a suitable alternative to *NMMDS*, with the advantage that the algorithm is faster and leads to a global optimum, instead that a local one.

3. A special case and a suggested solution

Empirical data are always subject to both measure error and rounding, so that a distance matrix may be only an estimate of the true one. In the following, we compare the additive constant method to a deeper insight to Euclidean approximation in the case of negative eigenvalues. This will be done in the special case in which *a)* individuals actual pattern is supposed to belong to an Euclidean space, *b)* factors accountable for individual scattering are supposed to exist, and *c)* distances are non-Euclidean due to measure errors. A *robust* analysis method not too much dependent on these errors would be precious. In this case, application of *NMMDS* would lose all metric information and additive constant technique axes and inertia would not be directly related to original pattern, since distances would be biased. In particular, the original true inertia would be overestimated.
Let us consider the eigenanalysis of a non-*psd* matrix, with eigenvectors constraint being

$$\sum_i u_{\alpha i} u_{\beta i} = \begin{cases} 1 & \text{if} \quad \alpha = \beta \quad \text{and} \quad \lambda > 0 \\ -1 & \text{if} \quad \alpha = \beta \quad \text{and} \quad \lambda < 0 \\ 0 & \text{otherwise} \end{cases} \tag{4}$$

and call it for simplicity *plain analysis*. Thus, eigenvectors corresponding to negative eigenvalues are imaginary, as well as individuals coordinates (Gower, 1985). Nevertheless, formula $\sum_i c_{i\alpha}^2 = \lambda_\alpha$ is still valid, but $\sum_\alpha \lambda_\alpha$ *is partitioned into two parts*: 1) a positive one, that overestimates the true inertia, with Euclidean real representation, i.e. an Euclidean approximation of individuals on the space spanned by real eigenvectors; and 2) a negative one, that fixes the overestimated part, and provides an imaginary representation (on the imaginary eigenvectors) of the bias introduced in the Euclidean approximation. Limited to this imaginary eigenspace, coordinates, contributions, distances, and inertia keep the same meaning of the real representation, with the difference that these are referred to the introduced bias. It is then *a principal coordinates analysis of introduced bias.* Considering a distance matrix D, if the corresponding S Torgerson's matrix is not *psd*, the equivalent tool is the S eigenanalysis with constraint $U'U = \Lambda$, including negative eigenvectors. In order to give a *geometrical meaning* to this technique, one must consider that in Euclidean spaces non-Euclidean geometric figures «*do not close*». In order to enable points representation, thus to close figures, some distances need to be enlarged. Consequently, the original inertia is increased and becomes overestimated. The added inertia does not exist in the data nor it may appear in trace (Λ) = trace (S). Then negative eigenvalues and imaginary axes are necessary for inertia balance: having negative contribution, they explain where and how much extra inertia was added, in order to allow the Euclidean representation.

4. Numerical examples

Let's start with a very simple example, considering a square of unit area, with all vertices lying on the axes. Their coordinates are thus either 0 or $\pm 1/\sqrt{2}$, each side length is 1, and diagonals length is $\sqrt{2}$. Inputting the computer $\sqrt{2}$ as a double precision number, the eigenvalues of the corresponding Torgerson matrix are correct and both points coordinates and interpoint distances are sufficiently well approximated. Giving the computer $\sqrt{2}$ with only 4 significant digits, say either 1.414 or 1.415 a third non-zero eigenvalue is obtained, negative in the latter case. In Tab. 1 are compared the essential results of *PCoA* on both the exact distance matrix and the one with rounding to 1.414 and both plain analysis and *PCoA* with additive constant on distance matrix with rounding to 1.415. In the first column each point distance from the centroid on the plane of first two axes is given; in the second the coordinates along the third, if existing; in the third column the Torgerson's matrix trace, in the fourth the points inertia on the first two eigenvectors plane, in the fifth the bias introduced in the side measures and the last the one introduced in the diagonals length. In case of rounding to 1.414, in order to keep the side length to 1, the points connected by a diagonal (1 and 4) are opposed on a third axis to the others (2 and 3), in order to fit the square sides length, since otherwise the reduced length diagonal would shrink the sides too (Fig.

Table 1: *Main results of four points analyses.*

Matrix	*distance from origin*	*III axis coordinate*	*trace*	*plane 1-2 inertia*	*bias*	
					side	*diagonal*
$1/\sqrt{2}$	.7071		2.0000	2.0000		
1.414	.7070	± .0086	1.9997	1.9994	-.0003	0
1.415 plain analysis	.7075	±i.0167	2.0022	2.0011	+.0011	0
1.415 additive constant	.7079		2.0045	2.0045	+.0022	+.0022

Figure 1: *Representation of four points on three-dimensional space, when diagonals are rounded.*

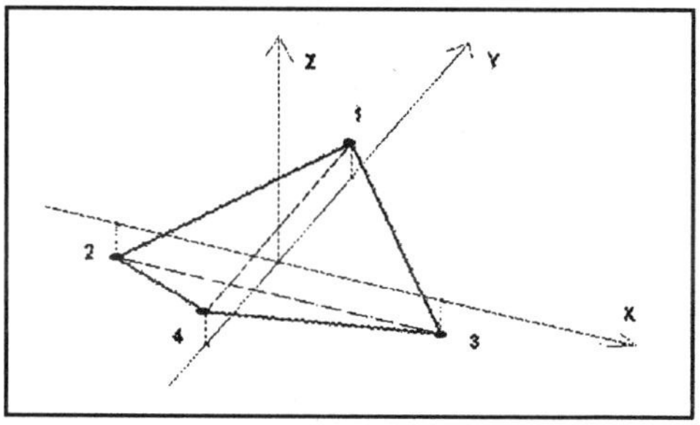

1). In case of rounding to 1.415, diagonals are too long to allow the side length 1, so that it becomes 1.00055. This biased distance is fixed on the fourth axis, an imaginary one with negative eigenvalue. Here the squared diffe-rence between imaginary points coordinates is negative and its value is exactly the one necessary to restore distance 1. Numerically, only sides lengths are biased and information is found on the fourth (imaginary) axis. If possible, a graphical representation, would be analogous to that in Fig. 1, with the difference that the imaginary vertical axis would show which couple of points distance was biased due to Euclidean representation needs. Using additive constant, all distances are biased, so that additive constant representation is more biased than plain solution, as Mardia (1978) proved.

Let us consider now a grid of 16 points in a 2-dimensional space, equally spaced, so that each grid's side is 2. *PCoA* returns the exact grid on the eigenspace, corresponding to two equal eigenvalues, that may be arbitrarily rotated. Modifying points 1-2 distance to $d_{12} = d_{21} = 2.1$, D matrix is no longer *psd*. Lingoes (1971) solution adds to each squared distance the same quantity, .4743589... and the correct representation of this biased structure is in 14 axes (i.e. the number of points - 2), where each biased distance is exactly represented. In Tab. 2 is shown the essential bias information: on the first column the matrix trace, on the second the inertia on the first two axes plane, on third and fourth the coordinates of points 1 and 2 on the third axis, and on the following maximum bias on plane 1-2 of both point position and distances considering points 1-2 and others separately.

In Fig. 2 the grid on plane spanned by axes 1 and 2 (*plane 1-2* in the following) is shown; rotation results from *PCoA*. On this plane, points 1 and 2 are those best represented, as well as their distance compared to all others. A third small axis opposes points 1 and 2 (the others remain close to the centroid), so that in three axes the (biased) distance 1-2 is completely represented. Other points and distances are adjusted on the following 11 axes, totally unuseful. Compared to the unbiased solution, all distances are biased by the same amount.

Using plain analysis, only four non-zero eigenvalues result. On Tab. 2 one may

Table 2: *Bias measures of grid representation through additive constant technique and plain analysis.*

Matrix	*trace*	*plane 1-2 inertia*	*axis 3 coord.*		*points bias*		*distances bias*	
			1	*2*	*1-2*	*others*	*1-2*	*others*
Unbiased grid	160.	160.						
Additive constant	160.7133	160.1636	.377	-.403	.0083	.005	.3282	
Plain analysis	160.0256	159.9386	-.302	.32	.0095	.004	.3338	.1857

Figure 2: *The 16 points grid on the plane of first two axes, with additive constant technique.*

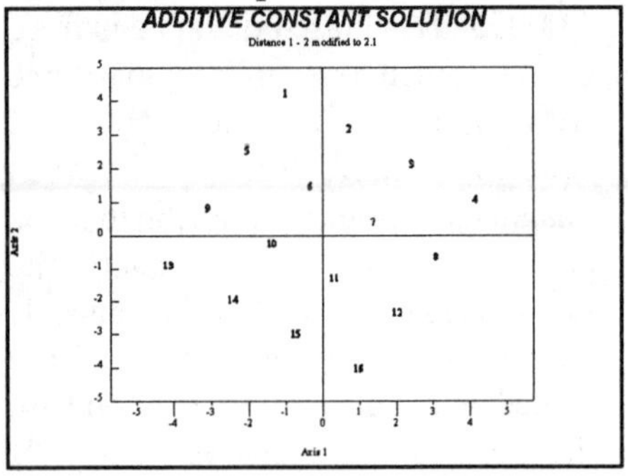

Figure 3: *The 16 points grid on the plane of axes 3 (real and 16 (imaginary).*

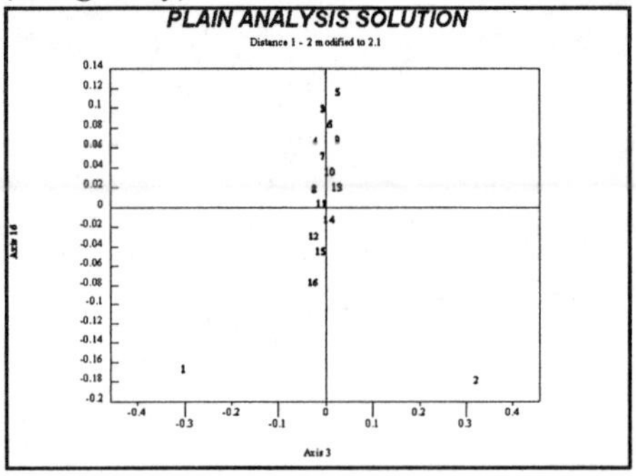

notice that all bias indicators are lower than those of additive constant. On the first two, the grid is again well represented, approximately with the same rotation than additive constant. On the third axis, points 1 and 2 are opposed and inertia on three axes overestimates the trace. The imaginary axis shows this overestimate. In fact, on this axis points 1 and 2 are opposed mostly to 3, 5, and 6, that is those closest to them, since their distances were mostly biased in order to adjust the biased 1-2 distance. Considering distances, on the plane 1-2 all those not involving points 1 and 2 are nearly exactly represented, whereas the others are somehow under-estimated. Summarizing, plain analysis gives less biased results than additive constant, in an essential number of axes, on plane 1-2 a better estimation of both points position and true distances is given, and, in addition, the bias introduced to distances is shown clearly by the imaginary axis (Fig. 3).

Let us now bias two distances, 1-2 and 15-16, to $d_{12} = d_{1516} = 2.1$. In this case, plain analysis gives 6 non-zero eigenvalues, two of which are negative. The pattern of points on the plane spanned by the first two axes is nearly the same as before. On the plane 3-4 (Fig. 4), points 2 and 15 are opposed to points 1 and 16, meaning that all these distances were originally biased. On the imaginary plane 15-16 (Fig. 5), opposition among points 1,2, 15, and 16 to all others, means that all these distances

Figure 4: *The 16-points grid with two biased distances on the plane of real axes 3 and 4.*

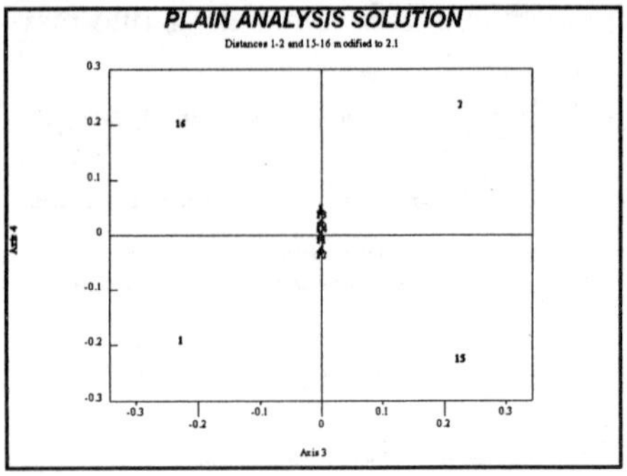

Figure 5: *The 16-points grid with two biased distances on the plane of imaginary axes 15 and 16.*

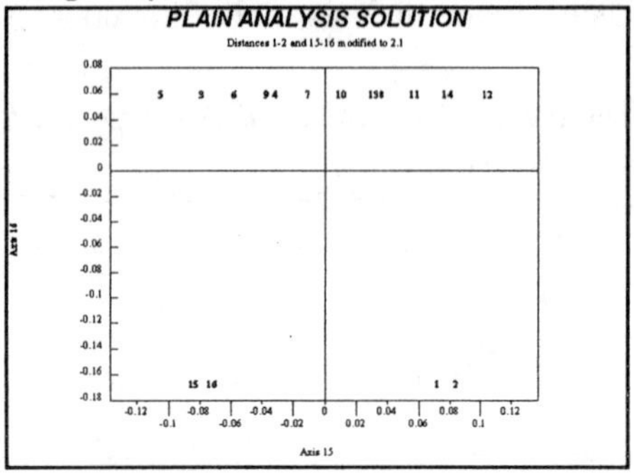

were biased for representation purposes, in particular those among points 1, 2 and the closest to them (on the second quadrant), and those among points 15-16 and the closest to them (on the first). In this way all bias introduced in order to fit Euclidean representation is visible.

5. Conclusions

The problem of the Euclidean approximation of points based on an association matrix, when necessary, may be solved in different ways. Both *PCA* and *PCoA* are exact solutions, provided that the association matrix is *psd*. Considering distance matrices, when this is not the case, an approximation may be done according to several techniques: *NMMDS* is always possible, but the algorithm is time consuming, initial configuration dependent, leading to local optima, and losing metric information, if present. The additive constant technique is competitive with *NMMDS*, since it is based on eigenanalysis, so it is faster and leads to a global solution, that may little differ from *NMMDS*. Nevertheless, no advantage comes from its metric structure, since it is biased in respect to the original one. Critchley (1980) proposes a general formula for non-Euclidean distances, consisting in transforming distances according to some monotonic function. A *psd* matrix is obtained through an additive constant-like transformation that is proved to have optimal properties. In this frame, Joly and Le Calvé (1986) proved that, given any non-Euclidean distance d_{ip}, there exists a maximum exponent ϵ, called *Euclidean Index*, such that the *Hadamard* power α of matrix D, D^{α}, is *psd* for every α such that $1/2 \leq \alpha \leq \epsilon < 1$. The index ϵ may be found by iteration and has optimal properties, since its bias of original distances is minimum. In addition, ϵ may be chosen as a good indicator of the considered distance nature, since it informs about the distance departure from Euclideanity. The use of an Euclidean index may lead to more interesting re-sults, both theoretical and practical: considering distances obtained through a for-mula, by including the found index in it, Euclidean distance would be get directly.

In the case of biased Euclidean distances, Joly and Le Calvé (1986) proposal seems of reduced interest, whereas the use of plain analysis leads to more natural and easily understandable results than additive constant. Compared to the latter, plain analysis less biases distances, seems simpler, more straightforward, and more *robust*, outlining the true structure and not the biased one. In addition, it shows exactly the extra bias, necessary to get the Euclidean approximation. For this reason, once the computed distance is suspected to be Euclidean, it may be a helpful tool for detecting distances measure errors or, if distance is computed in some way, which may be the couples of individuals responsible of the bias. In the spirit of exploratory analyses, this seems a very interesting opportunity.

As a final remark, since Bénasséni (1994) proposes a method aiming at adding a constant *only to some distances*, i.e. those suspected to non-Euclideanity, it may be interesting to compare it with plain analysis. It is likely that the results obtained through plain analysis may be used as an input for Bénasséni procedure.

Acknowledgements

This work was carried out with research grants n. 96.03830.CT12 and 96.03541.CT15 of Consiglio Nazionale delle Ricerche. I wish to thank Giuseppe Bove, Maurizio Vichi, and an anonymous referee for their friendly support and helpful critics and suggestions.

References

Bénasséni, J. (1984). Partial Additive Constant, *J. Stat. Comp. Simul.*, 49, 179-193.

Borg, J. & Lingoes, J. (1987). *Multidimensional Similarity Structure Analysis*, Springer Verlag, New York.

Cailliez, F. (1983). The Analytical Solution of the Additive Constant Problem, *Psychometrika*, 48, 2, 305-308.

Critchley, F. (1980). Optimal Norm Characterisations of Multidimensional Scaling Methods and some Related Data Analysis Problems, in: *Data Analysis and Informatics*, Diday, E. *et al.* (Eds.), Amsterdam, North-Holland, 209-229.

Gower, J.C. (1966). Some Distance Properties of Latent Root and Vector Methods used in Multivariate Analysis, *Biometrika*, 53, 325-338.

Gower, J.C. (1985). Properties of Euclidean and Non-Euclidean Distance Matrices, *Linear Algebra and its Applications,* 67, 81-95.

Joly, S. & Le Calvé, G. (1986). Étude des puissances d'une distance, *Statistique et Analyse des données,* 11, 3, 30-50.

Kruskal, J.B. (1964a). Multidimensional Scaling by Optimizing Goodness of Fit to a Nonmetric Hypothesis, *Psychometrika*, 29, 1-27.

Kruskal, J.B. (1964b). Nonmetric Multidimensional Scaling: a Numerical Method, *Psychometrika*, 29, 115-129.

Lang, S. (1972). *Linear Algebra,* Addison-Wesley, Reading, Mass.

Le Calvé, G. (1976). *Quelques remarques sur certains aspects de l'analyse factorielle*, Lab. Analyse des Données, Université de Rennes II, Cahier n. 2.

Lingoes, J.C. (1971). Some Boundary Conditions for a Monotone Analysis of Symmetric Matrices, *Psychometrika*, 36, 195-203.

Mardia, K.V. (1978). Some Properties of Classical Multi-Dimensional Scaling, *Comm. in Statistics - Theory and Methods*, Series A, 7, 13, 1233-1241.

Mardia, K.V., Kent, J.T. & Bibby, J.M. (1979). *Multivariate Analysis*, Academic Press, London.

Messick, S.J. & Abelson, R.P. (1956). The Additive Constant Problem in Multidimensional Scaling, *Psychometrika*, 21, 1-15.

Saito, T. (1978). The Problem of the Additive Constant and Eigenvalues in Metric Multidimensional Scaling, *Psychometrika*, 43, 2, 193-201.

Seber, G.A.F. (1984). *Multivariate Observations*, J. Wiley & Sons, New York.

Torgerson, W.S. (1958). *Theory and Methods of Scaling*, J.Wiley & Sons, NewYork.

Analysing Dissimilarities through Multigraphs

Angela Montanari Gabriele Soffritti
Dipartimento di Scienze Statistiche, Università di Bologna
Via Belle Arti 41, 40126 Bologna, Italy
montanar@stat.unibo.it soffritt@stat.unibo.it

Abstract: In this paper a very general way of modelling dissimilarities is proposed based on ideas derived from multigraph theory. The proposed method admits Gower's dissimilarity index as a special case and gives the possibility to cope with measurement errors and within subject variability when computing dissimilarities. The approach also allows to assign a different importance to the same dissimilarity value on different regions of the variable domain.

Keywords: Dissimilarities, Multigraphs, Cluster Analysis, Ordination Methods.

1. Introduction

Many of the classical multivariate statistical methods - such as cluster analysis and multidimensional scaling - require proximity matrices as input and their performances are heavily conditioned by how accurately the computed proximities describe the real differences among the observed units.
This is perhaps one reason for the large number of similarity or dissimilarity measures that can be found in statistical literature for numeric or binary variables. Lists and discussions of their properties are presented by, amongst others, Gordon (1981), Gower (1985), Gower & Legendre (1986), Snijders, Dormaar, van Schuur, Dijkman-Caes & Driessen (1990), Everitt (1993) and Cox & Cox (1994).
When one has to deal with mixed variables, the range of possible choices narrows down. The most widely - and perhaps the only - suggested solution is Gower's dissimilarity coefficient (Gower 1971) defined as $d_{ij}=1-s_{ij}$ (i and j being two generic units) where

$$s_{ij}=\Sigma_k s_{ijk}\delta_{ijk}/\Sigma_k\delta_{ijk}$$

and where $s_{ijk}=1-|x_{ik}-x_{jk}|/Range_k$ for numeric variables and $s_{ijk}=1$ or $s_{ijk}=0$ for binary and nominal variables according to whether the two units show the same or a different value on the k-th variable; δ_{ijk} is typically 1 or 0 depending on whether or not the comparison is valid for the k-th variable.
A different way of analysing mixed data has been proposed by Godehardt (1990) in the cluster analysis context and is based on graph-theoretic concepts. For each variable (or block of variables taking values on the same measurement

scale and phenomenally linked) a specific similarity or dissimilarity measure is calculated between every pair of units. Being m the number of variables (or of blocks), this gives m dissimilarities between any pair and defines a multigraph with the n objects as vertices. Every variable (or block) is thus considered as a layer of a related multigraph. Coherently with cluster analysis goals and with linkage methods philosophy, a pair of vertices in a layer is joined by an edge if the dissimilarity between the corresponding objects is less than a specified threshold d which is generally different for the different layers and is dependent on the dissimilarity measure that has been used to evaluate differences on the variable (or the block) corresponding to the layer (for blocks containing a single quantitative variable Godehardt suggests to choose d as a function of the variable standard deviation but no further advice is given neither as to which function has to be used nor as to what choice is suitable for nominal variables and for blocks comprising more than one variable). Cluster analysis is then performed on what Godehardt calls the s-projection of a multigraph, that is a graph having the same vertices as the multigraph and in which two vertices are linked if they result joined by at least s edges in the multigraph itself.
This approach to modelling dissimilarities gives interesting results as far as cluster analysis is concerned but may not be completely satisfactory when the same data have to be analysed by ordination methods too. In fact while cluster analysis aims at dissecting the data into homogeneous groups, ordination aims at graduating dissimilarities between units (Krzanowski & Marriott 1994) and therefore, in this context, the choice of a single threshold for each layer may sometimes conceal important features.

2. A weighting procedure for dissimilarities

Godehardt's model for dissimilarities can be modified to allow a more detailed unit description by weighting the edges of the multigraph, that is by giving edges a weight that is linked to the corresponding dissimilarity in a linear or non linear way, according to the amount of information one has about the units and the variables and wants to be conveyed into the analysis.
To put it into an operational perspective, for the k-th variable (or block of variables) t_k thresholds ${}_kd_1< \dots <{}_kd_l< \dots <{}_kd_{t_k}$ (i.e. a different number of thresholds can be used for different variables) and t_k weights ${}_kv_1< \dots <{}_kv_l< \dots <{}_kv_{t_k}$ are defined such that, denoted as d_{ijk} the dissimilarity between the units i and j on variable k, if $d_{ijk}<{}_kd_1$, then ${}_kv_1=0$; if $d_{ijk}\geq{}_kd_{t_k}$, then ${}_kv_{t_k}=1$; else ${}_kd_l\leq d_{ijk}<{}_kd_{(l+1)}$, ${}_kv_l$ is set equal to a user defined value that is generally different for different l's.
In so doing the dissimilarities (however computed) are mapped onto the 0-1 interval, with the effect of giving the same weight to all the dissimilarities falling between the same thresholds and of allowing non linear weighting of dissimilarities within the same variable. Different weights can also be given to different variables by simply changing the width of the interval onto which dissimi-

larities are mapped: for instance, a variable whose mapping interval has been set equal to 0-2 is given a double weight with respect to the remaining ones.
The suggested procedure is very general and can be applied to all the dissimilarity measures usually examined in the statistical literature (see Cox & Cox 1994) for quantitative, qualitative and dichotomous variables.
Once the multigraph has been constructed in the previously described way, proximity analysis can be performed on its projection in which a pair of vertices is joined by an edge whose weight equals the sum of the weights of the corresponding edges in the multigraph, divided by the sum of the weights given to the different variables (the sum equals m if all the variables are equally weighted).
Multigraphs represent an elegant model for dissimilarities, but are not the only way in which the procedure we are describing can be formulated. Alternatively one can consider m dissimilarity matrices (one for each variable or block of variables), transform them in as many matrices $\mathbf{V}_k$ whose entries are defined as

$$v_{ijk} = \begin{cases} 0 & \text{if } d_{ijk} <_k d_l \\ f_k(d_{ijk}) & \text{if } _k d_l \leq d_{ijk} <_k d_{(l+1)} \qquad l = 1,\ldots,t_k - 1 \\ 1 & \text{if } d_{ijk} \geq_k d_{t_k} \end{cases}$$

where f_k is a monotone increasing function of the observed dissimilarities, and synthesize them in a single matrix $\mathbf{A}$ whose entries are given by

$$a_{ij}=\Sigma_k v_{ijk} w_k / \Sigma_k w_k$$

where w_k is the k-th variable weight.
The crucial point of the method is the choice of the thresholds and of the weighting procedure. One can either devise a procedure for determining the thresholds (linearly or non linearly depending on the research aims) and then weight differences linearly, or choose equally spaced thresholds and devise a weighting criterion, which gives linear or non linear weights depending on the character nature and the exploratory goals. From the result point of view the two approaches are not dual to each other. For a given value of the maximum difference allowed between two units in order for them to be considered as identical, the first one produces a small number of intervals of increasing (or decreasing) width and all the differences within each interval are equally weighted; in so doing it highly flattens "between subjects" variability. If one wants to cope with measurement errors while preserving data variability the second approach is preferable; in the following we will describe it in detail.
The first step to be made is the choice of a constant ε_k, for each variable or block of variables, which just measures the maximum difference allowed between two units in order for them to be considered as identical ($\varepsilon_k = _k d_1$). After that, the number of equally spaced thresholds can be determined, for variables measured on any measurement scale as well as for blocks of variables, as

$$t_k = int({}_kd_{MAX}/\varepsilon_k)$$

where *int(a)* denotes the integer part of *a*, that is the largest integer less than or equal to *a* (e.g. *int*(2.7)=2), and ${}_kd_{MAX}$=MAX$\{d_{ijk}\}$. The weighting function f_k can then be chosen within the monotone increasing convex functions, if one wishes to give differences an importance which is less than proportional to the distance values, within the monotone increasing concave functions, if the importance is to be more than proportional, within straight functions, if difference importance is to be proportional to their values (fig. 1) (see Anderberg (1973) for a slightly different proposal). Once the weighting function has been defined, v_{ijk} is determined as

$$v_{ijk} = \gamma_k f_k(int(d_{ijk}/\varepsilon_k))$$

where γ_k is a normalizing constant which compels v_{ijk} to lie between 0 and 1[1].

Figure 1: *Weight behavior for three different choices of the weighting function:* $v_{ijk}=(int(d_{ijk}/\varepsilon_k))^2/t_k^2$ *(solid line),* $v_{ijk}=int(d_{ijk}/\varepsilon_k)/t_k$ *(dashed line),* $v_{ijk}=(int(d_{ijk}/\varepsilon_k))^{1/2}/t_k^{1/2}$ *(dotted line)*

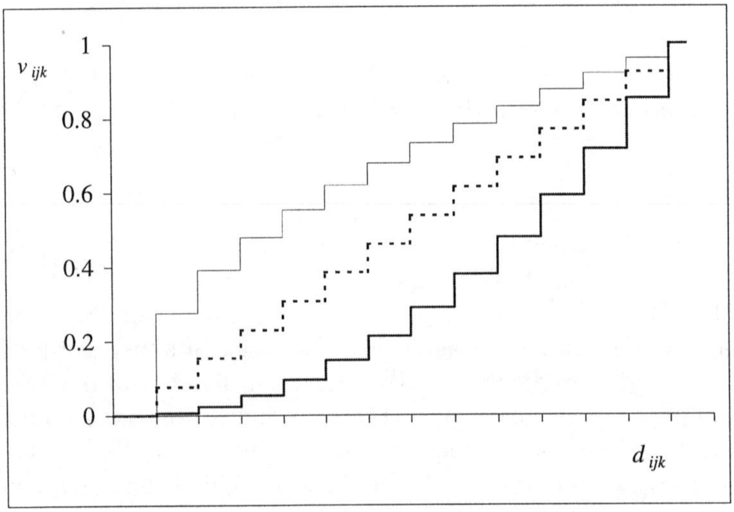

In order to fix ideas the following simple example may be of help. Let's consider five houses on which three variables have been observed: the surface area (X_1 in m^2), the presence of a garage (X_2, 1='yes', 0='no'), the location of the house (X_3, 1='town centre', 2='suburbs', 3='country'). The data matrix **X** is given in fig. 2a.

[1] If the chosen function f_k does not pass through the origin, it has to be translated to g_k so as $g_k(0)=0$.

For the three blocks (one for each variable) $\varepsilon_1=5$, $\varepsilon_2={}_2d_{MAX}$, $\varepsilon_3={}_3d_{MAX}$ have been chosen, thus giving $t_1=13$, $t_2=t_3=1$ (in these last two cases there is no need to choose any weighting function). The weighting function for X_1 has been defined as $f_1(int(d_{ij1}/\varepsilon_1))=(int(d_{ij1}/\varepsilon_1))^2$ where $d_{ij1}=|x_{i1}-x_{j1}|$, thus overweighting large differences, so $v_{ij1}=(int(d_{ij1}/\varepsilon_1))^2/t_1^2$. The multigraph in fig. 2b illustrates the weights v_{ijk} for each variable and each couple of units, and the weighted graph in fig. 2c the corresponding dissimilarities a_{ij} obtained by using $w_1=2$, $w_2=w_3=1$ as weights for the three variables. No edge has been drawn if $v_{ijk}=0$ or $a_{ij}=0$.

Figure 2: *A simple illustrative example of the proposed procedure*

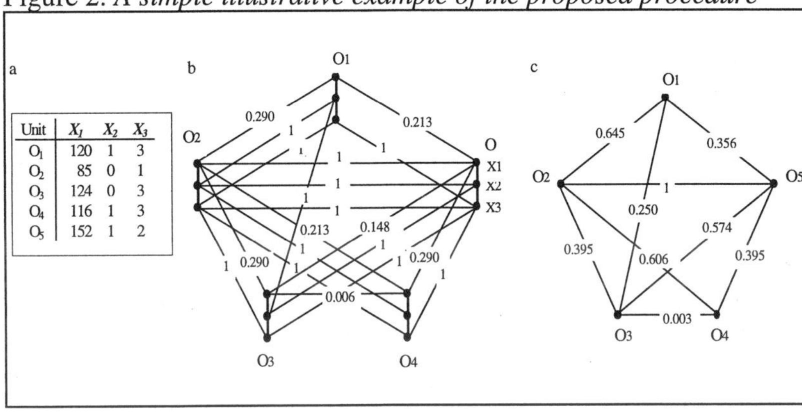

Unit	X_1	X_2	X_3
O_1	120	1	3
O_2	85	0	1
O_3	124	0	3
O_4	116	1	3
O_5	152	1	2

3. A further weighting procedure for single variables

The weighting procedure illustrated in the previous paragraph can be modified, for blocks composed by a single variable, to assign a different weight to the same dissimilarity value on different regions of the variable domain. This possibility (not contemplated in the usual dissimilarity measures) could be very useful for some variables. For example, the importance of a given difference (e.g. £ 100000) between the monthly incomes x_i and x_j of two individuals i and j when x_i and x_j are both high is not so high as when x_i and x_j are both low: a different weight should be assigned to the difference on the basis of the position of x_i and x_j in the income domain. This result can be obtained by means of the following weighting procedure.

The first step to be made is the construction of a partition $\{I_1,\ldots, I_h,\ldots, I_g\}$ of the domain $[{}_kx_{min}, {}_kx_{max}]$ of the variable X_k, where $I_h=[{}_kx_{h-1}, {}_kx_h)$ is the generic class of the partition, with ${}_kRange_h={}_kx_h-{}_kx_{h-1}$, and ${}_kx_0={}_kx_{min}$, ${}_kx_g={}_kx_{max}$. After that, a constant ${}_k\varepsilon_h$ for each class I_h has to be chosen, thus giving the corresponding number of equally spaced thresholds ${}_kt_h=int({}_kRange_h/{}_k\varepsilon_h)$, and an eventually different function ${}_kf_h$ for each class I_h, whose normalizing constant is ${}_k\gamma_h$, defined. The computation of the weight v_{ijk} then depends on the classes which the units i and j belong to:

$$v_{ijk} = \begin{cases} {}_k\beta_h\, {}_k\gamma_h\, {}_kf_h(int(\frac{|x_{ik}-x_{jk}|}{{}_k\varepsilon_h})) & \text{if } x_{ik}\in I_h \text{ and } x_{jk}\in I_h \\ \\ {}_k\beta_h\, {}_k\gamma_h\, {}_kf_h(int(\frac{|x_{ik}-{}_kx_h|}{{}_k\varepsilon_h}))+ & \\ +{}_k\beta_{(h+1)}\, {}_k\gamma_{(h+1)}\, {}_kf_{(h+1)}(int(\frac{|x_{jk}-{}_kx_h|}{{}_k\varepsilon_{(h+1)}})) & \text{if } x_{ik}\in I_h \text{ and } x_{jk}\in I_{h+1} \\ \\ {}_k\beta_h\, {}_k\gamma_h\, {}_kf_h(int(\frac{|x_{ik}-{}_kx_h|}{{}_k\varepsilon_h}))+ & \\ +\sum_{q=h+1}^{t-1} {}_k\beta_q + {}_k\beta_{(h+t)}\, {}_k\gamma_{(h+t)}\, {}_kf_{(h+t)}(int(\frac{|x_{jk}-{}_kx_{t-1}|}{{}_k\varepsilon_{(h+t)}})) & \text{if } x_{ik}\in I_h \text{ and } x_{jk}\in I_{h+t} \end{cases}$$

where $t>1$, ${}_k\beta_h={}_kRange_h/Range_k$, and it has been supposed, without loss of generality, $x_{ik}\leq x_{jk}$.

In this procedure, different constraints on the ${}_k\varepsilon_h$'s and the ${}_k\beta_h$'s have to be imposed for $h=1,\ldots, g$ according to whether one wishes to overweight or underweight large differences: in the first case ${}_k\varepsilon_h\leq{}_k\varepsilon_{h+1}$ and ${}_k\beta_h\leq{}_k\beta_{h+1}$ (thus ${}_kRange_h\leq {}_kRange_{h+1}$), in the second one ${}_k\varepsilon_h\geq{}_k\varepsilon_{h+1}$ and ${}_k\beta_h\geq{}_k\beta_{h+1}$ (thus ${}_kRange_h\geq{}_kRange_{h+1}$). The following simple example helps clarifying this further weighting procedure and illustrate its effectiveness. Let's consider six families belonging to the group of two-person families for which the monthly income has been observed: the six values, together with the group minimum and maximum, are illustrated in fig. 3.

Figure 3: *Six monthly incomes (x_{ik}, in thousand £) classified into three classes: I_1=[500, 2000), I_2=[2000, 5000), I_3=[5000, 10000]*

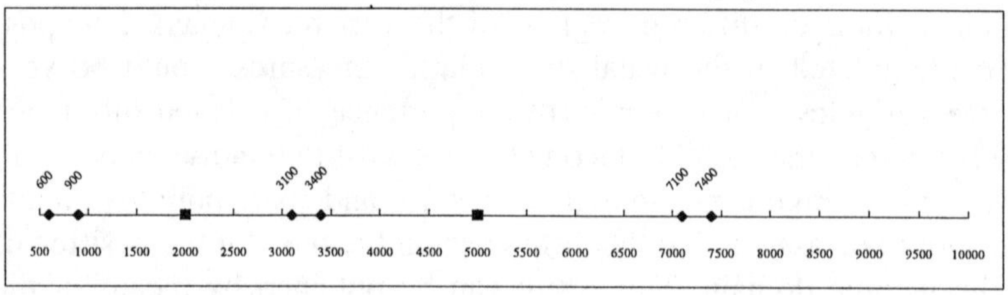

As it can be seen, two families are relatively poor, two relatively rich, and two intermediate ones, and the so formed three couples of families have exactly the same income difference. In order to give income differences an importance which is less than proportional to their values, three (one for each income situation) weighting functions ${}_kf_h$ have been chosen within the monotone increasing convex functions, and the three income classes I_h have been defined so as to satisfy the constraints ${}_kRange_h\leq{}_kRange_{h+1}$. All the parameters chosen in this simple example are listed in table 1.

Table 1: *Parameters used to compute the dissimilarities for the example of figure 3, where* $d_{ijk}=|x_{ik}-x_{jk}|$

I_h	${}_k\beta_h$	${}_k\varepsilon_h$	${}_kt_h$	${}_k\gamma_h\ {}_kf_h$
[500, 2000)	0.1579	50	30	$(int(d_{ijk}/50))^2/30^2$
[2000, 5000)	0.3158	100	30	$(int(d_{ijk}/100))^3/30^3$
[5000, 10000]	0.5263	250	20	$(int(d_{ijk}/250))^4/20^4$

Table 2 shows the dissimilarities v_{ijk} computed by means of the proposed procedure for each couple of the six families. By the inspection of the values it clearly emerges that the three previously described couples of families are not considered as equally dissimilar; the same thing occurs when the comparison is for instance between v_{13k} and v_{24k}.

Table 2: *Dissimilarity (upper triangular) matrix* $\mathbf{V}_k$ *for the example of figure 3*

Units	O_1	O_2	O_3	O_4	O_5	O_6
O_1	0	0.0063	0.1531	0.1696	0.4668	0.4749
O_2		0	0.1005	0.1170	0.4142	0.4223
O_3			0	0.0003	0.0937	0.1018
O_4				0	0.0614	0.0695
O_5					0	$3\cdot10^{-6}$
O_6						0

4. Conclusions

The suggested approach is very general as it admits Godehardt's model for dissimilarities and Gower's coefficient as special cases. In fact, if a single threshold is chosen for each variable, one obtains Godehardt's model; if on each layer dissimilarity is evaluated following Gower's suggestions (as many blocks as variables are considered), the number of different thresholds within each variable is set equal to the number of different dissimilarity values ($t_k=n(n-1)/2$ for $k=1, \ldots, m$), the weights are made to correspond to the dissimilarities themselves, sorted in ascending order (${}_kd_l$ is set equal to the l-th ordered d_{ijk} computed dissimilarity value, for $l=1, \ldots, t_k$, and the weighting function is simply the reverse of the variable range), and all the variable weights are set equal to one, Gower's solution is obtained.

The appeal of the suggested solution is its versatility. It allows to analyse mixed variables using for each of them the proximity measure one deems best, thus freeing the researcher from sticking to Gower's suggested indices and it doesn't require variable transformation in order to synthesize dissimilarities expressed in different units. Furthermore it can cope with measurement errors and treat characters with a high within-subject variability by equally weighting slightly different dissimilarities; a possibility which is not contemplated in either Gower's index or in the multivariate methods (such as INDSCAL) which can be used, from a different perspective, to synthesize the m dissimilarity matrices defined with respect to the m variables.

A further interesting aspect of the threshold based procedure lies in that it can also be successfully applied to single variables, for which distances of the same entity located at different positions of the variable domain assume a different importance.

Moreover a suitable choice of weights can prevent from considering equally dissimilar two units which differ in all layers or which significantly differ in only one of their variables and are similar in the remaining $m-1$.

References

Anderberg, M. R. (1973). *Cluster Analysis for Applications*, Academic Press, New York.

Cox, T. F. & Cox, M. A. A. (1994). *Multidimensional Scaling*, Chapman & Hall, London.

Everitt, B. S. (1993). *Cluster analysis* (3rd ed.), Arnold, London.

Godehardt, E. (1990). *Graphs as Structural Models*, Vieweg, Braunschweig-Wiesbaden.

Gordon, A. D. (1981). *Classification*, Chapman & Hall, London.

Gower, J. C. (1971). A general coefficient of similarity and some of its properties, *Biometrics*, 27, 857-871.

Gower, J. C. (1985). Measures of similarity, dissimilarity, and distance, in *Encyclopedia of Statistical Sciences, Volume 5*, Kotz, S., Johnson, N. L. & Read, C. B. (Eds.), Wiley, 397-405.

Gower, J. C. & Legendre, P. (1986). Metric and euclidean properties of dissimilarity coefficients, *Journal of Classification*, 3, 5-48.

Krzanowski, W. J. & Marriott, F. H. C. (1994). *Multivariate Analysis, Part I, Distributions, Ordination and Inference*, Edward Arnold, London.

Snijders, T. A. B., Dormaar, M., van Schuur, W. H., Dijkman-Caes, C. & Driessen, G. (1990). Distribution of some similarity coefficients for dyadic binary data in the case of associated attributes, *Journal of Classification*, 7, 5-31.

Professional Positioning based on Dominant Eigenvalue Scores (DES), Dimensional Scaling (DS) and Multidimensional Scaling (MDS) Synthesis of Binary Evaluations Matrix of Experts

Claudio Quintano
Institute Statistics and Mathematics, Naval University Institute of Naples
e-mail: quintano@nava3.uninav.it

Abstract: Wider research aimed to obtain timely evaluations of the Professional Market in the province of Naples through in-depth interviews with experts. This paper shows some results on one set of professions, according to DS and MDS procedures.

Keywords: Profession Market, Binary Evaluation Matrix, Dominant Eigenvalue Scores (DES), Dimensional Scaling (DS), Multidimensional Scaling (MDS).

1. Introduction

The purpose is to obtain forecasts ranking evaluations (year 2005) about given professional sets, through interviews with a group of experts. These subjective evaluations were preferred to accurate quantifications which employ traditional methods, require analytical and sophisticated procedures, are hard to carry out and are not always reliable.
The reference geographical area is the province of Naples.

2. Data Collected from Experts

This research was based on interviews with qualified persons in most of the professional fields examined, called *privileged interlocutors*. These are experts who have either studied or gained wide professional experience in the professions under study.
The size of the group cannot be too high because of the quali-quantitative nature

of the interview; empirical results show that 15-20 participants is enough in all situations. In this research the number of experts to interview was twenty.

3. The Construction of Item Evaluations

The questionnaire was constructed to get forecasts on professional classifications, in the near future (year 2005), in the province of Naples. About each homogeneous list, relative scores for each possible pair of professions were sought, that is, for each pair of professions, according to the Saaty scale, we asked experts to indicate how many times *one* profession is greater than or less than *another* one, regarding particular *aspects* (*difficulties to overcome* and *future development perspectives*). The Saaty scale is composed by all the whole numbers included in the range 1-9 and by the relating reciprocal numbers, inverting the object of evaluations reference. This scale derives from empirical studies which have demonstrated that the maximum ability of simultaneous comparisons of a person may range from five to nine objects comparison.

4. Binary Evaluation Matrices as Input for Multivariate Methods

The need to have evaluations for each pair of professions led to the compilation of a matrix for each group of professions being directly compared, having on the rows and in the columns the same professions to compare. This matrix can be defined as a binary evaluation matrix and it is very useful in surveys. In fact, we may sometimes prefer to obtain binary comparisons between each profession and the others, rather than asking respondents to grade a number of n professions according to their importance regarding the evaluation criterion. The generic element a_{ij} ($i<j$) was the value attributed to the profession of the ith row when it was compared with the jth column profession, that is how many times the ith profession is more or less than the jth one, regarding *particular aspects* (*difficulties to overcome* and *future development perspectives*). Of course, if the first profession is more than the second one, a_{ij} will be a number from 1 to 9 (Saaty scale), otherwise the reciprocal value, from 1 to 9, that will be in the range 0 ÷ 1. In other words, the matrix so defined is an n-rowed square matrix, called *reciprocal*, as it satisfies the following conditions:
$a_{ij}= 1/a_{ji}$, each element is equal to the reciprocity of its symmetric element;
$a_{ii}= 1$, the diagonal elements, resulting from the comparison of each element with itself, are equal to one.

5. Multivariate Methods

5.1. The Dominant Eigenvalue Scores (DES) Method

According to the Dominant Eigenvalue Theory, a matrix so defined has all the eigenvalues null, except one, equal to n, therefore it is called dominant. The elements of its related eigenvector, also called dominant, are the scores to attribute to each profession because they express the importance, in normalised terms, implicitly given to each profession in the comparison attribution of scores. This means that the scores forming the list are independent of any unity of measurement. So, the importance of DES derives from the possibility of obtaining a professional classification indirectly, that is, without asking interviewees, but extracting it from professional pair evaluations, which are codified judgements and are easier to obtain from the Saaty scale.
Thus, dominant eigenvector elements can be used in computations to give each profession a concise score.
The theoretical model, on which the DES is based, can be showed through a simple example. Considering a number of n professions, whose normalised scores are given by the vector $w = (w_1, w_2, ..., w_n)$, with $\sum w_i^2 = 1$, and supposing that only the score ratios $a_{ij}=w_i/w_j$ of the professions are known, we can construct a score ratio matrix:

$$A = \begin{bmatrix} w_1/w_1 & w_1/w_2 & \dots & w_1/w_n \\ w_2/w_1 & w_2/w_2 & \dots & w_2/w_n \\ \dots & \dots & \dots & \dots \\ w_n/w_1 & w_n/w_2 & \dots & w_n/w_n \end{bmatrix}$$

A is a reciprocal matrix, because $a_{ij}= 1/a_{ji}$. The positivity of its elements can be easily demonstrated. Then, the normalised scores of the professions compared can be obtained resolving the following equation system in the unknown w:

$$\mathrm{Aw} = \lambda \mathrm{w} \tag{1}$$

System (1) can be also written as follows, according to eigenvalue theory:

$$(\mathrm{A}-\lambda \mathrm{I})\ \mathrm{w}= 0. \tag{2}$$

It is easily recognisable as an eigenvalue problem, which has a valid solution if and only if n is an eigenvalue of A; w will be, in this case, the corresponding eigenvector.

Three results of positive matrix theory can be set out:

1. the theorem of Perron-Frobenius states that only one dominant eigenvalue exists for positive matrices;
2. as matrix A has a rank equal to one (because each row is a constant multiple of the first one), all its eigenvalues are null except one;
3. the eigenvalue sum of a positive matrix is equal to its trace (that is the sum of its diagonal elements).

With reference to the third point, this will be: $a_{ii}= w_i/w_i=1$ and, then:

$$\sum \lambda_i = tr(A) = \sum a_{ii} = n \tag{3}$$

In light of what has been said on positive matrix properties, only one of the eigenvalues λ_i, defined λ_{max}, will be equal to n, with $\lambda_i = 0$ for $\lambda_i \neq \lambda_{max}$. The related eigenvector w will be derived from the normalisation of any column of A. Thus, the solution of the equation system (1) or of the corresponding eigenvalue problem (2) enable us to calculate the vector of normalised scores from the matrix of binary evaluations. Therefore, according to these principles, from binary evaluation matrices we can derive important normalised scores to ascribe directly to the professions being compared.

5.2. Dimensional and Multidimensional Scaling (DS And MDS) Methods

Besides deriving scores, binary evaluation matrices are also the basis for projecting professions in a two-dimensional context, highlighting associations and contrapositions.

The techniques applied were DS, which uses as its input the entire single expert binary evaluation matrix and MDS, which uses as its input all of the experts' binary evaluation matrices, in order to synthesise them. The latter technique, in particular, enable us to have a comprehensive description of individual analyses, highlighting, at the same time, relations between professions which would be unlikely to emerge from individual evaluations. The choice of a multidimensional statistical analysis technique, which is very sophisticated, was influenced by the nature and the availability of data, besides by our objectives, which were:

a) to highlight the similarities and differences among professions according to the importance attributed to them by the group of experts;
b) to point out the (unknown) factors which most influence the evaluations

given by each expert because the available data are proximity judgements about couples of professions expressed in conformity with a quantitative type of scale.

The DS technique uses factorial methods, more precisely, it is included in individual difference techniques. Given the quantitative nature of the data, due to the application of a special evaluation scale, a method based on the sorting and not a metric method was chosen.

It consists of a representation of professional positions in a smaller space dimension (in general 2), showing the unknown factors on which evaluations were based. Besides synthesis of individual evaluations, carried out using the multidimensional-scaling technique, a separate representation of experts was made in the same space, showing the contributions given by each of them.

This type of analysis enable us to pick out the particular features and the common characteristics of the various professions, taking into account that reducing an *n* dimensional space into a simplified one with only *two* dimensions causes necessarily the loss of a part of the variability of the phenomenon analysed and that two elements appearing near to each other in the two dimensional space would not be the same as in the original space-dimension.

6. Some Comments

Tables 1 and 2, the first referring to *difficulties to overcome*, the second to *future development perspectives*, highlight, on the one hand, a comparison between DES analysis and similarity-dissimilarity analyses and, on the other hand, a comparison between expert n. 5's results and those deriving from the whole group of experts.

Before reading these tables, we need to make some observations. First, the DS and MDS techniques cause the loss of a certain degree of the original variability, caused by the choice of a smaller number of dimensions than the original one. Thus, from this point of view, the score analysis, which is not affected by this problem, is preferable.

Another distinction can also be made regarding similarity and dissimilarity analyses. In fact, the results obtained with the MDS technique are more reliable than the ones referring to the single expert, because MDS includes an iterative procedure whose stop criterion is connected to the stress value, which may not overcome a fixed threshold value. Furthermore, this technique gives the rate of original variability captured by the two factorial axes. In this research reference is made mainly to the first factorial axis, because of its greatest explicative power.

Table 1: *Synthesis of the most significant results for the Culture and Training professional set referring to expert n. 5 and to the whole group of experts with regard to the evaluations made according to the criterion, difficulties to overcome.*

Pro-Fes-sions	DES analysis (*)		DS analysis		MDS analysis	
	expert n.5	expert group	quadrant (**)	oppositions (***)	quadrant (**)	oppositions (***)
a	7	7	IV	h,g,b, e,d	IV	e,g,h,d,b
b	1	3	II	a,c,f,i	III	f,a,i,c
c	9	8	IV	h,g,b, e,d	IV	e,g,h,d,b
d	3,5	6	II	a,c,f,i	II	f,a,i,c
e	3,5	1	II	a,c,f,i	II	f,a,i,c
f	6	9	IV	h,g,b, e,d	I	e,g,h,d,b
g	5	4	II	a,c,f,i	III	f,a,i,c
h	2	2	II	a,c,f,i	III	f,a,i,c
i	8	5	IV	h,g,b, e,d	I	e,g,h,d,b

(*) The attribution of scores must be read in this way: 1 the most important profession, according to the criterion being used, 2 the second one in importance, ... up to 9, the least important; (**) The attribution of numbers to the four quadrants was carried out in an anticlockwise way with reference to both factorial axes, considering as first quadrant the one whose co-ordinates are both positive; (***) Contrapositions are considered only with reference to the first factorial axis. The lists follow a decreasing order in the absolute value of the co-ordinate on the first factorial axis, so that the first in the list are different, the last are similar.

Professional list: *a* Librarian; *b* Teacher-Training Co-ordinator; *c* Documentalist; *d* Elementary School Teacher; *e* Support Teacher; *f* Archives Operator; *g* Career Advisor; *h* Training Planner; *i* Restorer.

Table 2: *Synthesis of the most significant results for the Culture and Training professional set referring to expert n. 5 and to the whole group of experts with reference to the evaluations made according to future development perspectives.*

Profes-sions	DES analysis (*)		DS analysis		MDS analysis	
	expert n.5	expert group	quadrant (**)	oppositions (***)	quadrant (**)	oppositions (***)
a	4	7	IV	g,h,d,e	I	g,h,b,e
b	5	6	IV	g,h,d,e	II	i,f,a,c,d
c	7	9	IV	g,h,d,e	I	g,h,b,e
d	8	5	II	i,f,a,c,b	IV	g,h,b,e
e	9	4	II	i,f,a,c,b	III	i,f,a,c,d
f	6	8	IV	g,h,d,e	I	g,h,b,e
g	1	2	II	i,f,a,c,b	III	i,f,a,c,d
h	2	3	III	i,f,a,c,b	II	i,f,a,c,d
i	3	1	I	g,h,d,e	IV	g,h,b,e

For (*) - (**) - (***) and Professional list see Table 1.

As we can see from Tables 1 and 2, expert n. 5 evaluations are particularly close to those of the whole group of experts. In fact, for example, with reference to the DES analysis, both expert n. 5 and the group of experts attribute the greatest importance to Teacher-Training Co-ordinator, Support Teacher and Training Planner, with regard to *difficulties to overcome*, and to Career Advisor, Training Planner and Restorer, for future development perspectives.
Similar results also come from DS and MDS techniques. In fact, the contrapositions arising from professional projections in two-dimensional space are the same with reference to expert n. 5 and to the evaluations of the whole group of experts: the professions which the most demanding job is connected are Training Planner, Career Advisor, Teacher-Training Co-ordinator, Support Teacher and Elementary School Teacher, in antithesis to those with more operative functions, which are: Archives Operator, Restorer, Documentalist and Librarian. It also confirms what emerged from the Derived Subject Weights, about the primary role of expert n. 5 in the factorial axes construction. The results referring to future development perspectives are quite similar.
In Table 3 professions in contrapositions in DS and MDS analyses, referring only to the first factorial axis, were listed in each column. We can verify that the professional clusters in contraposition are exactly the same for expert n. 5 and the whole group of experts with reference to *difficulties to overcome*, while, for *future development perspectives*, the only professions that do not respect this principle are Teacher-Training Co-ordinator, on the one hand, and Elementary School Teacher, on the other.
Other important considerations could also be made on differences between couples of normalised scores, obtained from, on the one hand, DES analysis, that is by dominant eigenvector elements, and, on the other hand, DS and MDS analyses, using as scores the first factorial axis co-ordinates, although they refer only to a part of total phenomenon variability. These differences highlight the gap existing between each couple of professions in accordance to the experts' evaluations.
There is a positive correlation between single expert evaluations and whole group of experts evaluations, according both to difficulties to overcome and future development perspectives. This confirms the widely representative role of expert n. 5 in the whole group of experts.

Table 3: *Clusters of professions according to their positioning in DS and MDS applications, for (in order) expert n. 5 and the whole group of experts, with respect to the first factorial axis and with reference to both difficulties to overcome and future development perspectives.*

Aspects			
Difficulties to overcome		**Future development perspectives**	
Expert n. 5 - DS	Exp. group-MDS	Expert n. 5 - DS	Expert group-MDS
Archives Operat. Restorer Documentalist Librarian	Support Teacher Orientation Expert Training Planner Elem. School T. Teach. Training C.	Orientation Exp. Training Planner Elem. School T. Support Teacher	Orientation Expert Training Planner Teacher Training C. Support Teacher
⇳	⇳	⇳	⇳
Training Plann. Orientation Exp. T. Training C. Support Teacher Elem. School T.	Archives Operator Restorer Documentalist Librarian	Restorer Archives Operat. Librarian Documentalist T. Training C.	Restorer Archives Operator Librarian Documentalist Element. School T.

The procedure employed enable us to draw some conclusion: elaborations made through DES analysis, on the one hand, and DS and MDS analyses, on the other hand, converge to produce the same results, but highlight different aspects: the first gives a more complete picture, in one-dimensional space; DS and MDS analyses provide information only on a part of total phenomenon variability, highlighting associations and dissimilarities which the first one does not do.

References

ISFOL, (1987). *Repertorio delle professioni*, Istituto Poligrafico e Zecca dello Stato, Roma.

ISFOL, (1989-1995). *Osservatorio, Formazione-Orientamento-Occupazione-Nuove Tecnologie-Professionalità*, various numbers.

Marbach, G. (1988). *Le ricerche di mercato*, UTET, Torino.

Markpack, (1990). *Tutorial and Reference Manual*, PRISM Sarl, France.

Saaty, T. L. (1977). Scaling Method for Priorities in Hierarchical Structures, *Journal of Mathematical Psycology*, XV, 3, 333-336.

SPSS. (1994). *Categories 6.1*, SPSS Inc., United States of America.

Non-Metric Full-Multidimensional Scaling

Maurizio Vichi
Dipartimento di Metodi Quantitativi e Teoria Economica, Viale Pindaro 42, 65127, Pescara, Italy, e-mail: Vichi@DMQTE.unich.it

Abstact: This paper focuses on some solutions for non-metric full-Multidimensional Scaling (MDS), minimizing the STRESS and S-STRESS loss functions. In particular, the linear transformations of dissimilarities into Euclidean distances minimizing the two loss functions are given. A non trivial result for S-STRESS with a quadratic transformation of dissimilarities, constraining its coefficients, is also obtained.

keywords: Metric MDS, Non-metric MDS.

1. Introduction

Multidimensional scaling (MDS) refers to a class of techniques for approximating a (nxn) matrix of observed dissimilarities $\mathbf{D}\equiv[d_{ij}]$ between n objects, with a (nxn) Minkowski distance matrix $\mathbf{E}_m\equiv[e_{ij}=(\sum_{h=1}^{p}|x_{ih}-x_{jh}|^m)^{1/m}]$ having an associated *configuration of n points* $[x_{ij}]\equiv\mathbf{X}\in\Re^{n\times p}$ of a p (generally low) dimensional Minkowski space. Interest in defining "the best" approximation in an Euclidean space dates back at least as the seminal work of Torgerson (1958). While approximating $\mathbf{D}$ a MDS technique should preserve in $\mathbf{E}_m$ the *pattern* (*manifold*) observed in $\mathbf{D}$, thus requiring to: *i*) minimize a loss function between $\mathbf{D}$ and $\mathbf{E}_m$; *ii*) preserve a monotone relationship between $\mathbf{D}$ and $\mathbf{E}_m$. A metric MDS procedure satisfies *i*); while a non-metric MDS satisfies *i*) subject to *ii*). The two most popular loss functions in MDS are Kruskal's raw STRESS and the S-STRESS,

$$\Sigma_1(\mathbf{X}) = \frac{1}{2}||\mathbf{D}-\mathbf{E}(\mathbf{X})||^2, \tag{1a}$$

$$\Sigma_1(\mathbf{X}) = \frac{1}{2}||\mathbf{D}*\mathbf{D}-\mathbf{E}(\mathbf{X})*\mathbf{E}(\mathbf{X})||^2, \tag{1b}$$

where matrix $\mathbf{E}(\mathbf{X})$ is the Euclidean matrix written as a function of the configuration of points, and $\mathbf{A}*\mathbf{B}$ is the Hadamard (direct) product of two matrices with equal dimensions. Functions (1a) and (1b) are often normalized and weighted.

Therefore, a metric MDS technique involves minimizing (1a) or (1b) over the set $\Re^{n\times p}$, while a non-metric MDS method consists in minimizing (1a) or (1b)

between $\mathbf{E}(\mathbf{X})$ and $f(\mathbf{D})$, with f an arbitrary monotone transformation, over the set $\Re^{n\times p}$, as well as the monotone transformation f. In practice, this is classically done numerically interleaving steepest descent steps on the configuration with estimation of f via isotonic regression of the current distances $\mathbf{E}(\mathbf{X})$ on the dissimilarities $\mathbf{D}$.
Full-multidimensional scaling refers to the case where the dimensionality of the configuration is at most $p=n-1$, and therefore the number of dimensions is not constrained to be small.
Notice that the non-metric MDS problem is much more complex than the metric case. In this paper we study some analytical solutions for non-metric full-MDS, for the Euclidean case (m=2), when the monotone relation *ii*) is linear or quadratic. A numerical example is given to show the proposed solutions.
An outline of the material in this paper is as follows. Section 2 recalls non-metric MDS, and gives three solutions of non-metric MDS. Section 3 applies the proposed transformations. Section 4 reports a discussion on the use of these transformations and gives some conclusions.

2. Non-metric Full-Multidimensional Scaling

Let $\mathbf{Z} = \mathbf{I}-(1/n)\mathbf{1}\mathbf{1}'$ be the $(n\times n)$ idempotent matrix denoting the orthogonal projection onto the orthogonal complement of $\mathbf{1}$ in $\Re^n$, where $\mathbf{1}$ is a n-vector of unitary elements. Let $\mathbf{D}_n$ be the set of $(n\times n)$ dissimilarity matrices $\mathbf{D}\equiv[d_{ij} : d_{ji}=d_{ij}\geq 0,\ d_{ii}=0\ \forall\ 1\leq i,j\leq n]$. Let $\mathbf{E}_n$ and E_n^2 be the sets of $(n\times n)$ Euclidean matrices $\mathbf{E}\equiv[e_{ij}]$ and the corresponding $(n\times n)$ squared Euclidean matrices $\mathbf{E}*\mathbf{E}\equiv[e_{ij}^2]$. Let $t(\mathbf{D})=(-\frac{1}{2}\mathbf{ZDZ})$ denote the linear and one-to-one Young-Householder transform. The set $t(\mathrm{E}_n^2)$ is the closed convex cone of the positive semi-definite matrices of order n, with interior the positive definite matrices.
Let each matrix $\mathbf{E}$ be identified by the $n(n-1)/2$ component vector $\overline{vec}(\mathbf{E})$ obtained vectorizing elements of a triangle below (above) the diagonal of $\mathbf{E}$, row-by-row (i.e., $e_{12}, \ldots, e_{1n}, e_{23}, \ldots$). For n=3, the set $\overline{vec}(\mathrm{E}_3^2)$ has the conical form shown in figure 1.
The boundary of $\overline{vec}(\mathrm{E}_3^2)$ is given by matrices with $\det(t(\mathbf{D}*\mathbf{D}))=0$, i.e., $2d_{12}^2 d_{13}^2+2d_{12}^2 d_{23}^2+2d_{13}^2 d_{23}^2-d_{12}^4-d_{13}^4-d_{23}^4=0$.
For S-STRESS the global optimum of metric full-MDS is given by the solution of an eigen-problem, see for example Mathar (1985), while no analytical solution is known for STRESS. However, a local optimum found by an optimization method turns out to be the global optimum since $\mathbf{E}_n$ is a convex cone and STRESS is convex, so that there is only a single minimum.
Geometrically, metric full-MDS consists in finding the point $\overline{vec}(\mathbf{E})$ or $(\overline{vec}(\mathbf{E}*\mathbf{E}))$ of contact with the convex cones $\overline{vec}(\mathbf{E}_n)$ or $(\overline{vec}(\mathrm{E}_n^2))$ on the hyper-

sphere with centre in the point $\overline{vec}(\mathbf{D})$ or ($\overline{vec}(\mathbf{D*D})$); while for the non-metric case the solution is also bounded by $n(n-1)/2$ -1 hyper-planes. In figure 1 the solution of non-metric full-MDS using S-STRESS, for n=3, and $\overline{vec}(\mathbf{D*D})=(3,3,20)'$ is shown. Since we have $d_{12}\leq d_{13}\leq d_{23}$ it follows that $e_{12}\leq e_{13}\leq e_{23}$. The sphere with centre (3,3,20)′ and minimum radius (S-STRESS)$^{1/2}=2\sqrt{2}$, contacts $\overline{vec}(\mathrm{E}_3^2)$ in the point $\overline{vec}(\mathbf{E*E})$ = (5, 5, 20)′. The two planes $d_{12}-d_{13}=0$, $d_{13}-d_{23}=0$ bound the solution, as it is shown in figure 1.
After these preliminary remarks on the MDS, we are now in position to state some results for non-metric full-MDS. The first identifies the coefficients (positive) of the linear transformation of dissimilarities **D**, which give an Euclidean matrix **E** minimizing $tr(\mathbf{D}-\mathbf{E})^2$.

Theorem 1: Let $\mathbf{D} \in \mathbb{D}_n$. The linear transformation of **D**:

$$\mathbf{E}=a(\mathbf{11}'-\mathbf{I})+b\mathbf{D}, \tag{2}$$

is Euclidean and such that: $\min\{tr(\mathbf{D}-\mathbf{E})^2 \mid \mathbf{D}\in \mathbb{D}_n, -½\,\mathbf{Z}(\mathbf{E*E})\mathbf{Z} \geq 0\}$,
i.e., it minimizes the raw STRESS, when:

$$b=b^* = \frac{c\sum_{i=1}^{n}\sum_{j=1}^{n} d_{ij} + \sum_{i=1}^{n}\sum_{j=1}^{n} d_{ij}^2}{c^2(n-1)n + \sum_{i=1}^{n}\sum_{j=1}^{n} d_{ij}^2 + 2c\sum_{i=1}^{n}\sum_{j=1}^{n} d_{ij}} \quad ; \quad a=a^* = c\, b^*, \tag{3}$$

where c is the maximum real eigenvalue of $\begin{bmatrix} \mathbf{0} & 2t(\mathbf{D*D}) \\ -\mathbf{I} & -4t(\mathbf{D}) \end{bmatrix}$.
Note that c is the minimum additive constant of Cailliez (1983).

Proof. Matrix $a(\mathbf{11}'-\mathbf{I})+\mathbf{D}$ is Euclidean iff $a \geq c$ the minimum additive constant (Cailliez, 1983). Thus, matrix $\mathbf{E}=b[c(\mathbf{11}'-\mathbf{I})+\mathbf{D}]$ is Euclidean for every b>0, since $\mathbf{W} = -½\,\mathbf{Z}(\mathbf{E*E})\mathbf{Z}\geq 0$ is positive semi-definite.
Furthermore, $F=tr(\mathbf{D}-\mathbf{E})^2 = tr\mathbf{D}^2 + b^2\, tr[c(\mathbf{11}'-\mathbf{I})+\mathbf{D}]^2 - 2b\,tr\{\mathbf{D}[c(\mathbf{11}'-\mathbf{I})+\mathbf{D}]\}$ and the Normal equation is: $(dF/db) = b\, tr[c(\mathbf{11}'-\mathbf{I})+\mathbf{D}]^2 - tr\{\mathbf{D}[c(\mathbf{11}'-\mathbf{I})+\mathbf{D}]\}=0$, so that (3) follows. ■

The configuration of points is found in at most $p=n-2$ dimensions since **1** is an eigenvector with an induced zero root and an extra zero eigenvalue has to be found to have **E** lying on the boundary of $\mathbb{E}_n$.
The coordinates of $\mathbf{E}=a^*(\mathbf{11}'-\mathbf{I})+b^*\mathbf{D}$ are: $\mathbf{X}=[\mathbf{u}_1,..,\mathbf{u}_{n-2}]\mathrm{diag}(\lambda_{(1)},..,\lambda_{(n-2)})^{1/2}$, where $\lambda_{(1)},...,\lambda_{(n-2)}$ are the positive eigenvalues of $t(a^{*2}(\mathbf{11}'-\mathbf{I})+2a^*b^*\mathbf{D}+b^{*2}\mathbf{D*D})$ in decreasing order, and $\mathbf{u}_1,...,\mathbf{u}_{n-2}$ are the corresponding eigenvectors. A similar result can be stated for S-STRESS.

Figure 1: *Geometrical representation of the non-metric full-MDS*

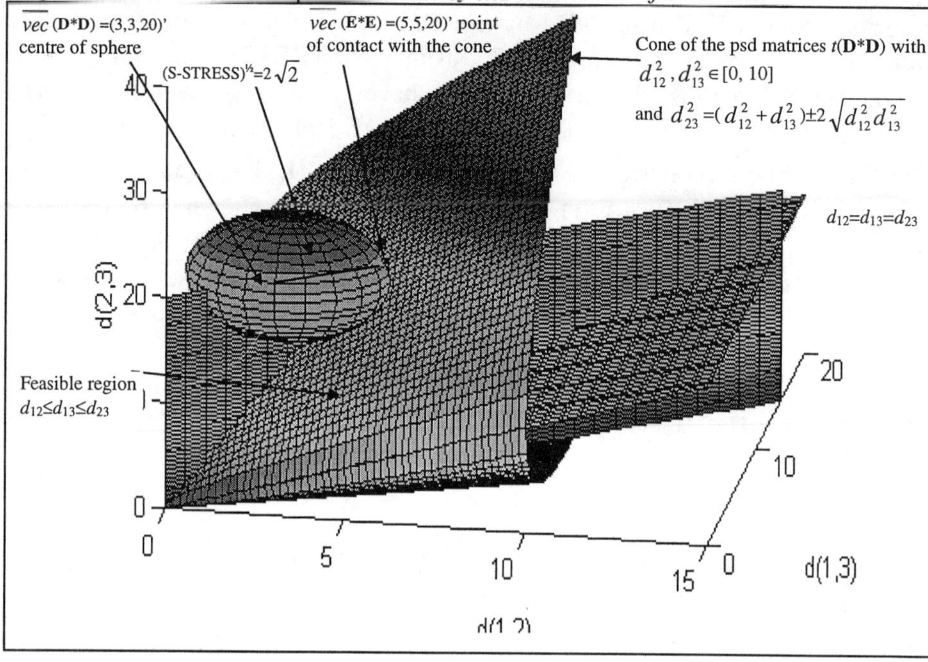

Theorem 2: Let $\mathbf{D} \in \mathbf{D}_n$, and let α be the minimum eigenvalue of $t(\mathbf{D}^*\mathbf{D})$. The linear transformation of $(\mathbf{D}^*\mathbf{D})$:

$$\mathbf{E}^*\mathbf{E} = d(\mathbf{11}'\text{-}\mathbf{I}) + f\,\mathbf{D}^*\mathbf{D} \tag{4}$$

is Euclidean and such that, $\min\{tr(\mathbf{D}^*\mathbf{D}\text{-}\mathbf{E}^*\mathbf{E})^2 \mid \mathbf{D}\in \mathbf{D}_n, \text{-}½\, \mathbf{Z}(\mathbf{E}^*\mathbf{E})\mathbf{Z} \geq 0\}$ i.e., it minimizes S-STRESS, when:

$$f=f^* = \frac{tr\{\mathbf{D}^*\mathbf{D}[-2\alpha(\mathbf{11}'-\mathbf{I})+\mathbf{D}^*\mathbf{D}]\}}{tr[-2\alpha(\mathbf{11}'-\mathbf{I})+\mathbf{D}^*\mathbf{D}]^2}; \quad d=d^* = -2\alpha f^*. \tag{5}$$

Proof. Matrix $a(\mathbf{11}'\text{-}\mathbf{I})+\mathbf{D}^*\mathbf{D}$ is Euclidean iff $d \geq -2\alpha$ (Lingoes, 1971). For every $f > 0$ matrix $\mathbf{E}^*\mathbf{E} = b[-2\alpha(\mathbf{11}'\text{-}\mathbf{I}) + \mathbf{D}^*\mathbf{D}]$ it is such that $\mathbf{E}$ is Euclidean. Furthermore, $\text{G} = tr(\mathbf{D}^*\mathbf{D}\text{-}\mathbf{E}^*\mathbf{E})^2 = tr(\mathbf{D}^*\mathbf{D})^2 + f^2\, tr[-2\alpha(\mathbf{11}'\text{-}\mathbf{I}) + \mathbf{D}^*\mathbf{D}]^2 +$ $-2f tr\{\mathbf{D}^*\mathbf{D}[-2\alpha(\mathbf{11}'\text{-}\mathbf{I})+\mathbf{D}^*\mathbf{D}]\}$, and the Normal equation is $(\text{dF}/\text{d}f)=f tr[-2\alpha(\mathbf{11}'\text{-}\mathbf{I})+\mathbf{D}^*\mathbf{D}]^2\text{-}tr\{\mathbf{D}^*\mathbf{D}[-2\alpha(\mathbf{11}'\text{-}\mathbf{I})+\mathbf{D}^*\mathbf{D}]\}=0$, so (5) follows. ■

Also in this case the configuration of points is found in at most $p=n\text{-}2$ dimensions with coordinates: $\mathbf{X}=[\mathbf{u}_1, ..., \mathbf{u}_{n-2}]\text{diag}(\lambda_{(1)},..., \lambda_{(n-2)})^{1/2}$, where $\lambda_{(1)},..., \lambda_{(n-2)}$ are the ordered eigenvalues of $t(d^*(\mathbf{11}'\text{-}\mathbf{I})+f^*\mathbf{D}^*\mathbf{D})$, and $\mathbf{u}_1,...,\mathbf{u}_{n-2}$ the corresponding eigenvectors.

The results given by theorem 1 and 2 are a non trivial consequence of the additive constant problem. More complex is the problem to find a non-linear and monotone transformation of the dissimilarities which is Euclidean and minimizing the (1a) or (1b). A first result is given for (1b), when a quadratic mapping is used.

Theorem 3: Let $\mathbf{D}\in \mathbf{D}_n$, and let α and β be minimum eigenvalues of $-\frac{1}{2}\mathbf{Z}(\mathbf{D}*\mathbf{D})\mathbf{Z}$ and $-\frac{1}{2}\,\mathbf{Z}(\mathbf{D}*\mathbf{D}*\mathbf{D}*\mathbf{D})\mathbf{Z}$ respectively. The quadratic transformation of $(\mathbf{D}*\mathbf{D})$:

$$\mathbf{E}*\mathbf{E} = g(\mathbf{11}'\text{-}\mathbf{I}) + h\,\mathbf{D}*\mathbf{D} + l\,\mathbf{D}*\mathbf{D}*\mathbf{D}*\mathbf{D}, \text{ with } g,h,l>0, \tag{6}$$

is such that $\mathbf{E}$ is Euclidean if,

$$g,h \geq 0 \text{ and } l \geq (-g/2 - \alpha h)/\,\beta > 0 \tag{7}$$

and $\min\{tr(\mathbf{D}*\mathbf{D}\text{-}\mathbf{E}*\mathbf{E})^2 \mid \mathbf{D}\in\mathbf{D}_n, -\frac{1}{2}\mathbf{Z}(\mathbf{E}*\mathbf{E})\mathbf{Z} \geq 0,\ g,h\geq 0 \text{ and } l=(-g/2-\alpha h)/\beta>0\}$, i.e., the S-STRESS subject to constraints (7) is minimized for

$$g=g^*=\frac{tr(\mathbf{D}_2\mathbf{F})tr\mathbf{G}^2 - tr(\mathbf{FG})tr(\mathbf{D}_2\mathbf{G})}{tr\mathbf{F}^2 tr\mathbf{G}^2 - (tr\mathbf{FG})^2} \quad h=h^*=\frac{tr\mathbf{F}^2 tr(\mathbf{D}_2\mathbf{G}) - tr(\mathbf{FG})tr(\mathbf{D}_2\mathbf{F})}{tr\mathbf{F}^2 tr\mathbf{G}^2 - (tr\mathbf{FG})^2}$$
$$l=l^*=(-g^*/2 - \alpha h^*)/\,\beta, \tag{8}$$

where $\mathbf{D}_2$=$\mathbf{D}*\mathbf{D}$; $\mathbf{F}=(\mathbf{11}'\text{-}\mathbf{I})-(1/2\beta)\mathbf{D}*\mathbf{D}*\mathbf{D}*\mathbf{D}$; $\mathbf{G}=\mathbf{D}*\mathbf{D}-(\alpha/\beta)\,\mathbf{D}*\mathbf{D}*\mathbf{D}*\mathbf{D}$.

Proof: Matrix $\mathbf{W}=-\frac{1}{2}\,\mathbf{Z}(\mathbf{E}*\mathbf{E})\mathbf{Z}$ has to be p.s.d or $\mathbf{x}'\mathbf{W}\mathbf{x}\geq 0\ \forall \mathbf{x}\in\Re^n$, with $\mathbf{x}'\mathbf{x}=1$.

$$\mathbf{x}'\mathbf{W}\mathbf{x} = \tfrac{1}{2}\,g\,\mathbf{x}'\mathbf{Z}\mathbf{x} + h\mathbf{x}'[-\tfrac{1}{2}\,\mathbf{Z}(\mathbf{D}*\mathbf{D})\mathbf{Z}]\mathbf{x} + l\mathbf{x}'[-\tfrac{1}{2}\,\mathbf{Z}(\mathbf{D}*\mathbf{D}*\mathbf{D}*\mathbf{D})\mathbf{Z}]\mathbf{x}. \tag{9}$$

Since $\mathbf{x}'[-\frac{1}{2}\,\mathbf{Z}(\mathbf{D}*\mathbf{D})\mathbf{Z}]\mathbf{x} \geq \alpha\mathbf{x}'\mathbf{x}$, $\mathbf{x}'[-\frac{1}{2}\,\mathbf{Z}(\mathbf{D}*\mathbf{D}*\mathbf{D}*\mathbf{D})\mathbf{Z}]\mathbf{x} \geq \beta\mathbf{x}'\mathbf{x}$,
$\mathbf{x}'\mathbf{Z}\mathbf{x}\geq 0$ for $g,h\geq 0$, $\mathbf{x}'\mathbf{W}\mathbf{x} = (\frac{1}{2}\,g + \alpha h + \beta l)\mathbf{x}'\mathbf{Z}\mathbf{x}\geq 0$ if (7) holds.
Furthermore,

$$H = tr(\mathbf{D}_2 - g(\mathbf{11}'\text{-}\mathbf{I}) - h\mathbf{D}_2 + [(-g/2 - \alpha h)/\beta]\mathbf{D}_4)^2 =$$
$$= tr\,\mathbf{D}_2^2 + g^2\,tr\mathbf{F}^2 + h^2\,tr\mathbf{G}^2 - 2gtr\,\mathbf{D}_2\mathbf{F} + 2h\,tr\mathbf{D}_2\mathbf{F} + 2gh\,tr\mathbf{FG}. \tag{10}$$

The system of Normal equations is:

$$\begin{cases} \partial H/\partial g = tr\mathbf{F}^2 + htr\mathbf{FG} = tr\mathbf{D}_2\mathbf{F} \\ \partial H/\partial h = tr\mathbf{F}\,\mathbf{G} + htr\mathbf{G}^2 = tr\mathbf{D}_2\mathbf{G} \end{cases} \tag{11}$$

from which (8) follows. ■

It has to be noted that theorem 3 finds the solution of a constrained problem on the coefficients and not of the unconstrained one. In this case the configuration of points is found in at most p=n-1 dimensions: $\mathbf{X}=[\mathbf{u}_1,\ldots,\mathbf{u}_{n-1}]\mathrm{diag}(\lambda_{(1)},\ldots,\lambda_{(n-1)})^{1/2}$, where $\lambda_{(1)},\ldots,\lambda_{(n-1)}$ are ranked eigenvalues of $t(g^*(\mathbf{11}'\text{-}\mathbf{I})+h^*\mathbf{D}*\mathbf{D}+l^*\mathbf{D}*\mathbf{D}*\mathbf{D}*\mathbf{D})$ and $\mathbf{u}_1,\ldots,\mathbf{u}_{n-1}$ are the corresponding eigenvectors. The optimal least squares approximation (6) of the dissimilarity matrix **D**, can be obtained solving the following quadratic constrained problem:

$$\begin{cases} \text{minimize } tr(\mathbf{D}*\mathbf{D}-\mathbf{E}*\mathbf{E})^2 \\ \text{subject to} \\ \mathbf{E}*\mathbf{E} = g(\mathbf{11}'-\mathbf{I}) + h(\mathbf{D}*\mathbf{D}) + l(\mathbf{D}*\mathbf{D}*\mathbf{D}*\mathbf{D}) \\ \lambda_{(n-1)}(t(\mathbf{E}*\mathbf{E})) = 0 \end{cases} \tag{12}$$

where $\lambda_{(n-1)}(t(\mathbf{E}*\mathbf{E}))$ is the second smallest eigenvalue of $t(\mathbf{E}*\mathbf{E})$. Problem (12) has been solved using a Sequential Quadratic Programming (SQP) algorithm. Comparative studies of non linear programming algorithms indicate that the SQP algorithm performs very well in terms of successful solutions, with a superlinear rate of convergence. An overview of SQP methods is given in Powell (1983). The analytical solution g^*, h^*, l^*, (8) can be used as initial guess for problem (12), and in the example we analyzed the rate of convergence was improved. Often the optimal solution of problem (12) is close to the analytical solution given by (8).

3. Numerical Example

The dissimilarity matrix **D** reported in Table 1 has been defined with an uniform distribution in [0, 10]. **D** is not Euclidean since $t(\mathbf{D}*\mathbf{D})$ has two large negative eigenvalues (-25.8640 , -9.5391).

Table 1: *Dissimilarity matrix* **D**

0	5.3685	0.5950	0.8896	2.7131	4.0907	4.7404
5.3685	0	3.2896	4.7819	5.9717	1.6145	8.2947
0.5950	3.2896	0	8.1212	6.1011	7.0149	0.9220
0.8896	4.7819	8.1212	0	8.3864	4.5161	9.5660
2.7131	5.9717	6.1011	8.3864	0	9.5169	6.4001
4.0907	1.6145	7.0149	4.5161	9.5169	0	6.1094
4.7404	8.2947	0.9220	9.5660	6.4001	6.1094	0

The best linear least squares approximation of matrix **D** in table 1, according to STRESS is yielded applying the result of Theorem 1, that gives:

$\mathbf{E} = 3.8720(\mathbf{11}'\text{-}\mathbf{I}) + 0.3098\mathbf{D}$,
with STRESS = 75.9819;

The best linear least squares approximation of matrix **D*D**, according to S-STRESS is obtained applying Theorem 2, that gives:

$$\mathbf{E*E} = 23.6578(\mathbf{11'}\text{-}\mathbf{I}) + 0.4573\mathbf{D*D},$$
with S-STRESS = 5443.2

The best least squares quadratic constrained by (7) approximation of **D*D**, according to S-STRESS is achieved applying Theorem 3, that gives:

$$\mathbf{E*E} = 25.2658(\mathbf{11'}\text{-}\mathbf{I}) + 0.3067\mathbf{D*D} + 0.0018\mathbf{D*D*D*D},$$
with S-STRESS = 5404.0.

Problem (12) was solves using the SQP algorithm after 153 function evaluations with a convergence constant equal 10^{-7}:

$$\mathbf{E*E} = 53.4127(\mathbf{11'}\text{-}\mathbf{I}) + 0.0102\mathbf{D*D*D*D},$$
with S-STRESS=3365.0.

This last Euclidean matrix has two zero eigenvalues.

4. Discussion

In this paper non-metric full-MDS is discussed. Three solutions are given in the cases where the relation between **D**, (**D*D**) and **E**, (**E*E**) is linear or quadratic. We suggest the use of these solutions: (*a*) as initial guesses for iterative global non-metric MDS procedures; or (*b*) before applying a metric MDS technique, because in this case often the configuration associated to **E** and obtained in reduced dimensions by a metric MDS is less distorted. This reduced distortion is confirmed, for example, when a clustering technique is applied on **D**, (**D*D**) and the solution is compared with those two obtained on **E**, (**E*E**) (identifying a configuration in 2 dimensions) using or not the above transformations before applying a metric MDS technique.

For example, in Figure 2 the dendrogram of the single linkage applied on the square of dissimilarities in Table 1 is shown. The dendrogram obtained applying the single linkage algorithm on the best least squares Euclidean approximation of matrix **D*D** is shown in Figure 2 (*b*). It can be noted that the two dendrograms (Figures 1 *a* and *b*) exhibit different classifications at different levels of fusion. The dendrogram in Figure 2 (*c*), obtained applying single linkage on the quadratic approximation given by Theorem 3, presents the same topology of the dendrogram in Figure 2 (*a*), with slightly different lengths of linkages. This confirms that the approximation in case (*b*) produces a larger distortion in the classification pattern observed in Table 1, with respect to a monotone approximation.

Figure 2: Comparison among the single linkage's dendrograms applied on the: (*a*) square of dissimilarities in table 1; (*b*) best least squares Euclidean approximation of **D*D**; (*c*) Quadratic least squares approximation given by theorem 3.

(*a*) Single linkage applied on the **D*D**, with **D** in Table 1

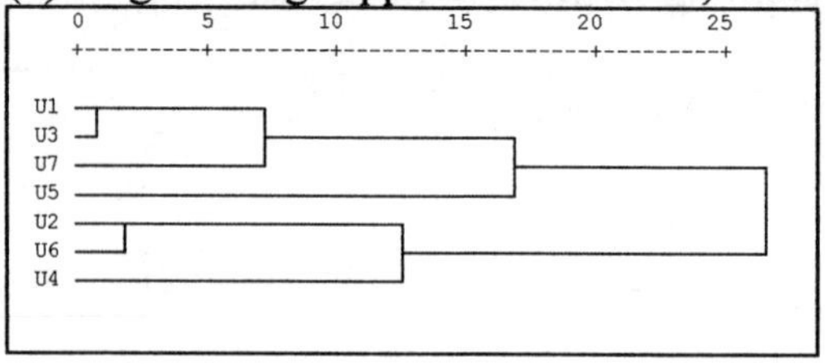

(*b*) Single linkage applied on the best least squares Euclidean approximation

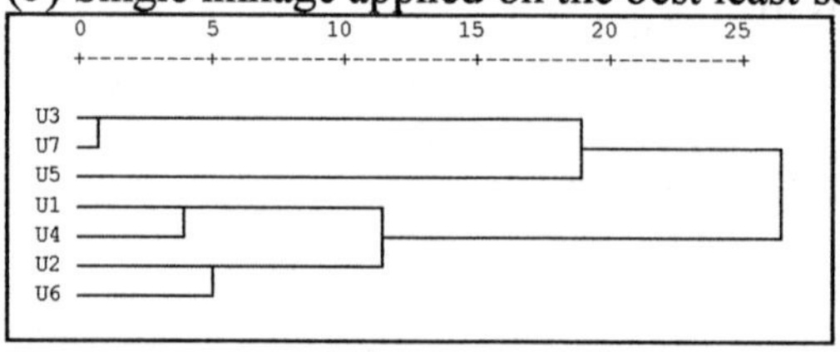

(*c*) Single linkage applied on the quadratic Euclidean approximation of **D*D**:

$$\mathbf{E*E} = 25.2658(\mathbf{11'}\text{-}\mathbf{I}) + 0.3067\mathbf{D*D} + 0.0018\mathbf{D*D*D*D}$$

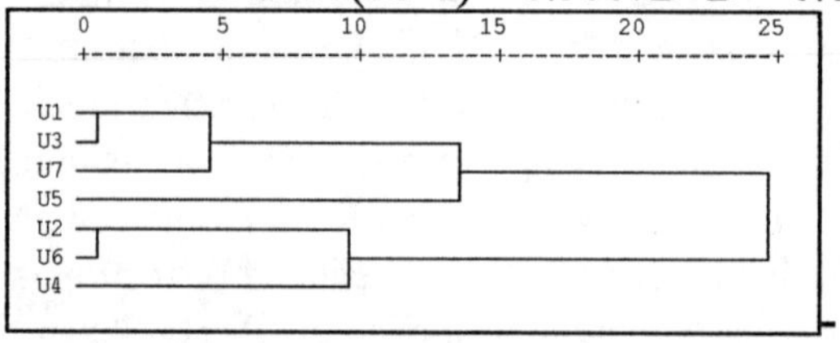

References

Cailliez F. (1983). The analytical solution of the additive constant problem, *Psychometrika*, 48, 2, 305-308.

Mathar, R. (1985). The Best Euclidean Fit to a Given Distance Matrix in Prescribed Dimensions, *Linear Algebra and Its Applications*, 67, 1-6.

Powell, M. J. D. (1983), Variable Metric Methods for Constrained Optimization, *Mathematical Programming: the State of the Art*, Eds., A. Bachem, M. Grötschel and B. Korte, Berlin: Springer-Verlag, 288-311.

Torgerson, W.S. (1958). Theory and Methods of Scaling, New York, Wiley.

DYNAMIC FACTOR ANALYSIS

Isabella Corazziari
Dipartimento di Matematica e Statistica
Università degli studi di Napoli Federico II

Abstract : This paper represents an extension of Dynamic Factor Analysis (AFD) models proposed in the '70s by Coppi and Zannella. AFD models are specific for data-array whose third dimension is time. They consider time as an explicit element which gives rise to part of the observed variability. AFD models integrate two different strategies. The first aims at studying the relationships between variables and units, averaged over time, by factorial analysis of specific covariance-matrices. The second aims at studying time evolution of both variables and units by time regression and autoregressive models.

Key words : array of data or cubic matrices, factorial analysis, time series, regression and autoregressive models.

1. Introduction

Since the years 1980's a growing interest has been given to the statistical treatment of data classified according to the following three criteria (or modes, cfr. Tucker 1966): statistical unit, quantitative variable and time of data collecting. This kind of data may be represented in a cubic matrix **X** (Law and others, 1984) whose generic element is x_{ijt}, where i=1,...,N is the unit index, j=1,...,J is the variable index and t=1,...,T is the time index, when we observe the same units and variables in each time (or occasion).

Different statistical models have been proposed for this kind of data. The Dynamic Factorial Analysis (AFD) models are one of these (Coppi, Zannella, 1979).

In this work, we will discuss the possible extensions of AFD regression models, used to describe the time variability of the array, to polynomial functions in t of order greater than 1 and non linear functions. The possibility of introducing autoregressive models will be considered. Finally parameter estimates will be discussed from the data reconstitution point of view for AFD model I. This allows us to join both the regression and the factorial strategies.

In this context a descriptive approach has been used instead of a probabilistic one.

2. Data preprocessing

The array **X** can be reduced to a bidimensional matrix, depending on which of the two index i or t is used as the external one. In fact **X** can be reduced to a matrix of dimensions IT x J, by overposing the matrices $\{\mathbf{X}_t, t=1,...,T\}$, where $\mathbf{X}_t$ is the matrix units by variables observed in time t, or **X** can be reduced to a matrix of dimensions TI x J, by overposing the matrices $\{\mathbf{X}_i, i=1,...,I\}$, where $\mathbf{X}_i$ is the matrix times by variables for the unit i.
We can introduce *weights* for each mode of the array **X**. As regards units, we consider the diagonal matrix $\mathbf{D}(IxI)= \{d_i, i=1,...I; \Sigma_i d_i=1\}$. For the variables we consider the diagonal matrix $\mathbf{M}(JxJ)=\{m_j, j=1,...,J\}$. In AFD we suggest as m_j the reciprocal of the mean of the IT observations for variable j. The main aim of weigthing variables is to eliminate differences in measurement units and character intensities, which could influence the analysis, by modifying comparisons between and/or within occasions. As regards time we consider the diagonal matrix $\mathbf{L}(TxT)= \{l_t, t=1,...,T; \Sigma_i l_t=1\}$. Greater weights can be attributed to central time values, and smaller weights to extreme time values of the period of interest, following an approach of moving average in time series analysis.
Each element x_{ijt} is weighted by the quantities $(d_i \cdot m_j \cdot l_t)$ which can be considered as elements of an array **P**, obtained as the following tensorial product $(\mathbf{dxm'})\otimes\underline{\mathbf{l}}$, where **d**, **m**, **l** are the main diagonals of the matrices **D**, **M**, **L** respectively. The total sum of the elements of **P** is the trace of **M**.

3. Sources of variation in X

AFD considers three sources of variation which describe the observed data in the period of interest.
The first one is the time evolution of each variable averaged over units. The corresponding covariance matrix is $^*\mathbf{S_t}$:

$$^*\mathbf{S_t} = [(\mathbf{I\text{-}1} \text{ x } \mathbf{l}')^*\mathbf{X}_t]' \, \mathbf{L} \, [(\mathbf{I\text{-}1} \text{ x } \mathbf{l}')^*\mathbf{X}_t]$$

where $^*\mathbf{X}_t = \{\bar{x}_{.jt} \; j=1, ...,J \; t=1, ...,T\}$, and $\bar{x}_{.jt} = \Sigma_i \, x_{ijt} \cdot d_i$.

The generic element of $^*\mathbf{S_t}$ is ${}_t s^*_{jj'} = \sum_t \left(\bar{x}_{.jt} - \bar{x}_{.j.}\right)\left(\bar{x}_{.j't} - \bar{x}_{.j'.}\right) \cdot l_t$.

The second source of variation considers the structural relationships between units and variables, averaged over time.
The corresponding covariance matrix is $^*\mathbf{S_i}$:

$$^*\mathbf{S_i} = [(\mathbf{I\text{-}\underline{1}} \text{ x } \underline{\mathbf{d}}')^*\mathbf{X}_i]' \, \mathbf{D} \, [(\mathbf{I\text{-}\underline{1}} \text{ x } \underline{\mathbf{d}}')^*\mathbf{X}_i]$$

where $^*\mathbf{X}_i = \{\bar{x}_{ij.}\ i=1,...,I\ j=1,...,J\}$, and $\bar{x}_{ij.} = \sum_t x_{ijt} \cdot l_t$.

The generic element of $^*\mathbf{S_i}$ is ${}^*_i s_{jj'} = \sum_i \left(\bar{x}_{ij.} - \bar{x}_{.j.}\right)\left(\bar{x}_{ij'.} - \bar{x}_{.j'.}\right) \cdot d_i$.

The third source of variation rises by the differential time evolution of the units, resulting from the interaction of the two modes unit and time, without the interactions of the other two-mode combinations, unit-variable and variable-time. This variability is described by the covariance matrix $\mathbf{S}_{it}$ of the values $(x_{ijt} - \bar{x}_{ij.} - \bar{x}_{.jt} + \bar{x}_{.j.})$.

The three covariance matrices rise from the decomposition of the total covariance matrix **S** of the array **X.**

Considering **X** as a two dimensional matrix, **S** can be obtained as follows:

$$\mathbf{S}=\{[\mathbf{I} - \underline{\mathbf{1}}\ (\underline{l} \otimes \underline{\mathbf{d}})']\ \mathbf{X}\}'\text{diag}(\underline{l} \otimes \underline{\mathbf{d}})\ \{\mathbf{X}\ [\mathbf{I} - \underline{\mathbf{1}}\ (\underline{l} \otimes \underline{\mathbf{d}})']\}$$

when **X** results from the overposition of $\mathbf{X}_t$ and $(\underline{l} \otimes \underline{\mathbf{d}}) = \begin{bmatrix} l_1\underline{\mathbf{d}} \\ l_2\underline{\mathbf{d}} \\ \cdots \\ l_T\underline{\mathbf{d}} \end{bmatrix}$

I is the identity matrix of order IT, and $\underline{\mathbf{1}}$ is a vector of IT 1's; or

$$\mathbf{S}=\{[\mathbf{I} - \underline{\mathbf{1}}\ (\underline{\mathbf{d}} \otimes \underline{l})']\ \mathbf{X}\}'\text{diag}(\underline{\mathbf{d}} \otimes \underline{l})\ \{\mathbf{X}\ [\mathbf{I} - \underline{\mathbf{1}}\ (\underline{\mathbf{d}} \otimes \underline{l})']\}$$

when **X** results from the overposition of the matrices $\mathbf{X}_i$ and $(\underline{\mathbf{d}} \otimes \underline{l}) = \begin{bmatrix} d_1\underline{\mathbf{l}} \\ d_2\underline{\mathbf{l}} \\ \cdots \\ d_I\underline{\mathbf{l}} \end{bmatrix}$

I is the identity matrix of order TI, and $\underline{\mathbf{1}}$ is a vector of TI elements.

It can be shown that $\mathbf{S} = {}^*\mathbf{S}_i + {}^*\mathbf{S}_t + \mathbf{S}_{it}$.

Finally we can define the following covariance matrices:

$\bar{\mathbf{S}}_t = \sum_t \mathbf{S}(t) \cdot l_t\ \ t=1,...,T$, where $\mathbf{S}(t)$ is the covariance matrix of the observations at time t;

$\bar{\mathbf{S}}_i = \sum_i \mathbf{S}(i) \cdot d_i\ \ i=1,...,I$, where $\mathbf{S}(i)$ is the covariance matrix for the unit i, considering times as observations.

It can be shown that $\bar{\mathbf{S}}_t = {}^*\mathbf{S}_i + \mathbf{S}_{it}$ and $\bar{\mathbf{S}}_i = {}^*\mathbf{S}_t + \mathbf{S}_{it}$. Looking at the decomposition of **S,** we can write **S** in the following two ways: $\mathbf{S} = \bar{\mathbf{S}}_t + {}^*\mathbf{S}_t$ or $\mathbf{S} = \bar{\mathbf{S}}_i + {}^*\mathbf{S}_i$.

In AFD the three sources of variations are described using factorial methods and time regression models.

4. AFD models

The first three AFD models consider different strategies to analyse the "static" variability and the differential time evolution of the units.
As regards time evolution of the centres $\overline{x}_{.jt}$ all of the three AFD models consider a linear regression model for each variable j, where the independent variable is time. The parameters are obtained by ordinary least squares. The assumptions about residuals are the classic ones: $cov[e_{.jt},e_{.j't'}]=w_j$, if j=j' e t=t', and 0 otherwise.
The variability of the centres $\overline{x}_{ij.}$ is analized by factorial analysis of specific covariance matrices in each of the three AFD models.
In the first model, factorial analysis is applied to the covariance matrix $\overline{\mathbf{S}}_t$. By projecting the matrices $\mathbf{X}_t$ centred in each time we obtain the factorial representation of each unit in each time. The representation of the centres $\overline{x}_{ij.}$ is obtained by projecting the matrix $^*\mathbf{X}_i$ centred, on the factorial plane, reminding the decomposition $\overline{\mathbf{S}}_t = {}^*\mathbf{S}_i + \mathbf{S}_{it}$.
In the second AFD model, factorial analysis is applied to the matrix $^*\mathbf{S}_i$. In this way we obtain the factorial representation of the centres $\overline{x}_{ij.}$.
In the third model, the factorial analysis is applied to the matrix ${}_y\overline{\mathbf{S}}_t = \Sigma_t\, {}_y\mathbf{S}(t)\cdot l_t$, where ${}_y\overline{\mathbf{S}}_t$ and ${}_y\mathbf{S}(t)$ are obtained as in the first model, but considering observed data processed as follows: $y_{ijt} = x_{ijt} - \hat{x}_{ijt}$, where $\hat{x}_{ijt}$ is the time regression value corresponding to x_{ijt} , describing the total time evolution of unit i for each variable j.
As regards differential time evolution of the units, in the first model it is described by comparing the projection of each unit in each time, with the projection of the corresponding centre $\overline{x}_{ij.}$:

$$F_{iht} - \overline{F}_{ih} = \sum_{j=1}^{J}\left[c_{jh}\cdot\left(x_{ijt} - \overline{x}_{.jt}\right) - c_{jh}\cdot\left(\overline{x}_{ij.} - \overline{x}_{.j.}\right)\right] = \sum_{j=1}^{J} c_{jh}\cdot\left(x_{ijt} - \overline{x}_{.jt} - \overline{x}_{ij.} - \overline{x}_{.j.}\right)$$

where F_{iht} and $\overline{F}_{ih}$ are the factorial scores corresponding to the unit i, on the factor h; $\underline{\mathbf{c}}_h$, h=1, ..., H is the eigenvector corresponding to the h° eigenvalue of $\overline{\mathbf{S}}_t$. This representation rises from the decomposition $\overline{\mathbf{S}}_t = {}^*\mathbf{S}_i + \mathbf{S}_{it}$.
Both the second and the third AFD models describe the differential time evolution of the units, starting from a time regression model for each unit, whose parameters are calculated by ordinary least squares:

$$x_{ijt} = a_{ij} + b_{ij}\cdot t + e_{ijt},\ j=1,\dots,J \text{ and } i=1,\dots,I$$

the assumptions are: $cov[e_{ijt},e_{ij't'}]=w_j$ if j=j' and t=t'; 0 otherwise.
Differential time evolution of each unit can be measured considering the differences of the two regression parameters b_i and b_{ij}, j=1,...,J.
The different strategies are based on the two possible decomposition of the matrix of total covariances **S**, $\mathbf{S}=\overline{\mathbf{S}}_t+{}^*\mathbf{S}_t$ and $\mathbf{S}=\overline{\mathbf{S}}_i+{}^*\mathbf{S}_i$. Model I considers the first decomposition; both model II and III consider the second decomposition, but the third model uses the following approximation: ${}_y\overline{\mathbf{S}}_t\cong{}^*\mathbf{S}_i$.

5. Indices of the quality of fitness of the model to data

For each source of variation, indices measuring the quality of fitness of the model to data are calculated. As models are linear, the indices are calculated considering the trace of both the observed and the theoretical cov matrices.
As regards the centres $\overline{x}_{.jt}$, the quality index is the same for the three models:

$$^*I_t=tr(^*\hat{\mathbf{S}}_t)/tr(^*\mathbf{S}_t)$$

where $^*\hat{\mathbf{S}}_t$ is the covariance matrix of the regression values of $\overline{x}_{.jt}$.
As regards factorial structures, the indices for the first model are the following:

$$\overline{I}_t=[\Sigma_h \underline{c}_h'\cdot\overline{\mathbf{S}}_t\cdot\underline{c}_h]/tr(\overline{\mathbf{S}}_t)$$

where $\underline{c}_h$ is the eigenvector corresponding to the eigenvalue h° of $\overline{\mathbf{S}}_t$.
The following index measures how much variability is described in each time-occasion by the factorial plane:

$$I(t)=[\Sigma_h \underline{c}_h'\cdot\mathbf{S}(t)\cdot\underline{c}_h]/tr[\mathbf{S}(t)] \text{ with } t=1,...,T$$

As regards the centres $\overline{x}_{ij.}$ the corresponding quality index is:

$$^*I_i=\{\Sigma_h \underline{c}_h'\cdot{}^*\mathbf{S}_i\cdot\underline{c}_h\}/tr(^*\mathbf{S}_i)$$

Finally the quality index of the differential time evolution of the units is :

$$I_{it}=\{\Sigma_h \underline{c}_h'\cdot\mathbf{S}_{it}\cdot\underline{c}_h\}/tr(\mathbf{S}_{it})$$

In the second model the quality index of the factorial structure is:

$$^{*}I_i=\{\Sigma_h {}^{*}\underline{c}_h' \cdot {}^{*}\mathbf{S}_i \cdot {}^{*}\underline{c}_h\}/tr({}^{*}\mathbf{S}_i)$$

In the third model the index is:

$$_{y}I_t=[\Sigma_h {}^{y}\underline{c}_h' \cdot {}_{y}\overline{\mathbf{S}}_t \cdot {}^{y}\underline{c}_h]/tr({}_{y}\overline{\mathbf{S}}_t)$$

As regards differential time evolution of the units in the second and the third model, the quality index is:

$$I_{it}=tr(\hat{\mathbf{S}}_{it})/tr(\mathbf{S}_{it})$$

where $\hat{\mathbf{S}}_{it}$ is the covariance matrix of the values $\hat{x}_{ijt}-\hat{x}_{.jt}$ resulting from regressions. The index corresponding to the total time evolution of the units is:

$$\overline{I}_i=tr(\hat{\mathbf{S}}_i)/tr(\overline{\mathbf{S}}_i)$$

where $\hat{\mathbf{S}}_i$ is the theoretical covariance matrix, calculated as the mean over units of the theoretical covariance matrices of the regression values of each unit. We can consider the same kind of index for each unit:

$$I(i)=tr\hat{\mathbf{S}}(\mathbf{i})/tr[\mathbf{S}(i)] \quad i=1,...,I$$

where $\hat{\mathbf{S}}(\mathbf{i})$ is the covariance matrix of the regression values of each unit, considering times as the observations.

The summary indices of the total fitness of models to data are:

Model I: $I=[\overline{I}_t \cdot tr(\overline{\mathbf{S}}_t) + {}^{*}I_t \cdot tr({}^{*}\mathbf{S}_t)]/ tr(\mathbf{S})$

Modello II: $I=[\overline{I}_i \cdot tr(\overline{\mathbf{S}}_i) + {}^{*}I_i \cdot tr({}^{*}\mathbf{S}_i)]/ tr(\mathbf{S})$

Modello III: $I=[{}^{*}I_t \cdot tr({}^{*}\mathbf{S}_t)+ {}_{y}I_t \cdot tr({}_{y}\overline{\mathbf{S}}_t) + I_{it} \cdot tr(\mathbf{S}_{it})]/tr(\mathbf{S})$

6. Possible extensions of time regression models

It can be easily proved that:

1. $\overline{I}_t = \max_{(b_j \in B_j)}$, and $\overline{I}_i = \max_{(b_{ij} \in B_{ij})}$, where $\mathbf{B_j}$ and $\mathbf{B_{ij}}$ represent the classes of all linear estimators respectively for the regression models $\overline{x}_{.jt}=a_j+b_j \cdot t+e_{.jt}$, and $x_{ijt}=a_{ij}+b_{ij} \cdot t+e_{ijt}$ $i=1,...,I$, $j=1,...,J$;

2. $b_j=\Sigma_i b_{ij} \cdot d_i$

As regards the time evolution of the centres $\overline{x}_{.jt}$ and, for the models II and III time evolution of each unit, we can consider more general models, as $\overline{x}_{.jt}={}_jf(t)+e_{.jt}$ and $x_{ijt}={}_{ij}f(t)+e_{ijt}$, where ${}_jf(t)$ and ${}_{ij}f(t)$ are generic polinomial or non linear functions of time, corresponding to variable j. The assumptions about $e_{.jt}$ e_{ijt} are the same as in the case of the linear regression. In a descriptive context, the parameters can be calculated by ordinary least squares, whose solution can be obtained by numerical methods.

Considering ${}_jf(t)$ e ${}_{ij}f(t)$ as polinomial functions in t of the same order, the vectors of the parameters have the same properties we saw in the case of the simple linear regression:

1. the indices of fitness are maximus;
2. parameters for $\overline{x}_{.jt}$ are obtained as the mean over i of the parameters of x_{ijt}.

The differential time evolution of the units can be measured by the parameters ${}_{ij}{}^{d}\underline{\hat{b}}={}_{ij}\underline{\hat{b}}-{}_{j}\underline{\hat{b}}$, easily calculated considering centred data.

These considerations are valid with respect to more general functions, linear in the parameters:

$$\overline{x}_{.jt} = {}_jb_0+{}_jb_1\cdot g_1(t)+{}_jb_2\cdot g_2(t) +...+{}_jb_p\cdot g_k(t) +e_{.jt}$$

$$x_{ijt}= {}_{ij}b_0+ {}_{ij}b_1\cdot g_1(t)+ {}_{ij}b_2\cdot g_2(t) +...+ {}_{ij}b_p\cdot g_k(t) +e_{ijt}$$

When there are enough occasions, it is possible to introduce an autoregressive model for each variable, directly fitted to raw data if they exhibit stationarity, or to data after having removed the trend, which could be estimated by a linear time regression. The important aspect is the study of the time structure of the model as a supplementary source of information (Piccolo, 1974).

7. Reconstitution of the array X

We can reconstitute initial data when the model parameters are found by minimizing a lost function, which measures the fitness of the model to observed data. Let us consider the following lost function: $\Sigma_{ijt} (x_{ijt} - \hat{x}_{ijt})^2$ (2)

x_{ijt} can be written as $x_{ijt}= (x_{ijt} - \overline{x}_{.jt} - \overline{x}_{ij.} + \overline{x}_{.j.}) + (\overline{x}_{ij.} - \overline{x}_{.j.}) + \overline{x}_{.jt}$ (3)

where each part represents a different source of variation, and is parameterized in different ways in each AFD model.

In AFD model I the first two elements of (3) are described by a factorial model, the third one by a time regression. It can be proved that minimizing (2) by considering the appropriate expression for $\hat{x}_{ijt}$, gives us the same parameters for the factorial representation and the regression model as indicated above for the model itsself. We can reconstitute the observed data as $x_{ijt}\cong\Sigma_h F_{ith}\, a_{jh} + \hat{x}_{.jt}$, where $\sum_h F_{ith} a_{jh}$ is the factorial representation of $\mathbf{S}_t$, and $\hat{x}_{.jt}$ is the time regression value for $\bar{x}_{.jt}$. Similar considerations can be made for the other two AFD models, with more complications due to the more complex model structures of the various sources of variation.

8. Conclusions

AFD models represent an alternative to three-mode data analysis models such as the STATIS and the TUCKERs' models. AFD models seem to be more convenient for three-mode data whose third dimension is time, because they consider time as an explicit element which give rise to part of the observed variability in the array **X**. In this sense, time is considered as a dimension of different nature with respect to the other two dimensions, the unit and the variable.
Parameters from time regression models and considerations about time structure from autoregressive models considerably enrich the information about the relationships between units and variables given by factorial models.

References

Coppi R., Zannella F. (1979). L'analisi fattoriale di una serie temporale multipla relativa allo stesso insieme di unita' statistiche, *Atti della XXIX Riunione della SIS*.

Harshman R.A. e M.Lundy (1984). Data Preprocessing/Extended Parafac Model, in *Research Method For Multimode Data Analysis*, di H.G.Law ed altri.

Kroonenberg P.M. (1992). Three-mode component models. A survey of the literature, *Statistica Applicata*, vol.4 n°4.

Lavit C., Escoufier Y., Sabatier R., Traissac P. (1994), The ACT (Statis Method), *Computational Statistics & Data Analysis*, vol.18

Law H.G. (1984). *Research Method For Multimode Data Analysis*,

Piccolo D. (1974). *Analisi Delle Serie Temporali, I Processi Autoregressivi del Secondo Ordine*, Centro di Specializzazione e Ricerche Economico-Agrarie per il Mezzogiorno, Napoli.

Tucker L.R. (1966). Some Mathematical Notes On Three-Mode Factor Analysis, *Psychometrika*, vol.31, n°3.

A Non Symmetrical Generalised Co-Structure Analysis for Inspecting Quality Control Data

Vincenzo Esposito, Germana Scepi
Dipartimento di Matematica e Statistica, Università di Napoli "Federico II"
Via Cintia, Complesso Monte Sant'Angelo, 80126 Napoli
e-mail: binci@dms.unina.it

Abstract: The paper provides a contribution to factorial methods in multidimensional data analysis covering the gap of graphical representations of statistical units on which a multiple set of response variables as well as a common set of explanatory variables are observed. By joining the features of multiple Co-Inertia analysis with those of a geometrical non-symmetrical approach, the proposed technique gains remarkable advantages in identifying a typology of statistical units generated by the mentioned dependence structure.

Keywords: Multiple Sets, Co-Inertia, Orthogonal Projections, Graphical Displays.

1. Introduction

The prediction of multivariate responses by multivariate predictors is a relevant problem in applied Statistics. In real applications, a "true model" for the link between responses and predictors is seldom available and, in any case, does not provide a graphical insight for a better understanding of the data structure under study. Therefore, a geometrical approach seems more reasonable and helpful.
In this paper, we aim at visually inspecting the dependence structure between K sets of response variables with respect to a common set of explanatory variables. This approach helps in graphically identifying a typology of the statistical units induced by those sets of response variables with the highest discrimination power among them.
The usefulness of this technique in real applications is of primary concern to quality control problems, as it will be shown by an example on water pollution.

2. The Data Structure

The data structure at hand is similar to the typical one of Generalised Canonical Correlation Analysis (GCCA, Carroll 1968). In fact, it comprises K sets of q_j ($j=1,...,K$) quantitative response variables observed on the same n statistical units, represented by the K matrices $\boldsymbol{Y}_j$ of dimensions (nxq_j). Therefore, the $\boldsymbol{Y}_j$'s

are row-wise paired matrices whose rows belong to different multidimensional spaces.
Moreover, as we refer to a dependence structure, i.e. we work in a non-symmetrical context, we have one set of p quantitative explanatory variables observed on the same n units, represented by the $\boldsymbol{X}$ matrix of dimensions (nxp).
All variables are considered preliminarily centred with respect to their own arithmetic means as well as standardised to unit variance. As we work on pre-processed data, in the analysis we can refer to classical Euclidean metrics for interpreting graphical representations.
In such a context, three different strategies of analysis are possible:
1) a *global analysis* searching for a common plane where to compare the different matrices;
2) a *synthesis analysis* searching for a compromise matrix whose factorial representation aims at defining homogeneous classes of the statistical units;
3) a *detailed analysis* which, by means of the projections in supplementary of each matrix on the factorial planes of the synthesis analysis, allows to compare both statistical units and variables in the different matrices.
In literature, several techniques (e.g. STATIS, Lavit 1988; Analyse Factorielle Multiple, Escofier & Pagès 1984; Multiple Co-Inertia Analysis, Chessel & Hanafi 1996) have been proposed for the analysis of multiple tables but they refer only to the symmetrical structure of interdependence.
In the following, our technique is developed in the direction of Multiple Co-Inertia Analysis but in a non-symmetrical context.

3. The Analysis

Geometrically speaking, Multiple Co-Inertia Analysis studies the response variables structure as K clouds of n points in the spaces R^{q_j} regardless of any explanatory variables. On the other hand, Non Symmetrical Comparative Analysis of Co-Inertia (Esposito 1997; Balbi & Esposito, 1997) takes into account the explanatory variables but is confined to deal with just 2 tables which must also be totally paired, i.e. same response variables observed under 2 different conditions. Hereinafter, we aim at dealing with multiple sets $(K > 2)$ of response variables as well as with row-wise paired tables.
The analysis is set in the geometrical framework of Principal Component Analysis onto a Reference subspace (PCAR, D'Ambra & Lauro, 1982). PCAR looks for the principal components of the image of the response variables on the subspace (in our case, R^p) spanned by the explanatory ones in order to take into account the non-symmetrical relationship between two sets of variables. Thus, we project the $\boldsymbol{Y}_j$'s on the space R^p through the common orthogonal operator $\boldsymbol{P_x} = \boldsymbol{X}(\boldsymbol{X}'\boldsymbol{X})^{-1}\boldsymbol{X}'$ so as to define the new matrices $\boldsymbol{Y}_j^* = \boldsymbol{P_x}\boldsymbol{Y}_j$.

In the core of the analysis we first search for K block-normalised vectors w_j^1's, one for each subspace R^{q_j} spanned by the variables in Y_j^*, as well as for one normalised auxiliary variable z^1 in the statistical units subspace R^n. After that, vectors w_j^2's and z^2 are searched to be orthogonal with, respectively, w_j^1's and z^1. The choice of these vectors is based on the maximisation of:

$$\sum_{j=1}^{K} \pi_j \left(\boldsymbol{P_x Y_j w_j} | z \right)^2 \tag{1}$$

where each π_j represents the weight assigned to each Y_j^*. Such a weighting system is necessary in order to take into account the different number of response variables in each $\mathbf{Y}_j$ (i.e., $\pi_j = q_j / \sum_j q_j$ used in the application) or the different variability of each response variables set (i.e., $\pi_j = Var\,\mathbf{Y}_j / \sum_j Var\mathbf{Y}_j$).

The quantity in (1), as $Var(z) = 1$, may be differently expressed as:

$$\sum_{j=1}^{K} \pi_j Var(\mathbf{P_x Y}_j \mathbf{w}_j) \rho^2 (\mathbf{P_x Y}_j \mathbf{w}_j, \mathbf{z}) \tag{2}$$

where $Var(\cdot)$ stands for the variance operator and $\rho(\cdot)$ for the correlation coefficient operator.

The equivalence in (2) points out that the proposed analysis jointly maximises the criteria of, respectively, PCAR (the term $Var(\mathbf{P_x Y}_j \mathbf{w}_j)$) and GCCA (the term $\rho^2(\mathbf{P_x Y}_j \mathbf{w}_j, \mathbf{z})$). Consequently, it succeeds in performing K separate analyses together with a global one of the Y_j^*'s.

By referring to the Cauchy-Schwartz inequality (Chessel & Hanafi, 1996), it can be easily shown that the majoring quantity of the expression in (2) is defined by:

$$\sum_{j=1}^{K} \pi_j \left\| \mathbf{Y}_j^{*'} \mathbf{z} \right\|^2 = \mathbf{z}' \left(\sum_{j=1}^{K} \pi_j \mathbf{P_x Y}_j \mathbf{Y}_j' \mathbf{P_x} \right) \mathbf{z}. \tag{3}$$

As the Y_j's are paired by rows, the first order solutions w_j^1's and z^1 are derived from a PCA performed on the pooled matrix $\mathbf{Y} = \left[Y_1^* | Y_2^* | \ldots | Y_K^* \right]$.

Namely, the axes w_j^1's are the q_j-dimensional vectors of the principal axis associated to the highest eigenvalue of a PCAR on the pooled *K* sets Y_j's with respect to *X*, while the variable z^1 is given by the relative principal component.

With respect to the second order solution w_j^2's and z^2, we work on the residual matrices $\mathbf{E}_j = \mathbf{Y}_j^* - \mathbf{Y}_j^* \mathbf{P}_{w_j}^1$ where $\mathbf{P}_{w_j}^1$ is the orthogonal projection operator on the subspace spanned by the vector w_j^1.

By juxtaposing all E_j's, we define a pooled matrix $\mathbf{E} = \left[E_1 | E_2 | ... | E_K \right]$. Finally, the first order solutions of $\mathbf{E}$ represent the second order solutions of $\mathbf{Y}$. Similarly, by iterating the procedure, the generic r-th order solution may be found.

The proposed technique provides one system of Co-Inertia axes where to project the *K* clouds of statistical units relative to each matrix Y_j^* in R^{q_j}. These representations, according with the chosen maximisation criterion, allow a global comparison of *K* configurations of the statistical units. Consequently, peculiar configurations of the units may be inspected.

On the same system, we can display the axes of *K* separate PCAR's with the objective of representing the inertia structure of each table itself.

More importantly, aiming at enhancing a typology of the units, we normalise all sets of co-ordinates to 1. In this way, we are allowed to simultaneously project, on a common display, both the *K* representations of each unit and the relative components of the auxiliary variables. Thereafter, *star-plots* are drawn which characterise the *K* behaviours of each unit with respect to a synthesis of theirs.

With regards to the variables, both explanatory and response ones are represented on the plane spanned by the auxiliary variables in R^n in order to show the links between the *K* sets Y_j^* as well as the influence of the variables in *X* on them.

4. An Application on Water Quality Control Multivariate Data

The proposed technique is very useful for visualising quality control multivariate problems (Lauro, Scepi & Balbi, 1996, for a review). In particular, it helps in identifying both a control typology and the causes of eventual out of controls. The latter is a very important topic in the analysis of multivariate control process in which it is very difficult to detect the variables actually determining out of control situations.

In the following example, we refer to the well-known data by Verneaux (1973) relative to a study in hydrobiology. The aim is to investigate the water quality of the French River Doubs taking into account both its ichtyologic fauna and its chemical, physical and biological components.

Therefore, two data matrices are available. The first one crosses 35 sites observed along the river with 11 variables (explanatory ones) referring to morphological features of the river (distance from the source, altitude, slope, minimum mean flow) and to water quality indicators (pH, total hardness, phosphates, nitrates, ammoniacal azote, oxygen, Biological Request of Oxygen). The second one crosses the same sites with an index of presence of 27 species (response variables) of fishes in each site. These data have been analysed by Chessel et al. (1987) in the framework of Canonical Correspondence Analysis. They result in partitioning the 27 species into 8 groups. In our analysis, we take into account their partition. However, as there are two groups each formed by just one species, we first aggregate these groups to the nearest groups on the factorial plane therein obtained. We thus finally consider a 6 groups' partition of the species.

By applying our technique to the same data, taking into account the results by Chessel et al. (1987), our added value consists in identifying a common structure of the sites so as to detect "anomalous" situations.

Figure 1 represents the discrimination power of each group of species, that is its capacity in yielding a typology of the sites. This capacity is computed, for each axis, as the squared covariance between each system of co-ordinates and the auxiliary variable relative to the same order.

It is clearly shown that group 3 is the one with the highest overall discrimination power. For this reason, in the following, we show some of the graphical representations relative to group 3 in order to explain the interpretative power of the technique we propose.

Figure 1: *Representation of Discrimination Power of Species Groups*

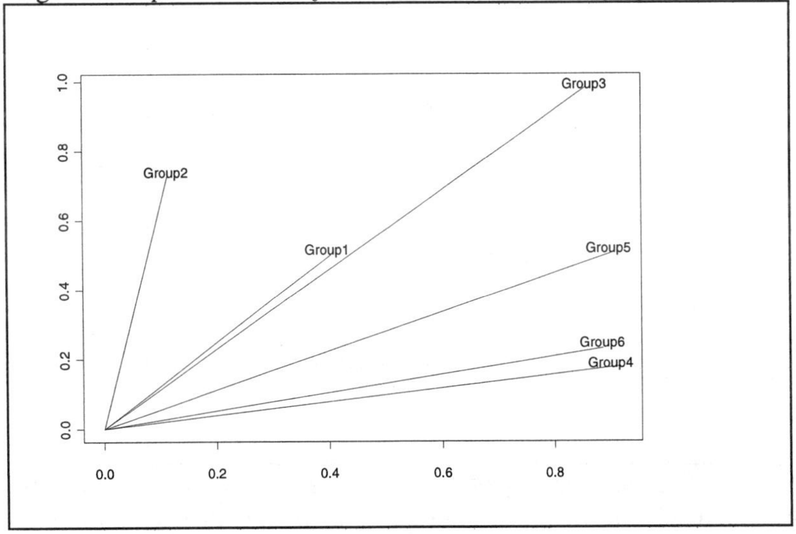

Figure 2 refers to the representation of sites in the species space in group 3. The axes, displayed as full lines, are the ones obtained by performing a single PCAR on group 3. They almost match with the principal axes of GCCA (dotted lines).

Figure 2: *Representation of Sites in the space relative to Group 3*

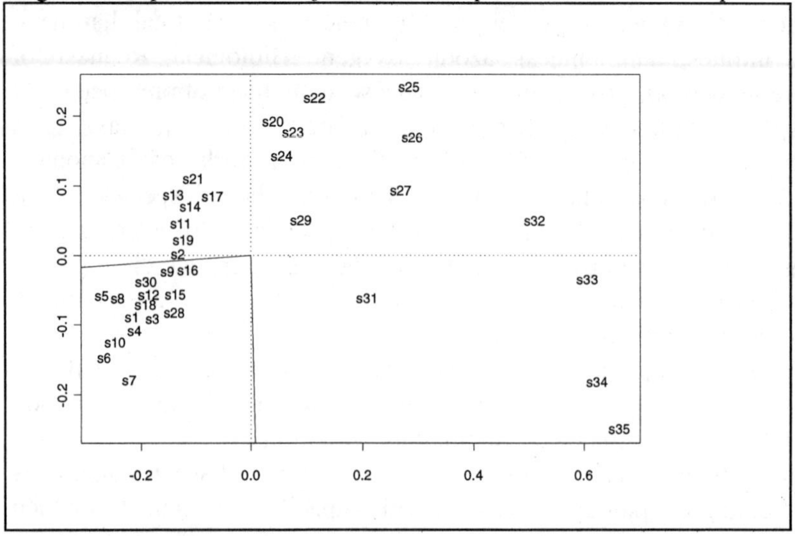

The first axis (the horizontal one) discriminates between sites nearest to the source (on the left) and those farest from the source itself (on the right). In fact, the sites are numbered from s1 to s35 on the basis of their distance from the source.

Figure 3: *Explanatory Variables and Group 3 Species*

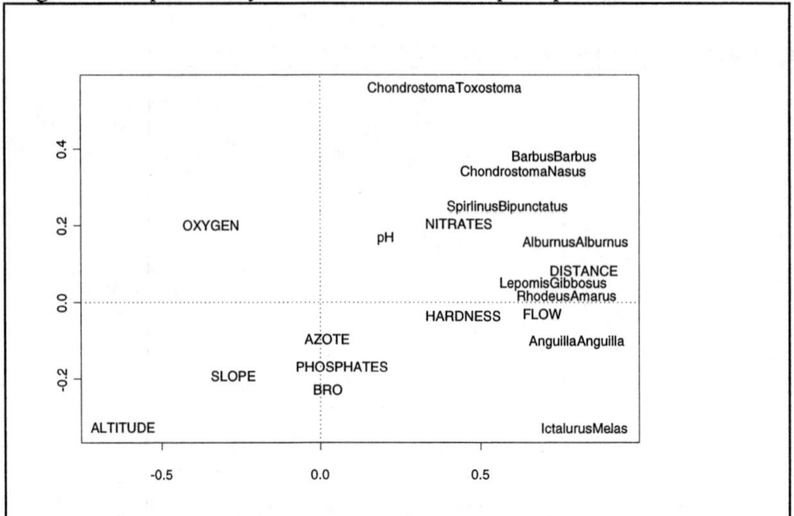

In order to understand the meaning of the second axis (the vertical one), we must interpret the display in Figure 3 that represents both explanatory variables and the 9 species (long words in lower-case) in Group 3.

The first axis opposes altitude to distance, as they are logically inversely related, and discriminates among the other morphological features (slope, hardness, and flow) relative to the position of the observed sites. The grouping of the species on the right hand side of this axis implies that their presence is very much influenced by the distance from the source. They are actually species usually living very far from the source.

The second axis, instead, discriminates among the chemical components of the water thus representing a synthesis of its quality. In particular, species suited to live in waters rich of azote, phosphates and BRO oppose to those who prefer oxygen, pH and nitrates.

In Figure 4, the *star-plots* of the behaviour of just the 'anomalous' sites in the different groups are represented. The centre of each star relates to the synthesis representation while its edges relate to the behaviours in the different groups. Moreover, by 'anomalous' sites we mean those who do not strictly conform themselves to a common behaviour around the origin as most of the sites do.

Figure 4: *Synthesis Representation of 'Anomalous' Sites*

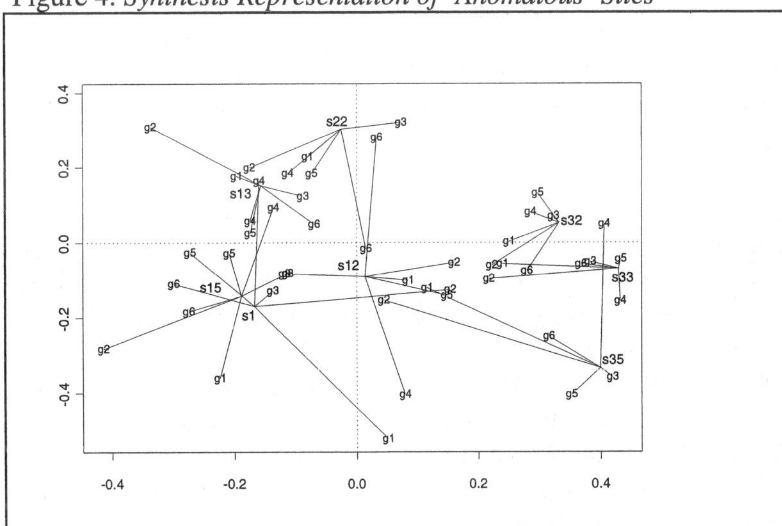

The sites very far from the source have a synthesis configuration that, with respect to the shape, are very similar to each other and altogether form a configuration substantially different from the other sites. However, each of them has a very different variability, e.g. s32 has a low variability since its representations are very close to each other, while s35 has a very high variability due to its different behaviours in groups 1, 2 and 4.

On the other side of the first axis, it is worth noting that s1 has the highest variability among all sites, and its synthesis may be considered quite anomalous due to its behaviour in group 1.

The positioning of the other sites on the factorial plane may be similarly commented.

5. Conclusions

The proposed technique represents a contribution to the non-symmetrical approach to the simultaneous analysis of multiple data sets. This area has not yet received a proper attention with respect to its relevance in real applications. In this perspective, it is important to note the awareness of the technique for the role played by the statistical units. Graphical representations of their configurations, both in each set and as a whole, allow better understanding of data at hand.

The approach we follow is entirely geometrical. The enrichment of the results by means of inferential tools represents a future task of research to be accomplished. Moreover, the idea behind this paper may turn out to be very profitable for studying complex covariance structures via a modelling approach.

References

Balbi, S. & Esposito, V. (1997). Rotated canonical analysis onto a reference subspace, *Invited lecture at second IASC World Conference*, Pasadena, USA.

Carroll, J.D. (1968). A generalization of canonical correlation analysis to three or more sets of variables, *Proceedings of the 76th Convention of the American Psychological Association*, 3, 227-228.

Chessel, D. & Hanafi, M. (1996). Analyses de la co-inertie de K nuages de points, *Revue de Statistique Appliquée*, 44, 2, 35-60.

Chessel, D., Lebreton, J.D. & Yoccoz, N. (1987). Propriétés de l'analyse canonique des correspondances: une illustration en hydrobiologie, *Revue de Statistique Appliquée*, 35, 4, 55-72.

D'Ambra, L. & Lauro, C. (1982). Analisi in componenti principali in rapporto ad un sottospazio di riferimento, *Rivista di Statistica Applicata*, 15, 1, 51-67.

Escofier, B. & Pagès, J. (1984). L'analyse factorielle multiple: une méthode de comparaison de groupes de variables, in: *Data Analysis and Informatics III*, Diday, E. et al. (Eds.), North Holland, 41-55.

Esposito, V. (1997). Un'analisi non simmetrica comparativa con osservazioni stratificate, *Convegno SIS "La Statistica per le Imprese"*, Torino.

Lauro, N., Scepi, G. & Balbi, S. (1996). Differenti approcci nella costruzione di carte di controllo multivariato, *Studi in Onore di G. Landenna*, CEDAM.

Lavit, C. (1988). *Analyse conjointe de tableaux quantitatifs*, Masson, Paris.

Verneaux, J. (1973). Cours d'eau de Franche-Comté (Massif du Jura). Recherches écologiques sur le réseau hydrographique du Doubs. *Essai de biotypologie. Thèse d'état*, Besançon.

Acknowledgement: *This research was supported by the CNR grant 93.01789.CT12 (resp. Prof. C. Lauro). This paper is the result of a joint work between the authors. However, Vincenzo Esposito was mainly responsible for redacting Sections 1, 3, and 5, while Germana Scepi for redacting Sections 2. and 4. Vincenzo Esposito developed the S-Plus algorithm for computations and graphics in Section 4.*

Principal Surfaces Constrained Analysis

Lombardo R.* & Tessitore G.**
*Institute of management and quantitative research, S.U.N.-Naples-Italy-
e.mail: lombardo@dmsna.dms.unina.it
**Department of Mathematics and Statistics, University of Naples "Federico II"-Italy-e.mail: gt@dmsna.dms.unina.it

Abstract: In this paper we present a non parametric adaptive procedure for the principal components non linear representation (*Principal Surfaces*, *PS*, LeBlanc and Tibshirani, 1994) in *Constrained Principal Component Analysis* (*CPCA*, D'Ambra & Lauro 1982, 1992 see also Principal Component Analysis with Instrumental Variables, Rao 1964; Redundancy Analysis, van der Wollenberg, 1977).

Keywords: Constrained Principal Component Analysis, Principal Surfaces, Multivariate Adaptive Splines.

1. Introduction

In applied multivariate statistical analysis, dimension reduction is an important problem related to the realization of an appropriate and easy representation.
For the attainment of data visualization and interpretation simplicity, in this paper we develop a generalization of *Constrained Principal Component Analysis* (*CPCA*), where we use the *Principal Surfaces* (*PS*) relaxing the linearity assumption of the final reprentation. This means that the non-linearity does not concern the original data matrix as in *Non-linear PCA* (Gifi 1990), but the component scores matrix (LeBlanc and Tibshirani, 1994).
The principal components are transformed by multivariate B-splines (*PS*) of zero or one degree than higher degree B-splines, as the last are heavy to compute and difficult to interpret. In particular crisp coding often generates well separated and contiguous representations easy to interpret (van Rijckevorsel 1987).
In the next sections, we first introduce the *PS* construction in the *CPCA* context; afterwards we present the computational procedure for the performance of *Principal Surface Constrained Analysis* (*PSCA*). This is based on a forward phase for the choice of the univariate spline optimal knot sequence (fixing the knots number) and on a backward phase for the optimal detection of the space dimension.

2. Adaptive Principal Constrained Surfaces

Studying the dependence structure between two sets of quantitative variables $\mathbf{X}_{(n,q)}$ (predictors) and $\mathbf{Y}_{(n,p)}$ (responses), with n observations, q predictors and p response variables ($p<q$); the non-linear generalization of *CPCA*, is here developed by estimating the generalized factor model:

$$\hat{\mathbf{y}}_j = f(\mathbf{c}_j) + \mathbf{e}_j \qquad \forall j = 1,\ldots,p \tag{1}$$

where is: $\hat{\mathbf{y}}_j$ the jth column of the response matrix projected onto the subspace of the predictors; $\mathbf{c}_j$ (the jth column of the component score matrix $\mathbf{C}$) the parameter vector, which takes values into R^p; and each $f(\mathbf{c}_j)$ a non linear function mapping R^p into R^n.
We minimize the expected square error:

$$E\left[\sum_{j=1}^{m}(\hat{\mathbf{y}}_j - f(\mathbf{c}_j))^2\right] \tag{2}$$

Hastie & Stuezle (1989) describe the Principal Curve as a non linear transformation of a principal component. Furthermore they show that a principal curve is a critical point of the squared distance function (2), in this sense it generalizes the minimum distance property of linear principal components. Le Blanc & Tibshirani use the tensorial product among principal curves and construct Principal Surfaces as multivariate splines.
In coherence we use B-spline functions for the $f(\mathbf{c}_j)$ for their known property of flexibility and prefer zero or one degree B-splines (*PS*), because they are easier to interpret and compute.
So our problem comes out to be the optimal parameter estimation of the following transformation:

$$f(\mathbf{C}) = \beta_0 + \sum_{m=1}^{M}\beta_m\prod_{k=1}^{K_m}\mathbf{b}_{(k,m)}\left(\mathbf{c}_{j(k,m)}\right) \tag{3}$$

where: β are the coefficients for the construction of the spline transformation; M is the number of the basis functions of the multivariate spline; K_m is the basis number of the univariate spline to construct tensors and $\mathbf{b}$ is the basis function of the generic component $\mathbf{c}_{j(k,m)}$, on the partition of the multivariate domain, used in the kth term of the mth product.
In our *CPCA* context we obtain for each response variable a *PS* on which units are projected. These *PS* allow a suitable non-linear representation of the non-symmetric relationship among responses and predictors, generating well

separated and contiguous representations. In the classical representation it might be more difficult to inspect and interpret similarities-dissimilarities among units. For example, units, apparently located near each other in the classical bi-dimensional plot, could belong to different interval grid areas. Using zero degree B-spline, the *PS* are parallelepipeda. The grid levels are the coefficients, computed by regressing $\hat{\mathbf{Y}}$ with respect to the non-linear component tensor (i.e. *PS*, the multivariate spline). Different units groupes appear on different grid levels for each variable surface. In a way we classify units according to a specified variable; the higher the grid level, the stronger the variable influence on units.

3. The Computational Procedure

The algorithm we propose for the optimal selection of the principal surface and for the dimension reduction, uses the recursive partitioning algorithm proposed by Friedman (1991), in the optimal knot detection phase.
It can be synthetized as follows:

Preliminary phase:
The procedure starts from the computation of the linear principal components $\mathbf{c}_1, \ldots, \mathbf{c}_m$.

Forward phase:
Fixing the degree of spline and the number of knots, (*n*-2) possible transformations are computed in correspondence with all the possible unit values, excluded the minimal and the maximum ones (being, respectively, values on which the left and right external knot are computed). The algorithm, performed for each component, chooses the knot which minimizes equation (2). The first knot partitions the units into two sets. For the choice of the second internal knot the algorithm tries each observation internal to the two selected sets and detects the optimal one minimizing the criterion. Analogously, the subsequent knots are located until the fixed number is reached.

Backward phase:
It chooses the space dimension for the construction of the *PS*. This is a trade-off problem between the accuracy of the summarization and the risk of overfitting related to an increasing space number. The adopted criterion is the *Generalized Cross-Validation* (GCV, Craven and Wahba, 1979):

$$GCV(R) = \frac{1}{\mathrm{n}} \frac{\sum_{i=1}^{\mathrm{n}} \left[\hat{\mathbf{y}}_{ij} - f\,(\mathbf{c}_{ij})\right]^2}{1 - \hat{C}(R)} \tag{4}$$

(for R=p,...,1) where $\hat{C}(R)=\frac{1}{n}\sum_{i=1}^{n}\left[\hat{\mathbf{y}}_{ij}-f\left(\mathbf{c}_{ij}\right)\right]^{2}+dR+1$ represents a suitable increasing cost function of the space dimension. In the regression context, Friedman (1991) motivates the choice of the constant: $2 \leq d \leq 4$. Larger values for d will lead to fewer knots.

The final phase:
The optimal PS is computed on the univariate transformation of the single components choosen by the forward phase and on the base of the univariate spline number (space dimension) suggested by the GCV criterion. The tensorial product among the transformed components defines the multivariate spline.

4. Application

In this section we compare an example of classical *CPCA* and *PSCA*. The aim is to study how criminality depends on some population and economic indicators (data source: Sole24 ore, italian newspaper of 30-12-1996).

Figure 1: *The classic representation*

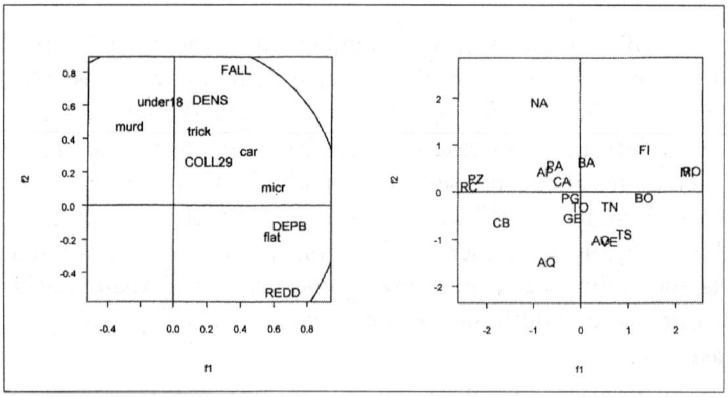

We investigate on the dependence structure among six predictors and six responses. The predictor variables are: income per capita **(REDD)**, bank deposit **(DEPB),** population density **(DENS)**, registered number of employment office, under-twentynine age **(COLL29)** and bankrupt enterprises number **(FALL)**. The response variables are: murderes number **(murd)**, car thefts number **(car)**, micro-criminality **(micr)**, flat thefts number **(flat)**, minority criminality **(under18)** and tricks number **(trick).** These variables are observed on 20 italian chief towns of province (Trieste **TS**, Bologna **BO**, Aosta **AO**, Trento **TN**, Roma **RM**, Campobasso **CB**, Bari **BA**, Ascoli-Piceno **AP**, Venezia **VE**,

Cagliari **CA**, Firenze **FI**, Genova **GE**, Reggio-Calabria **RC**, Aquila **AQ**, Milano **MI**, Napoli **NA**, Palermo **PA**, Perugia **PG**, Potenza **PZ**, Torino **TO**).
From the classical *CPCA* results, we see that the first two eigenvalues furnish the 84,3% of explained variance. In coherence the correlation circle shows (fig. 1) the positive correlation between the response variables **under18, murd,** and the positive direction of the second axis, while the right side of the first axis is strictly related to **micr** and **flat** variables. We also see that the predictor variables **TRICK, DENS** and **COLL29** have great importance for the explanation of **under18** and **murd,** while **DEPB** and **REDD** for **micr** and **flat** variables. The higher bank deposit, the higher microcriminality**.** As a matter of fact we observe that big towns like **MI, RO, BO** are principally characterized by **micr** and **flat** crimes and by **DEPB**. As well we notice the opposition between **NA** and **AQ** with respect to the second axis correlated to **under18** and **murd** variables.
With the *PSCA* analysis we would like to show that this interpretation does not allow to value the different locations of all towns and does not permit to understand clearly which is the particular variable that influence the towns proximities (similarities).
Differently we will point out the observations on *PS* representations.

Representation on Adaptive Principal Surface

PSCA allows a suitable representation of the dependence structure among variables. We construct surface graphs for each of the six responses.
After computing the principal components (tab. 1), to transform them we use zero degree B-splines with 3 different knots. The knots sequence per component (tab. 2) is defined on the basis of the criterion (2) and defines the grids. The number of components retained for the construction of surfaces (this solves the space dimension problem) is two as suggested by the *GCV* criterion. The tensor product of the transformed components computes the multivariate spline, i.e. the principal surface.

Table 1: *Principal Components*

	TS	BO	MI	AO	TN	RO	GE	AQ	FI	TO	VE	PG	AP	BA	CB	PA	NA	PZ	RC	CA
f1	,91	1,33	2,25	0,43	0,62	2,35	–,16	–,72	1,34	–,01	0,61	–,21	–,75	0,13	–1,7	–,55	–,87	–2,3	–2,4	–,39
f2	–,86	–,09	0,45	–0,9	–0,3	0,47	–0,5	–1,4	0,92	–,28	–1,0	–,09	0,46	0,66	–,69	0,59	1,93	0,31	0,15	0,26

Table 2*: Knots sequence*

	F1	F2
[1]	–2,4	–1,4
[2]	2,35	1,93
[3]	–0,72	–0,28

Regressing the projected variables $\hat{\mathbf{Y}}$ by the tensor matrix we compute the coefficients which are the grids levels.

For each response, we construct a *PS* where towns are located and get the following groupes:

1) **PZ, RC, NA, AP**
2) **PA, CA, PG, TO, TN, BO, MI, RO, FI, BA**
3) **CB**
4) **GE, AQ, AO, VE, TS**

Figure 2: under 18 *variable*

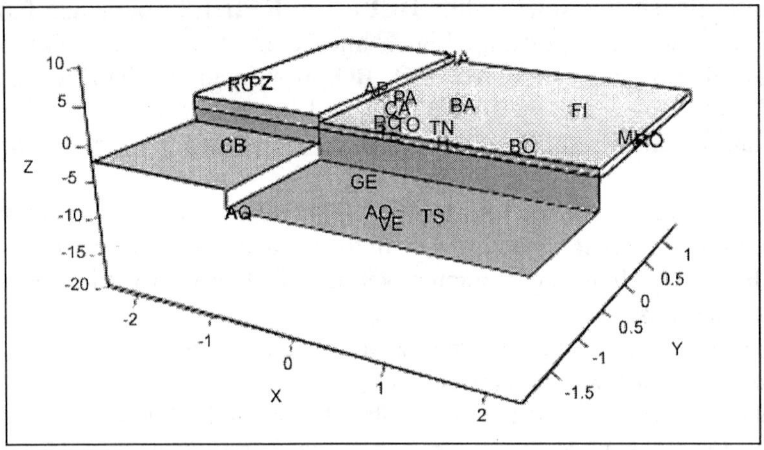

Figure 3: *Flat variable*

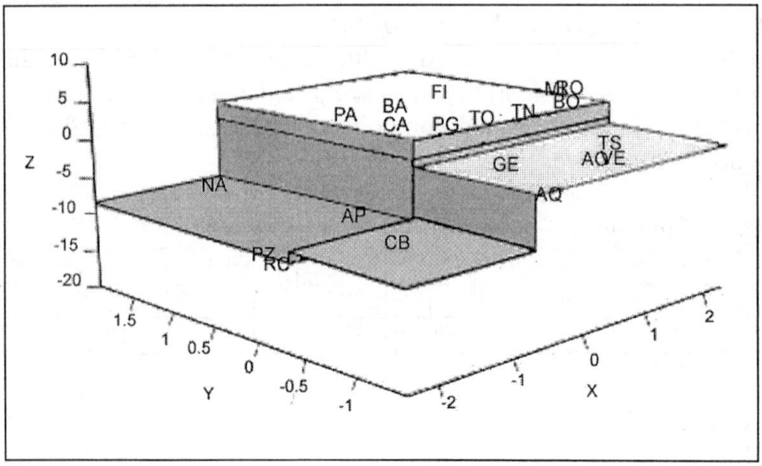

We notice the originality of group 3), which results looking at the *PS* of **under 18** (see fig.2)**,** where **CB** is clearly distinct from all the other chief provinces.

Furthermore for example the group 1) is detected mostly from the variable **flat** (see fig. 3). In this way, we can evaluate the negative or positive similarities of towns on the basis of their location at the grids.

5. Conclusions

In this paper we propose a non linear generalization of *CPCA*. The technique called *PSCA* has been implemented by an algorithm which appears to be an useful tool for data reduction and multivariate approximation function reconstruction. The example shows how principal surface representation allows to take into account the structural variables interactions; in particular the resulting "tri-dimensional" representation permits to investigate in depth the similarities among units due to a particular variable (classification variable).
An interesting extension of the procedure may concern a priori non linear transformation of both original variables and their components (Tessitore, Lombardo, van Rijckevorsel 1998) or the knots number optimal detection (Lombardo R., D'Ambra L., Tessitore G. 1997) or the robustness of the criterion.

References

Craven P. & Wahba G. (1979). Smoothing Noisy Data with Spline functions. Estimating the Correct Degree of Smoothing by the Method of Generalised Cross-Validation. *Numerische Mathematik,* 31: 377-403.

D'Ambra L. & Lauro N.C. (1982). Alcune estensioni dell'analisi in componenti principali in rapporto a sottospazi di riferimento. *Alcuni lavori di analisi statistica multivariata,* SIS Firenze. Ed. R. Leoni.

D'Ambra L. & Lauro N.C. (1992). Non Symmetrical exploratory data analysis. *Statistica Applicata,* 4,pag.511-529.

Gifi A. (1990). *Non Linear multivariate analysis*. Chichester: Wiley

Friedman J.H (1991). Multivariate Adaptive Regression Splines. *The Annals of Statistics,* 19,1:1-141.

Hastie T. & Stuezle W. (1989). Principal Curves. *J.A.S.A.*

LeBlanc M. & Tibshirani R. (1994). Adaptive Principal Surfaces. *The Journal of the American Statisical Association,* vol.89, No. 425.

Lombardo R. & Tessitore G. (1997) Adaptive Principal Surface in Constrained Principal Component Analysis. *Contributed papers IFCS classification and data analysis group,* Pescara 3 luglio.

Lombardo R., D'Ambra L., Tessitore G. (1997) An algorithm for detecting the number of knots in Constrained Principal Component Analysis. *Contributed papers,* vol. ISI 1997

Rao C.R. (1964). The use and interpretation of princi pal component analysis in applied research. *Sankhya, 26*.

Rijckevorsel J. L.A. van (1987). *The application of horseshoes and fuzzy coding in multiple correspondence analysis*. Leiden: DSWO Press.

Rijckevorsel J. L.A. van & de Leeuw J (1988) *Component and Correspondence Analysis*. Chicester: Wiley.

Rijckevorsel J. L.A. van & Tessitore G. (1993) An algorithm for Multivariate Adaptive Component and Correspondence Analysis (MACCA). *49th ISI session, Contributed Papers 2, pp.* 513-514

Tessitore G. , Lombardo R., Rijckevorsel J. L.A. van (1998) Non Linear Principal Surfaces Analysis. *VI conference IFCS, Short Papers, pp.*303-307.

Wollenberg A. van den (1977). Redundancy Analysis: an alternative for canonical correlation analysis. *Psychometrika,* 2.

Acknowledgments

The authors thank Prof. L. D'Ambra and the referees for the helpful comments. This paper was supported by CNR fund (1996), *responsible Prof. L. D'Ambra.*

Generalised Canonical Analysis on Symbolic Objects[(*)]

Rosanna Verde
Dip. di Matematica e Statistica, Università di Napoli "Federico II"
Monte S.Angelo, Via Cinthia, I-80126 Napoli, e-mail: verde@dms.unina.it

Abstract: In this paper we propose an extension of the Generalised Canonical Analysis to the study of *symbolic objects*. The aim is to analyse symbolic objects on a factorial plan. In the reduced sub-space, the symbolic objects are represented by polygons instead of points as in the classical data analysis. This kind of representation seems consistent with their original meaning of *complex information*. Furthermore we propose a symbolic interpretation of the factorial axes, and an evaluation of the quality of the images of the symbolic objects on the factorial plan.

Key words: Symbolic objects; Generalised Canonical Analysis; *fuzzy* coding.

1. Introduction

The main part of the real phenomena can be interpreted starting from the relations among the variables by which they are characterized. The symbolic analysis has been introduced by Diday (1989, 1991) in order to study complex data, which are often more representative of real phenomena than classical data (structured by individuals×variables).

In the last years, the development of symbolic data analysis has delivered many methods for the synthesis and the representation of complex information. The techniques proposed for the study of the symbolic objects are both symbolic and numerical. The symbolic techniques are typical of the Artificial Intelligence field and are based on a matching between symbolic objects of several orders. This kind of strategy allows to carry out generalization and specialization of *classes*, here represented by symbolic objects. On the other hand, the most part of numerical techniques are derived from classical statistical methods.

Several methods of the Data Analysis are suitably extended to the study of this particular kind of data.

Nevertheless, the application of numerical techniques to symbolic objects requires a numerical transformation of the variables that describe the objects in classical data matrices. The complex information expressed by the symbolic objects is recovered by a symbolic interpretation of the results.

(*) The paper has been supported by Esprit Project n. 20821 grant: *Symbolic Official Data Analysis System.* The author is grateful to prof. Carlo Lauro for helpful comments.

The generalization of factorial methods to symbolic objects is one of the research fields in the SODAS project; in this context, works by Gettler-Summa (1992); Chouakria *et al.* (1995, 1996); Lauro *et al.* (1997) should be mentioned. In the context of factorial approaches to symbolic objects, the present work aims at providing a particular visualization of symbolic objects on factorial plans and a symbolic interpretation of the factorial axes.

The main features of this work concern: symbolic objects definition; numerical transformation of the descriptors in *fuzzy* and binary coding matrices; extension of the Generalised Canonical Analysis to *fuzzy* coding data; graphical representation of the symbolic objects on a factorial plan; symbolic interpretation of the factorial axes; an application to real data.

2. *Symbolic object* definition

Let Ω be a set of individuals w called *elementary objects* and $Y=\{y_1,y_2,...,y_p\}$ be a set of numerical and nominal descriptors, with domains O_j's.

The most elementary symbolic object, called *event*, is denoted by $e_j=[y_j \in V_j]$, where $V_j \subseteq O_j$ is the set of values of y_j describing the symbolic object e_j. A conjunction of events is defined *symbolic assertion object*:

$$a = \wedge_j e_j = \wedge_{j=1,...,p}[y_j \in V_j]$$

A *logical function* defined on a set Ω is associated to a so that $a(w)$=true if $y_j(w) \in V_j$, $\forall j$. The *extension* of a, denoted $ext(a/\Omega)$, is the set of elements $w \in \Omega$ satisfying the condition $a(w)$=true.

Each assertion is described by multi-nominal, ordinal and *at intervals* continuous variables. Such variables are related among them.

Further information, given by probabilities, occurrences or beliefs can be associated to the categories of nominal variables. This kind of descriptors are called *modal* and the *mode* is the probability, the occurrence or the belief.

Moreover, the space of the symbolic objects descriptors can be reduced by logical constraints on the variables. Logical rules can induce the no-applicability (NA) of a variable, or of some categories, if another variable takes some values. Constraints can also expressed by taxonomic structures on the categories of nominal variables.

3. **Numerical coding of the symbolic object descriptors**

Factorial techniques have been proposed in order to analyse the relationships among the symbolic objects in reduced dimension sub-spaces.

The application of a numerical approach to symbolic objects requires their transformation in classical data and a successive reconstruction of the global information expressed by the objects, interpreting the results of the analysis in a symbolic way. The first phase consists of a suitable coding of the p variables, according to their nature, in classical matrices $\mathbf{Z}_j$ (j=1,...,p). In particular, the nominal variables are coded into binary matrices (0/1). More categories presented by an object are coded into more rows of the coding table $\mathbf{Z}_j$. If frequencies or probabilities are associated to the categories of a nominal descriptor, these values constitute the coding values of the descriptor.

We suggest a *fuzzy* coding of the quantitative descriptors (discrete and *at interval*) in order to retain the numerical information after the categorization of the numerical variables, and to take into account the variability of *at interval* variables. In this context, we propose piece-wise polynomial functions, the *Basic splines*, for a *fuzzy* coding of numerical variables. The most used coding functions, B-splines of degree 1, or semi-linear functions, identify three categories of each numerical character (e.g.: low-medium-high), and assign each value that can be taken by an object to these categories, with values in [0,1]. According to this kind of coding, the two bound-values of a continuous variable are coded into two different rows.

In order to take into account the original relationships among the variables, we propose to relate the rows of the coding matrices, corresponding to each symbolic object, by means of their Cartesian product.

In this way, some rows of the coding matrices $\mathbf{Z}_j$ (j=1,...,p) are duplicated and their increment depends on the number of symbolic objects, the number of continuous *at intervals* variables and the number of categories of nominal descriptors. Therefore, the total number of the rows of the coding matrices $\mathbf{Z}_j$ (j=1,...,p) is equal to $N=S\times 2^q \times \prod_{j=1}^{h} k_j$, where S is the number of symbolic objects considered in the analysis, q is the number of *at interval* variables, and k_j is the number of categories of the j-th multi-nominal variable (j=1,...,h). With $K=3\times q + \sum_{j=1}^{H} k_j$ we indicate the total number of the categories of the coded descriptors, which are given by the sum of the three fuzzy coding categories of the quantitative variables and the k_j categories of the multi-nominal variables as well as of the modal variables (j=1,...,H; with $H\geq h$).

From the juxtaposition of the binary and *fuzzy* coding tables: $[\mathbf{Z}_1|\ldots|\mathbf{Z}_j|\ldots|\mathbf{Z}_p]$ we obtain the global coding table $\mathbf{Z}$, of dimension (N×K).

Furthermore, in presence of dependence rules, the constrained symbolic objects can be reduced to a set of sub-objects, which are described by the values of the original variables consistent with the rules (Verde, 1997). Thus, the symbolic sub-objects can be interpreted as specialized symbolic objects.

They are treated in the analysis in the same way than others symbolic objects. In the coding phase, the category NA (*non-applicable*) is added up to the other coding categories of the variables, which are made unfeasible by logical conditions.

4. Generalised Canonical Analysis on symbolic objects

From a geometrical point of view, the rows of the coding table **Z**, relative to each symbolic object, correspond to the vertices of an hypercube in the space R^K of the categories of the coded variables.

A factorial approach based on the Generalised Canonical Analysis (GCA) is performed in order to visualize the relations among the hypercubes in a sub-space of reduced dimensions. In particular, we consider an extension of the classical GCA to the binary and *fuzzy* coded data matrices.

The proposed factorial analysis aims at decomposing the total inertia of the symbolic objects, i.e. the total inertia of the relative hypercube vertices, on factorial axes. Thus, as it is known, it searches for the axes of synthesis ν_j in each sub-space E_j, spanned by the column vectors of the coding matrices $\mathbf{Z}_j$ (j=1,...,p), and the orthogonal vectors ξ_α, as global synthesis of all ν_j. The criterion optimized is the average multiple correlation ratio: $\sum_{j=1}^{p}(\xi_\alpha|\nu_j)^2$, where ν_j and ξ_α are normalised to 1.

Let $\mathbf{P}_j=\mathbf{Z}_j(\mathbf{Z}_j'\mathbf{Z}_j)^{-1}\mathbf{Z}_j$ be the projection operator associated to the sub-space E_j and $\mathbf{P}=\Sigma_j\,\mathbf{P}_j$. The maximum inertia axes ξ_α are obtained maximizing:

$$\frac{1}{N\cdot p}\sum_{j=1}^{p}(\xi_\alpha|\nu_j)^2=\frac{1}{N\cdot p}\xi_\alpha'\mathbf{P}\xi_\alpha\,, \tag{1}$$

that is equivalent to solve the following characteristic equation, under the usual orthonormality constraints $\xi_\alpha'\xi_\alpha=1$ and $\xi_\alpha'\xi_{\alpha'}=0$ with $\alpha\neq\alpha'$:

$$\frac{1}{N\cdot p}\mathbf{Z}\Sigma^{-1}\mathbf{Z}'\xi_\alpha=\lambda_\alpha\xi_\alpha,\quad (\alpha=1,...,M\,;\,M=min[N\text{-}1,K\text{-}1]) \tag{2}$$

where: Σ^{-1} is a block diagonal matrix of elements $(\mathbf{Z}_j'\mathbf{Z}_j)^{-1}$ (j=1,...,p).

All vertices v_i (s=1,...,N) of all hypercubes are projected on the factorial plan, according to the classical formula of the coordinates of the individuals on the factorial axes:

$$\text{coor}_\alpha(v_s)=\mathbf{Z}\Sigma^{-1}\mathbf{Z}'\xi_\alpha\,, \tag{3}$$

5. Visualization of the symbolic objects on factorial plan

The analysis of symbolic objects by means of an extention of a classical factorial method requires an interpretation of the results in symbolic terms. Therefore, in order to display the objects consistently with their theoretical definition of *global and unitary* information, the projected vertices of each hypercube should be collected in an unique geometrical figure. Different kind of geometrical images can be proposed. The choice of the peculiar figure depends on two main considerations: an easy comparability among objects and the best recognition of the original symbolic objects by their deformed projections on the factorial plan. The first problem leads to visualize a symbolic object by means of the maximum covering rectangle of the projected hypercube vertices, according to the representation already proposed in other factorial approaches on these complex data (Chouakria *et al.*, 1995; 1996).

The length of the rectangles sides, representing the symbolic objects on factorial plans, is proportional to the variability of the descriptors that have highly contributed to the orientation of the factorial axes.

Nevertheless, this kind of representation is not always suitable to recognize the original symbolic objects. In fact, the vertices enveloping takes into account merely the vertices with minimum and maximum factorial coordinates, and it does not consider the spread of all its vertices on the plan. Therefore, the rectangles furnish over-sized representations of hypercubes with respect to the real surface recovered by the point-vertices. The first consequence deriving from this visualization way is to represent objects of different original dimensions by rectangles of the same dimensions.

An alternative, more suitable symbolic objects visualization seems to be a *convex enveloping* of the external vertices of each hypercube, which allows to consider the spread of the relative vertices.

The main advantage of this kind of representation is to allow a better recognition of different symbolic objects. Furthermore it furnishes an interpretation of the symbolic objects considering the variability of the vertices of the hypercubes along the directions of the factorial axes.

According to a symbolic objects visualization by rectangles, we propose to evaluate the quality of the projected hypercubes, on the factorial plan, as the average of the quality of the single segments linking the rectangle generator vertices. On the other hand, in the representations of the symbolic objects by convex envelops, the quality of the objects images can be evaluated by the quality of the internal diagonals of the hypercubes.

Moreover, in this context, an interesting interpretation of the factorial axes in symbolic terms can be given by symbolic assertions associated to the axes. The descriptors of the axes are the categories of the original p variables that have most contributed to the determination of the factorial axes (see e.g. in section 6).

6. A numerical example

An application of the proposed approach is performed on a set of data about the life quality in Italian cities (font: *Il sole 24 ore*, December 18, 1995). From these data, we have obtained 9 symbolic objects which are representative of the cities having different dimensions (L-"large", M-"medium", S-"small") in the

North, the Center, the South (including the Islands) of Italy (areas indicated by N,C, and S). The cities are characterized by 5 economic-demographics indicators, considered as *at interval* variables - I: income *per capita*; U: unemployment rate; M: micro-criminality; T: road traffic; C: cinema and theater expenditure *per capita.* The coding of each descriptor is realized by means of three B-splines of degree 1, corresponding to semi-linear functions. The coding functions identify the categories: low-medium-high of each indicator.

Each object is codified into two rows of the coding table associated to each descriptor. The vertices of the hypercubes associated to each object are equal to 2^5, which correspond to the number of rows of the coding table **Z** related to each object. The total number of **Z** rows is equal to $N=8\times2^5$ and the number of columns is $3\times5=15$, having been coded the descriptors with respect to 3 categories.

In the fig.1 the eight hypercubes are visualized as the maximum covering rectangles. The coordinates of the categories of the variables are represented too on the factorial plan. We observe that the first axis opposes the cities of the South area to the Northern ones, while the second axis opposes the large to the small dimensions cities. The dimensions of the rectangles, representing the symbolic objects on the factorial plan, are proportional to the variability of their descriptors.

Figure 1: *Rectangular images of the symbolic objects on the first factorial plan*

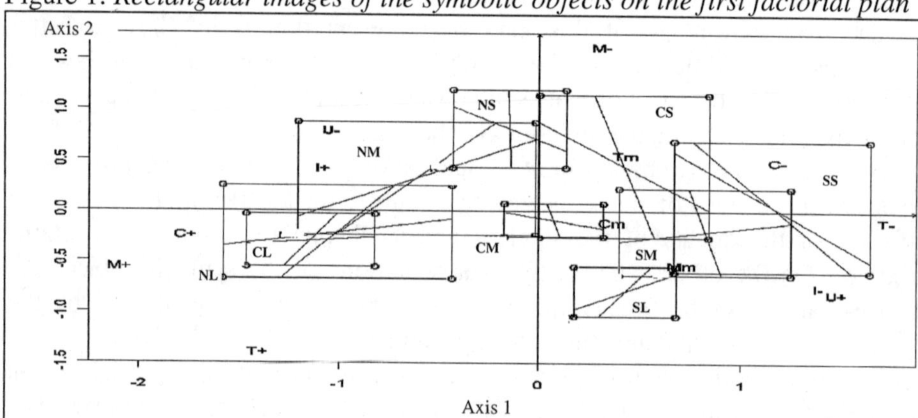

A suitable interpretation of the factorial plan (explained inertia=62%) is given by a description of the axes in terms of symbolic assertions, where the values of the descriptors are chosen on the basis of the contributions of the categories to the orientation to the factors. Therefore, the symbolic assertions associated to the axes 1 and 2 (*positive "+" and negative "-" side*) are the following:

a_{1+}=[income *per capita*={low}]∧[unemployment rate={high}]

a_{1-}=[micro-criminality={high}]∧[cinema/theater expenditure *p.c.*={high}]

a_{2+}=[micro-criminality={low}]∧[road traffic={medium}]

a_{2-}=[road traffic={high}]∧[micro-criminality={high}]

According to this interpretation, the small cities of the South Italy (*SS*) are strongly characterized by low income per capita and high unemployment rate (descriptors of a_{1+}), they are opposite to the large cities of the North (*NL*) where the expenditure *p.c* for cinema and theater is higher (descriptors of a_{1-}). The small cities of the North (*NS*) are well described by low micro-criminality and low road traffic (descriptors of a_{2+}), while the large cities both the North (*NL*) and the South (*SL*) are characterized by high road traffic and high micro-criminality (descriptors of a_{2-}).

It is worth noting that the rectangles representing the cities of the Center area (*CS*, *CM* and *CL*) are located more in proximity of the origin of the axes than the images of the others cities. This means that such cities assume some values of their descriptors near to the average of the values taken from the other objects. A particular case is given by the image of the large cities of the Center represented by a rectangle completely included in the image of the large cities of the North area (*NL*), presenting, at the same time, a little overlap with the image of medium cities of the North (*NM*) area. This means that the large cities of the Center (*CL*) area are characterized by the same values of the variables describing the large cities of the North Italy (*NL*), even if the last ones have an higher variability of the descriptors than the Center cities.

Figure 2: *Visualization of the symbolic objects by convex envelops*

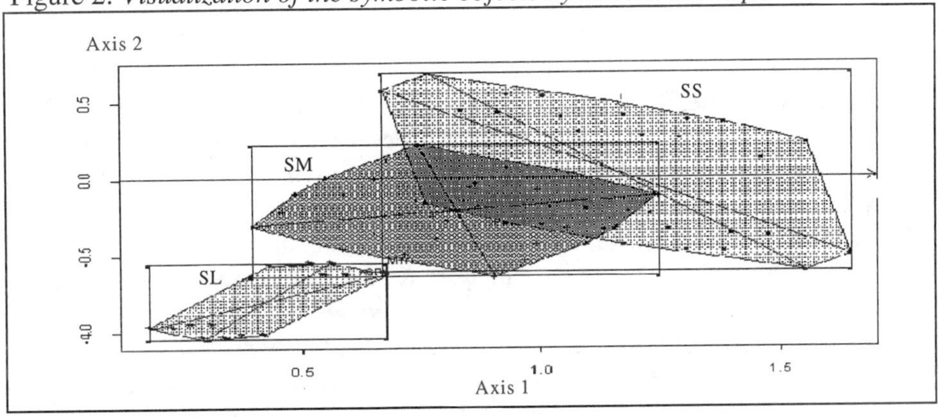

An alternative description of the symbolic objects is given by convex envelops. In the figure 2, it is shown a section of the factorial plan in which only the cities of the South area are represented. In particular, we observe that the images of the large and the medium cities are overlapped when rectangular representations are used while they are distinct using convex enveloping visualizations. The large overlapping between the medium and small cities images has similarly been reduced by visualizing the objects with convex envelops. Thus, this representation allows a better discrimination among objects than the rectangular one.

7. Conclusion

In conclusion, classical factorial approaches seem to be suitable techniques to visualize the relationships among the objects on reduced sub-spaces. They allow to consider the variability of the descriptors, proportionally to the dimensions of the symbolic objects images and they furnish an interpretation of the objects according to the symbolic meaning assumed by the factorial axes.

Furthermore, the visualization of the objects by convex envelops permits a better discrimination of different symbolic objects on the reduced sub-spaces. Alternative suitable ways to visualize symbolic objects, as ellipses or principal diagonals, can be proposed, in order to recognize the symbolic objects with respect to their original dimensions.

Another aspect, to be considered, concerns the global interpretation of the sub-objects in which the symbolic objects are decomposed by logical rules. In fact, on the factorial plan, these sub-objects are visualized independently of their original belonging to an unique object, either if a rectangular or convex enveloping visualization is used.

References

Chouakria, A., Diday, E., Cazes, P. (1995). Extention of Principal Component Analysis to interval data. In *New Techniques and Tecnologies for Statistics*.

Chouakria, A., Verde, R., Diday, E., Cazes, P. (1996) Généralisation de l'analyse factorielle des correspondances multiple à des objects symboliques. In Proc. *Quatriemes Journées de la Societé Francophone de Classification,* Vannes.

Diday, E. (1989). Knowledge representation and Symbolic Data Analysis. In Proc. *2nd Inter. Workshop on Data, Expert Knowledge, and Decision*. Hamburg.

Diday, E. (1991). Des objects de l'analyses de données à ceux de l'analyse de connaissances. In *Induction Symbolique et Numérique à partir de Données*, Eds: Diday and Kodratoff. Cépadues Editions, Toulouse, pp. 9-75.

Gettler-Summa, M. (1992). Factorial Axis Interpretation by Symbolic Objects. In *Journées Symbolique-Numerique*. Eds: Diday & Kodratoff, Pinson. Paris.

Kodratoff, Y. (1992). Symbolic or Numeric Induction?. In Proc. *3èmes Journées Symbolic Numérique: Analyse des Connaissances, Apprentissage, Raisonnement.* Lise-Ceremade, Paris IX *Dauphine*, Paris.

Nagabhushan, P., Gowda, C. K., Diday, E. (1995). Dimensionality reduction of symbolic data. *Pattern Recognition Letters,* **16**, 219-223.

Lauro, N.C., Palumbo, F., Verde, R. (1997). Factorial Discriminant Analysis Extended to Symbolic Object. In Proc. of the International Conf. *Ordinal and Symbolic Data Analysis*, Darmstadt (Germany).

Verde, R., De Angelis, P. (1997). Symbolic objects recognition on a factorial plan. In Proc. *NGUS'97-IV Inter. Meeting of Multidimensional Data Analysis*. Bilbao.

Analysis of Qualitative Variables in Structural Models with Unique Solutions

Giorgio Vittadini

Università degli Studi di Milano,
20126 – Viale Sarca 202
Milano, Italia, e-mail: vittadin@imiucca.csi.unimi.it

Abstract: A new method based on the Multidimensional Scaling and the Restricted Regression Component Decomposition is proposed in order to obtain solutions for structural models with mixed variables.

Keywords: Structural Models with Mixed Variables, Normality Hypothesis, Multidimensional Scaling, Alternating Least Squares, Restricted Regression Component Decomposition.

1. Structural Model with Qualitative Variables

The structural model for the study of causal relationship among latent variables is composed of one structural and two measurement equations:

$$H=HB+\Xi\Gamma+E=\Xi\Gamma(I-B)^{-1}+E(I-B)^{-1};\ Y=H\Lambda_y+U;\ X=\Xi\Lambda_x+U \tag{1}$$

where $Y'=(y'_{(1)},\ldots,y'_{(t)})(n_y,t)$, $X'=(x'_{(1)},\ldots,x'_{(t)})(n_x,t)$ are the observed mixed variables; $\Xi'=(\xi'_{(1)},\ldots,\xi'_{(t)})(n_\xi,t)$, $H'=(\eta'_{(1)},\ldots,\eta'_{(t)})(n_\eta,t)$ are the latent variables; $E'=(\varepsilon'_{(1)},\ldots,\varepsilon'_{(t)})\ (n_\varepsilon,t)$ are the errors in equations; $\Delta'=(\delta'_{(1)},\ldots,\delta'_{(t)})$ $(n_\delta t)$, $U'=(u'_{(1)},\ldots,u'_{(t)})\ (n_u,t)$, are the errors in variables. It is assumed that: all the random variables have zero mean and finite variance, B is a low matrix with zero on the main diagonal, (Y, X, H) are identically distributed and (Ξ, E, Δ, U) are identically and independently distributed. The model is usually proposed with restrictions on parameters and on covariances.
The solutions are reached starting from the variance covariance matrix of the reduced model where the variables H in the measurement models are substituted by the value of $(\Xi\Gamma(I-B)^{-1}+(I-B)^{-1}E)$ obtained from the structural equations.

2. Methods for Obtaining Solutions

First of all given that a continuous bivariate normal variable (J_l^*, J_m^*) with distribution $\Phi\left(j_l^*, j_m^*, \rho_{\left(j_l^*, j_m^*\right)}\right)$ underlies every pair of ordinal bivariate observed variables (J_l, J_m), the polychoric correlation between the two components of them is calculated. When there is one only ordinal variable, the polyserial correlation coefficient between an observed quantitative variable and the normal underlying variable is defined. The observed frequencies of the qualitative variables $n_{j_{l_q}, j_{m_r}}$ $(q = 1,\ldots n_l; r = 1,\ldots,n_r)$ given, the polychoric coefficients are reached in different ways. Jöreskog (1994) hypothesizes that the marginal probability of the normal variables are equal to the marginal frequencies of the two-way table $(\pi_{j_{l_q}\bullet} = n_{j_{l_q}\bullet})$ $(\pi_{j_{m_r}\bullet} = n_{j_{m_r}\bullet})$ of the ordinal variables. Therefore, first of all, he reaches the thresholds, using the marginal distributions of the normal bivariate distribution; for given thresholds using such distribution, he obtains the correlation coefficient maximizing the log likelihood of the sample respect to $\rho_{\left(j_l^*, j_m^*\right)}$. Lee et al. (1990) estimate the thresholds by means of Partition Maximum Likelihood (PML) which is simpler from the computational point of view.

Lee et al. (1995) estimate simultaneously the correlation coefficients and the thresholds concerning pairs of variables maximizing the log likelihood of the sample respect to $\rho_{\left(j_l^*, j_m^*\right)}$ and respect to every threshold $j_{l_q}^*, j_{m_r}^*$, by means of PML (but Jöreskog (1994) observes that "different estimates of thresholds for one variable may be obtained from different pairs of variables").

By means of Full Maximum Likelihood, in one case Lee et al. (1992) simultaneously reach all the thresholds of the polychoric correlations; in another case (Lee et al. (1990)), they reach also the parameters, the variances and the covariances of the latent variables. Moreover these two last methods take up too much computer time (Lee et al. (1990) Lee et al. (1995)).

In the first three methods, the parameters and the covariances of the latent variables and errors are obtained from the polychoric or polyserial correlations. There is not a unique understanding about the method of obtaining the parameters and the latent variables. Jöreskog (1990), (1994), Rigdon and Ferguson (1991) propose the Weighted Least Squares method and criticise the Maximum Likelihood method because the standard error parameter estimates are asymptotically incorrect, but Lee et al. (1995) continue to prefer the General Least Squares method and criticize the proposal of Jöreskog because it requires sample sizes larger than 200 and more computer time.

3. Some Critical Observations

Comparing, in some Montecarlo studies, different correlation coefficients when the underlying bivariate distribution is normal, the polychoric correlation is shown to be "the best in the sense of being closest to the true correlation" (Quiroga (1992)) [even if in a Montecarlo study Babakus et al. (1987) show that the polychoric coefficient provides the best estimates of model parameters and the worst fit statistics]. However, Muthen (1984), Aish and Jöreskog (1990) and Quiroga (1992) say also that "the assumption of underlying bivariate normality is too strong for most ordinary variables used in social sciences". There are not many studies on the use of polychoric coefficient with underlying not normal distribution. Lee and Lam (1988) study the robustness of polychoric coefficient only when the underlying distribution is elliptical (containing multivariate normal, platicentric and leptocentric distributions). Lee et al. (1995) even though obtaining quite satisfactory results about such robustness with moderate size random samples, say that "to draw a non definite conclusion, a longer simulation is needed". Moreover in literature only Quiroga (1992) tries, in a systematic way, to extend the distributive hypothesis of the continuous variables underlying the qualitative variables. First of all, she says that when you leave the normality assumptions, such variables are surely not consistent and slightly biased (Quiroga (1992)). Moreover, in a Montecarlo study she verifies that, when the underlying distribution is a skew-normal bivariate, the polychoric coefficient is the best choice only for sample size of 200-400 and large number of categories (5-9) and underestimates the true correlation coefficient. Then, by means of a measure of not normality, she verifies that when the underlying distributions are generated by the Fleishman-Vale-Maurelli polynomial transformation (with departure from normality due to skewness and kurtosis) the polychoric coefficient is robust, but overestimates the true correlation. Finally she proposes an extended polychoric coefficient with distribution given by a mixture of a normal and univariate skew-normal density function but she does not give empirical verifications. Therefore, until now, there are neither theoretical demonstrations nor empirical simulations which give satisfactory and generally valid reasons for using polychoric coefficient with underlying not normal distribution.

Moreover the solutions based on polychoric coefficients: are not sensible to different scale of qualitative variables because they generally deal only with ordinal variables, reach solutions from variance covariance-covariance matrix of the reduced model different from the solutions obtained from the observed variables of the original measurement models and have the same problems of non identification of parameters and indeterminacy of latent variables of structural models with quantitative variables (Vittadini 1989).

4. An Alternative Proposal

In the quantitative case the problems of not uniqueness of the solutions are resolved by using lincar combinations instead of causal latent variables of the observed variables (Wold (1982), Haagen and Vittadini (1991)). In this paper we propose to obtain the latent variables of the model as linear combinations of the observed mixed variables simultaneously quantified, by means of methods of multidimensional scaling using simultaneously optimal scaling and ordinary least squares method. In fact such methods resolve the problem of not normality of the variables because are distribution free (Young (1981)) and give unique solutions once chosen the method of multidimensional scaling. In order to avoid subjective choices about methodologies of multidimensional scaling (and therefore subjective solutions), we propose: to quantify the qualitative variable and to obtain the linear combinations of them by means of a unique objective function; to reach flexible solutions as regards to different kinds of linear combinations requested by the problems; to take into account the different scale of the qualitative variable. Therefore, among the variety of multidimensional scale methods we choose the family of Alsos method (Young (1981), and Keller and Wansbeek (1983)). These methods are based on alternating optimal scaling which quantifies qualitative variables and ordinary least squares which reach linear combinations of them in a iterative way. So they obtain solutions in a different way along the scale of the variables (ordinal-nominal, continuous-discrete), and the aim of analysis (e.g. Principal components, canonical correlation) giving answers to previous problems. Among the family of Alsos methods we avoid methods such as OSMOD (Saito and Otsu (1988)) or INDOMIX-CAMIX (Kiers (1991)) which obtain solutions in two stages. Instead we choose methods that simultaneously obtain quantifications of qualitative variables and their linear combinations (ADDALS (De Leeuw, Young, Takane (1976)) MORALS CORALS (Young, De Leeuw Takane, (1976)), PRINCALS (De Leeuw and Van Rijckevorsel (1980)) OVERALS (Van Der Burg and De Leeuw (1988)) respectively from the perspective of variance analysis, canonical correlation, principal components, multiple correspondence analysis. Moreover in order to obtain the latent variables from their real indicators as in Wold (1982), we apply the chosen Alsos methods to the subsets of mixed variables Y_β, X_δ characterized by submatrices $l_{y(\beta,\bullet)}, l_{x(\delta,\bullet)}$ with coefficients all different from zero. So we simultaneously obtain the quantification Y_β^*, X_δ^* of such mixed variables Y_β, X_δ and their linear transformations $\tilde{\eta}_\beta, \tilde{\xi}_\delta$ according to different aim of the analysis (e.g. canonical correlation, principal analysis etc.). Then in order to take into account the restrictions:

$$cov\left(\eta_\beta,\eta_\pi\right)=0;cov\left(\xi_\delta,\xi_\gamma\right)=0;cov\left(\delta_{\beta_k},\delta_{\beta_9}\right);cov\left(u_{\delta_\lambda},u_{\delta_\mu}\right)=0;$$
$$b_{(\beta,\mu)}=0;\gamma_{(\delta,\varphi)}=0;\ l_{y(\beta,\alpha)}=0;l_{x(\delta,v)}=0; \qquad (2)$$

we obtain by means of the Restricted Regression Component Decomposition (RRCD) of quantified variables Y^*, X^* (Haagen and Vittadini (1998)) by means of an iterative process:

$$\eta_\beta^+=Q_{*\eta_\mu\cup *y_\alpha}\eta_\beta^o\ (\eta_\beta^0=Q_{\tilde{\eta}_\pi}\tilde{\eta}_\beta\left(\beta\neq\pi\right);*\eta_\mu=Q_{H^o_{(\beta,\mu)}}\eta_\mu^o;*y_\alpha=Q^o_{H_{(\beta)}}y_\alpha\ )$$
$$\xi_\delta^+=Q_{*\eta_\varphi\cup *x_v}\xi_\delta^o\ (\ \xi_\delta^0=Q_{\tilde{\xi}_\gamma}\tilde{\xi}_\delta(\gamma\neq\delta);*\eta_\varphi=Q_{H^o_{(\varphi)}\cup\Xi^o_{(\delta)}}\eta_\varphi^o;*x_v=Q_{\Xi^o_{(\delta)}}x_v\ )$$
$$\tilde{\varepsilon}_j=Q_{\tilde{H}_{(j)}\cup\tilde{\Xi}}\tilde{\eta}_j((j)=1,...,j\text{-}1);\ \delta^0_{\beta_k}=Q_{\left(H^0\cup\Xi^0\cup y_{\beta_g}\right)}y_{\beta_k};u^0_{\delta_\lambda}=Q_{\left(X\cup\Xi^0\cup x_{\delta_\mu}\right)}x_{\delta_\lambda} \qquad (3)$$

where $H^\circ_{(\mu,\beta)}$ are the H° without $\eta^\circ_\mu,\ \eta^\circ_\beta,\ Q_{\tilde{\eta}_\pi}$ is the complement orthogonal to the orthogonal projector on the space generated by $\tilde{\eta}_\pi$ and the other symbols are defined in a similar way. So we have the following RRCD of $y_{\beta_k}, x_{\delta_\lambda}, \eta^o_\beta$:

$$Q_{\Xi^0\cup y^0_{\beta_\theta}/H^0}y_{\beta_k}=P_{H^0}y_{\beta_k}+Q_{H^0\cup\Xi^0\cup y_{\beta_\theta}}y_{\beta_k};Q_{Y\cup x_{\delta_\mu}/\Xi^0}x_{\delta_\lambda}=P_{\Xi^0}x_{\delta_\lambda}+Q_{Y\cup x_{\delta_\mu}\cup\Xi^0}x_{\delta_\lambda};$$
$$\eta^o_\beta=P_{H^o_{(\beta)}}\eta^o_\beta+P_{\Xi^0/H^o_{(\beta)}}\eta^o_\beta+Q_{H^o_{(\beta)}\cup\Xi^0}\eta^o_\beta \qquad (4)$$

5. Numerical Example

The following variables are observed on a sample of 150 families casually chosen from 4103 american families that have been codified in the Federal Reserve Board research regarding National Income and Wealth of 1983.
$\mathbf{Y_1}$ Job contract household (y_{11}), spouse (y_{12}); Occupation kind household (y_{13}), spouse (y_{14}); Occupation sector household (y_{15}), spouse (y_{16}). $\mathbf{Y_2}$ Total health (y_{21}); Incomc (y_{22}); Debt (y_{23}). $\mathbf{X_1}$ Age household (x_{11}), spouse (x_{12}); Sex household (x_{13}), spouse (x_{14}); Number of children (x_{15}); Race (x_{16}); Residence region (x_{17}); Civil Status (x_{18}); $\mathbf{X_2}$ Educational Level household (x_{21}), spouse (x_{22}); Full time job years household (x_{23}), spouse (x_{24}) ; Part time job years household (x_{25}), spouse (x_{26}); Latent variables: labour force (η_1); Health and income (η_2), Civil status (ξ_1), Instruction grade (ξ_2).
In order to verify the causal dependence of the latent variables H from the latent variables Ξ we use the alternative proposal shown in paragraph 4 with the following restrictions: $l_{y(Y_1,\eta_2)}=0, l_{y(Y_2,\eta_1)}=0, l_{x(X_1,\xi_2)}=0, l_{x(X_2,\xi_1)}=0$, $\text{cov}(\Delta_1,\Delta_2)=0$, $\text{cov}(U_1,U_2)=0$. The qualitative variables are quantified and the latent variables

are obtained as principal components with Princals method, the restrictions are then taken into account by means of RRCD. The variance-covariance of the observed variables and the results are shown in table 1.

Table 1: *The alternative proposal for the sample of American families.*

$$S_Y = \begin{bmatrix} 0.0293 & -0.0284 & -0.0085 & 0.0248 & -0.0049 & 0.0580 & 10.853 & -0.0165 & -0.0785 \\ -0.0284 & 8.2791 & -3.2858 & 0.0917 & 0.8078 & -0.3585 & -79.107 & 12.671 & 0.5563 \\ -0.0085 & -3.2858 & 17.0984 & -0.0313 & -0.6348 & 4.1994 & -325.60 & -17.426 & -1.1561 \\ 0.0248 & 0.0917 & -0.0313 & 0.1490 & 0.0107 & -0.0858 & 80.249 & 8.5699 & -0.0821 \\ -0.0049 & 0.8078 & -0.6348 & 0.0107 & 2.5869 & -0.7803 & -162.35 & 16.439 & 0.4694 \\ 0.0580 & -0.3585 & 4.1994 & -0.0858 & -0.7803 & 13.785 & -65.588 & -57.510 & 0.1070 \\ 10.853 & -79.107 & -325.606 & 80.249 & -0.0162 & -65.588 & 320981 & 40682 & -397.51 \\ -0.0165 & 12.671 & -17.426 & 8.5699 & 16.439 & -57.510 & 40682.1 & 17114 & -21.178 \\ -0.0785 & 0.5563 & -1.1561 & -0.0821 & 0.4694 & 0.1070 & -397.51 & -21.178 & 7.4188 \end{bmatrix}$$

$$S_X = \begin{bmatrix} 1003.4 & -16.291 & 5.9262 & 12.573 & -16.997 & -3.0333 & 37.238 & -11.617 & 9.0494 & -3.1692 & 1.8904 & -2.0307 & 34.181 & -69.770 \\ -16.291 & 109.04 & -5.5373 & 8.0773 & 0.7844 & -1.1737 & 7.0739 & 0.8466 & -2.5367 & -2.8791 & 107.29 & -2.1834 & 45.104 & 14.060 \\ 0.5926 & -0.5537 & 9.8888 & -0.1390 & -0.0330 & -1.1111 & 1.2672 & -0.0332 & -9.2626 & -0.1628 & -0.6510 & 2.4767 & 0.0669 & -0.0971 \\ 12.573 & 8.0773 & -1.3909 & 3.2374 & -0.1312 & -7.0707 & -4.0200 & -0.2396 & -3.5646 & -0.3074 & 7.7005 & -9.1676 & 3.8415 & 0.3945 \\ -16.997 & 0.7844 & -3.3030 & -0.1312 & 1.1162 & -5.7575 & -2.3432 & 0.3512 & 2.0414 & 0.0965 & -0.2843 & 6.1858 & -0.7796 & 1.2531 \\ -0.3033 & -0.1173 & -1.1111 & -0.0007 & -0.0057 & 1.0000 & 0.3016 & -0.0039 & 1.5454 & 0.0109 & -0.1712 & -1.6969 & -0.0370 & -0.0547 \\ 37.238 & 7.0739 & 1.2672 & -4.0200 & -2.3432 & 3.0161 & 197.33 & -1.3076 & -2.5917 & -8.4634 & 10.037 & 4.1361 & 14.046 & 0.2234 \\ -11.617 & 0.8466 & -3.3232 & -0.2396 & 0.3512 & -3.9393 & -1.3076 & 0.2403 & -3.3333 & 0.0719 & 0.4641 & 1.9676 & -0.2841 & 0.8446 \\ 0.9049 & -2.5367 & -9.2626 & -0.3564 & 0.2041 & 1.5454 & -2.5917 & -0.0033 & 4.0091 & 2.4420 & -5.7035 & 1.2327 & -0.6716 & -0.3408 \\ -3.1692 & -2.8791 & -1.6282 & 0.3074 & 0.0965 & 1.0909 & -8.4634 & 0.0719 & 2.4420 & 4.2763 & -4.3171 & 2.7717 & 0.4076 & 0.2460 \\ 1.8904 & 107.29 & -6.5101 & 7.7005 & -0.2843 & -1.7121 & 10.037 & 0.4641 & -5.7035 & -4.3171 & 127.84 & -1.2349 & 51.225 & 13.152 \\ -2.0307 & -2.1834 & 2.4767 & -0.9167 & 0.6185 & -1.6969 & 4.1361 & 0.1967 & 1.2327 & 0.2771 & -12.849 & 1.8704 & -2.8137 & 2.2771 \\ 34.181 & 45.014 & 6.6969 & 3.8415 & -0.7796 & -3.7070 & 14.046 & -0.2841 & -6.7161 & 0.4076 & 51.225 & -2.8137 & 84.526 & -14.738 \\ -69.770 & 14.060 & -9.7171 & 0.3945 & 1.2531 & -5.4747 & 0.2234 & 0.8446 & -3.4080 & 0.246 & 13.152 & 2.2771 & -14.738 & 47.498 \end{bmatrix}$$

$$B = \begin{bmatrix} 0 & -0.0962 \\ -0.0957 & 0 \end{bmatrix} \qquad l_{y(Y_1,\eta_1)} = \begin{bmatrix} 0.0089 & -1.9813 & 2.2994 & -0.0765 & -0.5419 & 1.7726 \end{bmatrix} \qquad l_{y(Y_2,\eta_2)} = \begin{bmatrix} 340.5859 & 98.4431 & -1.3412 \end{bmatrix}$$

$$\Gamma = \begin{bmatrix} 0.0171 & 0 \\ 0 & 0.0042 \end{bmatrix}$$

$$l_{x(X_1,\xi_1)} = \begin{bmatrix} 22.6810 & -1.2115 & 0.0568 & 0.5079 & -0.7008 & 0.0106 & 3.7964 & -0.4775 \end{bmatrix} \qquad l_{x(X_2,\xi_2)} = \begin{bmatrix} 1.7588 & 1.6581 & -3.5028 & 2.9609 & -3.9295 & 3.3142 \end{bmatrix}$$

$$S_H = \begin{bmatrix} 0.7220 & -0.0767 \\ -0.0767 & 0.4585 \end{bmatrix} \qquad S_{\Delta_1} = \begin{bmatrix} 0.0275 & -0.0053 & -0.0707 & 0.0207 & 0.0137 & 0.0483 \\ -0.0053 & 4.5237 & 1.1799 & -0.0355 & -0.2292 & 2.9146 \\ -0.0707 & 1.1799 & 11.0363 & 0.0180 & 0.8952 & 0.4736 \\ 0.0207 & -0.0355 & 0.0180 & 0.1306 & 0.0123 & 0.0542 \\ 0.0137 & -0.2292 & 0.8952 & 0.0123 & 2.1494 & 0.1236 \\ 0.0483 & 2.9146 & 0.4736 & 0.0542 & 0.1236 & 10.7678 \end{bmatrix} \qquad S_{\Delta_2} = \begin{bmatrix} 143417.7 & 6456.7 & 118.6 \\ 6456.7 & 7047.8 & 81.98 \\ 118.6 & 81.98 & 3.442 \end{bmatrix}$$

$$S_\Xi = \begin{bmatrix} 0.9393 & -0.0899 \\ -0.0899 & 0.9344 \end{bmatrix}$$

$$S_E = \begin{bmatrix} 0.9305 & 0.0872 \\ 0.0872 & 0.9258 \end{bmatrix}$$

$$S_{U_1} = \begin{bmatrix} 450.7216 & -2.2195 & -0.4068 & 0.9215 & -2.2092 & -0.4451 & -14.7014 & -1.4334 \\ -2.2195 & 95.1067 & -0.4428 & 7.4965 & -0.2513 & -0.0884 & 9.7258 & 0.3422 \\ -0.4068 & -0.4428 & 0.0841 & -0.1741 & 0.0188 & -0.0036 & 0.7438 & -0.0034 \\ 0.9215 & 7.4965 & -0.1741 & 2.7618 & 0.1694 & -0.0132 & -5.5328 & -0.0032 \\ -2.2092 & -0.2513 & 0.0188 & 0.1694 & 0.5023 & -0.0043 & -0.4254 & 0.0202 \\ -0.4451 & -0.0884 & -0.0036 & -0.0132 & -0.0043 & 0.0081 & 0.0983 & 0.0007 \\ -14.7014 & 9.7258 & 0.7438 & -5.5328 & -0.4254 & 0.0983 & 152.6526 & 0.0117 \\ -1.4334 & 0.3422 & -0.0034 & -0.0032 & 0.0202 & 0.0007 & 0.0117 & 0.0289 \end{bmatrix} \qquad S_{U_2} = \begin{bmatrix} 1.0689 & 0.1470 & 0.8455 & -1.0963 & 1.9265 & -0.9186 \\ 0.1470 & 1.2709 & 1.3589 & -1.5341 & 3.1266 & -0.5184 \\ 0.8455 & 1.3589 & 10.5682 & -2.8286 & 1.1425 & 0.8277 \\ -1.0963 & -1.5341 & 2.8286 & 10.3653 & 2.9974 & -1.6261 \\ 1.9265 & 3.1266 & 1.1425 & 2.9974 & 46.6684 & -10.1886 \\ -0.9186 & -0.5184 & 0.8277 & -1.6261 & -10.1886 & 23.8754 \end{bmatrix}$$

With the example we can verify that the alternative proposal respects all the properties of the structural model described in paragraph 1 and the restrictions indicated in this paragraph. But the alternative proposal obtains unique solutions solving all the problems of non-identification of parameters and indeterminacy of latent variables of the structural models with qualitative variables.

References

Aish, A. M., Jöreskog, K. G. (1990). A panel model for political efficacy and responsiveness: an application of LISREL7 with weighted least squares, *Quality and Quantity,* 24, 405-426.

Babakus, E., Ferguson, C.E. Jr, Jöreskog K.G. (1987). The sensitivity of confirmatory maximum likelihood factor analysis to violations of measurement scale and distributional assumptions, *Journal of Marketing Research*, 24, 222-228.

Dc Leeuw, J., Van Rijckevorsel, J.(1980). *Homals & princals, some generalizations of principal components analysis*, In E. Diday et al. (Eds.), Data Analysis and Informatics, North-Holland Publishing Company, 231-241.

De Leeuw, J., Young, F.W., Takane Y. (1976). Additive structure in qualitative data: an alternating least squares method with optimal scaling features, *Psychometrika,* 41, 471-503.

Haagen, K., Vittadini, G. (1991). Regression Component Decomposition in Structural Analysis, *Communications in Statistics,* 20, 1153-1161.

Haagen, K., Vittadini, G. (1998). Regression Component Decomposition Restricted. Un'alternativa al Lisrel model, *Metron*, 56, 1-2, in corso di pubblicazione.

Jöreskog, K.G. (1990). New developments in Lisrel: analysis of ordinal variables using polychoric correlations and weighted least squares, *Quality and Quantity*, 24, 387-404.

Jöreskog, K.G. (1994). On the estimation of polychoric correlations and their asymptotic covariance matrix, *Psychometrika,* 59, 3, 381-389.

Kiers, H. A. L. (1991). Simple structure in component analysis techniques for mixtures of qualitative and quantitative variables, *Psychometrika 56, 2, 197-212.*

Keller, W. J., Wansbeek, T. (1983). Multivariate methods for quantitative and qualitative data, *Journal of Econometrics*, 22, 91-111.

Lee, S.Y, Lam, M. L.(1988). Estimation of polychoric correlation with elliptical latent variables. *Journal of statistic Computation and Simulation*, 30, 173-188.

Lee, S.Y., Poon, W.Y., Bentler, P.M. (1990). Full maximum likelihood analysis of structural equation models with polytomous variables, *Statistics and Probability Letters,* 9, 91-97.

Lee, S.Y., Poon, W.Y., Bentler, P.M. (1995). A two-stage estimation of structural equation models with continuous and polytomous variables, *British Journal of Mathematical and Statistical Psychology*, 48, 339-358.

Muthen, B. (1984). A general structural equation model with dichotomous, ordered categorical, and continuous latent variable indicators, *Psychometrika,* 49, 1, 115-132.

Quiroga, A.M. (1992). *Studies of the Polychoric Correlation and other Correlation Measures for Ordinal Variables*, PhD thesis, Uppsala University.

Rigdon, E.E., Ferguson, C.E. Jr. (1991) The performance of the polychoric correlation coefficient and selected fitting functions in confirmatory factor analysis with ordinal data, *Journal of Marketing Research*, 28, 491-497.

Saito, T., Otsu, T. (1988). A method of optimal scaling for multivariate ordinal data and its extensions, *Psychometrika*, 53, 1, 5-25.

Van Der Burg, E., De Leeuw, J.(1988). Homogeneity analysis with k sets of variables: an alternating least squares method with optimal scaling features, *Psychometrika*, 53,2, 177-197.

Vittadini, G. (1989). *Indeterminacy Problems in the LISREL Model,* in Multivariate Behavioral Research, Forth Worth (Texas), 24, 4, 397-414.

Wold, H. (1982). *Soft Modelling: the basic design and some extensions,* in Jöreskog K.G., Wold H., *Systems under indirect observation: casuality, structure, prediction,* North - Holland, Amsterdam, 2, 1-54.

Young, F.W. (1981). Quantitative analysis of qualitative data, *Psychometrika,* 46, 357-388.

Young, F.W., De Leeuw, J., Takane, Y. (1976). Regression with qualitative and quantitative variables: an alternating least squares method with optimal scaling features, *Psychometrika,* 41, 4, 505-529.

Exploring Multivariate Spatial Data: Line Transect Data

Alessandra Capobianchi - Giovanna Jona-Lasinio
Dipartimento di Statistica, Probabilità e Statistiche Applicate – University of Rome «La Sapienza»
e-mail: jona@pow2.sta.uniroma1.it

Abstract: In this paper we describe an exploratory technique based on the diagonalization of cross-variogram matrices. Our aim is to describe the behavior of a multivariate set of spatial data in a dimensionally reduced space in such a way that the information on the spatial variation is preserved. Furthermore we propose a definition for the range of «variograms» in the multivariate case. Simulation studies and an application to botanical data collected on line transects are reported.

Keywords: Cross-variogram, Principal component analysis, singular value decomposition, spatial data.

1. Introduction

In the analysis of multivariate spatial data several difficulties arise. Because of the nature itself of the phenomena under study, we cannot assume independence between observations; instead we often want to model the specific type of dependence we observe. We usually model these data-sets as generated from a multivariate spatial random field (MSRF) under given assumptions on the type of relationship existing between locations and variables in each location. However before any serious attempt to model building can be made, we have to explore as deeply as possible the available data-set. This is very difficult because of the high dimensionality of the problem. Standard multivariate exploratory techniques like principal components analysis (PCA) and multidimensional scaling, usually fail to give us a good representation of this kind of multivariate data-set as they do not allow the information on the spatial arrangement of observations to be included.
The literature on the treatment of multivariate spatial data mostly deals with geo-statistical techniques (Borgman and Frahme, 1976, Journel and Huijbregts, 1978, Wackernagel, 1995, Christakos 1992). In Davis and Greenes (1983) PCA is applied to the sample correlation matrix in order to produce co-kriging results by computing a series of univariate variogram on a ``weight" matrix, whose rows are uncorrelated, their purpose being to provide an orthogonal basis for the variables, so that orthogonality holds spatially as well as in each considered location. In Di Bella and Jona Lasinio (1996 and references therein) some ordination methods in which the spatial information is included are described,

most of them oriented to pattern recognition in botanical studies, where great attention is given to reconsider *a posteriori* the dimension of sample units through the detection of the real scale of the data.

In Tailiang Xie and Myers (1995a, 1995b) a very interesting proposal is developed. The authors illustrate a technique to determine an orthonormal matrix which simultaneously diagonalizes, at selected lags, the cross-variogram matrices (or nearly diagonalizes it in the sense that the sum of squares of off-diagonal elements is small compared to the sum of squares of diagonal elements). In this paper, following the same line of thoughts, we describe an exploratory technique, based on the spectral decomposition of cross-variogram matrices, that allows us to represent our multivariate set of data in a more parsimonious manner keeping all the available spatial information along the analysis. As a tool for exploring the global behavior of the spatial variability, we propose a multivariate extension of the idea of the variogram *range*.

The choice of cross-variogram matrices is motivated by the fact that cross-variograms are symmetrical and non-negative definite matrices that can always be defined even when the usual assumptions of second order stationarity cannot be made.

2. Principal Components Analysis of MSRF

Formally our multivariate data-set is thought as generated by a k dimensional (second order stationary or intrinsic stationary of some given order (Matheron, 1973)) multivariate isotropic spatial random field (MSRF) $\mathbf{X}(s)=(X_1(s),\ldots, X_k(s))^T$, for all $s\in D$, with $s\in D$ where D is a finite partition of, say, n sites (or locations) of a given region. We assume that for each value of $s\in D$ observations of the whole MSRF are available. We denote by $\mathbf{C}(0)$ the kxk covariance matrix between the process components evaluated in each site and by $\mathbf{C}(h)$ the cross-covariance matrix of the field evaluated between sites $s_i,s_j\in D$ such that $d(s_i,s_j)=h$ is the value of the distance between them. Under second order stationarity assumption (we'll deal with the intrinsic case later on) we can define the set of *cross-variogram* matrices (see for instance Cressie, 1991 and references therein) as:

$$2\Gamma(h)=\mathrm{Var}(\mathbf{X}(s_i)-\mathbf{X}(s_j)), \qquad (1)$$

For all pairs $s_i,s_j\in D$, $h=d(s_i,s_j)$ with $h=h_1,\ldots,h_m$.

In $\Gamma(h)$, information on the spatial variation of the MSRF at distance h is represented. As it is well known through PCA we are able to reduce a set of correlated random variables into an uncorrelated set by an orthogonal transformation. The aim of the transformation is to reconstruct a k dimensional random variable $\mathbf{U}$ by $p<k$ linear functions without much loss of information; then we seek the best linear predictor based on that p functions. The efficiency of prediction may be measured by the residual variance and an over all measure of the predictive efficiency is the sum of such variances. If we want to include the information available on the spatial behavior of the RF in these p linear

functions, we have to build them starting from a measure of the spatial variation itself, *i.e.* the set of matrices given in (1). We can proceed as in Tailiang Xie and Myers (1995a, 1995b) by finding a matrix that *nearly* diagonalize the m cross-variogram matrices. Then, by linearly transforming the original data into *almost* spatially un-correlated vectors, we can treat each variable separately. However as this procedure involves some complex and computationally expensive calculations, we propose to proceed in a rougher, but informative way.

Being our main interest the study of the joint variation in space of the SRF for all value of the distance between sites, we have to synthesize the information contained in the m cross-variogram matrices. The most natural choice in this direction is to build a *synthesis* matrix and apply it PCA. We build the following synthesis matrix:

$$\Gamma=\sum_{h=1}^{h_m}\Gamma(h) \tag{2}$$

Other choices of *synthesis* matrix are possible. For example $\Gamma=\sum_{h=1}^{h_m}\Gamma(h)/m$ (which is equivalent to (2)) or, using some previous knowledge about the considered phenomena, we can build a weight system $\{w(h)\colon \Sigma_h\, w(h)=1,\ h=h_1,\ldots,h_m\}$ and compute $\Gamma=\Sigma_h\, w(h)\Gamma(h)$. Our method remains valid for all choices of Γ.

Through the spectral decomposition of Γ we find the orthogonal transformation $\mathbf{B}$ that minimizes

$$\text{Trace}\,(\Gamma\text{-}\Gamma\mathbf{B}(\mathbf{B}^T\Gamma\mathbf{B})^{-1}\mathbf{B}^T\Gamma)=\sum_{h=1}^{h_m}\text{trace}\,(\Gamma(h)\text{-}\,\Gamma(h)\mathbf{B}(\mathbf{B}^T\Gamma\mathbf{B})^{-1}\mathbf{B}^T\Gamma(h))\,-$$

$$-2\sum_{h\neq j}^{h_m}\text{trace}\,(\Gamma(h)\text{-}\,\Gamma(h)\mathbf{B}(\mathbf{B}^T\Gamma\mathbf{B})^{-1}\mathbf{B}^T\Gamma(j)) \tag{3}$$

As Γ can be seen as a global measure of spatial variation, (3) can be seen as an over all spatial residual prediction variance.

Further considerations have to be made. In the univariate setting it is of great interest to evaluate the range (h^*) of the variogram, as we can consider observations taken at locations, say, s_i, s_j, such that $d(s_i, s_j)>h^*$ almost uncorrelated. Using Γ we can identify a multivariate equivalent of the variogram's range . More precisely we write (2) in terms of its eigenvalues and eigenvectors, *i.e.* $\Gamma=\mathbf{Q}\Lambda\mathbf{Q}^T$ where $\mathbf{Q}$ is the k x k matrix of its normalized eigenvectors and Λ is the diagonal matrix of its eigenvalues. Then we can decompose each eigenvalue of Γ in terms of *contributions* given to its value by each chosen lag h:

$$\lambda_i=\mathbf{q}_i^T\Gamma\,\mathbf{q}_i=\mathbf{q}_i^T\sum_{h=1}^{hm}\Gamma(h)\,\mathbf{q}_i=\sum_{h=1}^{hm}\mathbf{q}_i^T(\mathbf{C}(0)-\mathbf{C}(h))\mathbf{q}_i \quad (4)$$

then we can define:

$$h_i=\arg\max_h \mathbf{q}_i^T(\mathbf{C}(0)-\mathbf{C}(h))\mathbf{q}_i \quad (5)$$

The first eigenvalue takes into account the largest amount of the variability of the SRF, h_1* may be seen as a global range for the cross-variogram of the SRF, *i.e.* the distance such that our observations are *maximally* uncorrelated. In fact under second order stationarity assumptions we have $\Gamma(h)=\mathbf{C}(0)-\mathbf{C}(h)$ then

$$h_1=\arg\max_h \mathbf{q}_1^T(\mathbf{C}(0)-\mathbf{C}(h))\mathbf{q}_1\equiv\arg\min_h \mathbf{q}_1^T\mathbf{C}(h)\mathbf{q}_1$$

In other words we define the range of the cross-variogram to be the range of the univariate variogram of the first principal component obtained from the singular value decomposition of Γ. In general we can define as many ranges as many components we compute. More precisely, let $\chi_i(s_j)=\mathbf{q}_i^T\mathbf{X}(s_j)$ be the i-th principal component, its variogram is given by:

$$\gamma_\chi(h)=E(\chi_i(s)-\chi_i(s+h))^2=E(\mathbf{q}_i^T\mathbf{X}(s)-\mathbf{q}_i^T\mathbf{X}(s+h))^2=\mathbf{q}_i^T\Gamma(h)\mathbf{q}_i \quad (6)$$

Notice that the cross-variogram of the χ's are given by $\gamma_{\chi\chi}(h)=\mathbf{q}_i^T\Gamma(h)\mathbf{q}_j$ and $\Sigma_h\gamma_{\chi\chi}(h)=\lambda_i\delta_{ij}$ being principal components orthogonal to each other.
In what follows we usually choose h_1* as global range, as the first eigenvalue accounts for the largest part of the whole spatial variability. However other choices are possible, we can compute $p<k$ components and take the largest value of h_i* (i=1,...,p) to be the global range. But this choice may be misleading, as the corresponding h* may be due to, say, some secondary spatial pattern in the field. Let us clarify this last remark with an example. We consider a data-set of botanical data where, along a transect of 30 sites, the coverage of 5 species of plants are observed. The data are organized in a 5x30 matrix and the environment suggests that a second order stationarity assumption is reasonable. From previous study, we already know that the association of three of the five species induces a primary pattern in the vegetation which influences blocks of 3 sites at a time (one site is a square of $1m^2$), a secondary pattern is induced by the antagonistic behavior of two species (the presence of one species nearly excludes the presence of the other one) and a third pattern is due to the spread presence of a third rare species and it influences blocks of 8 sites. Applying the proposed method and considering the first 3 principal components (the eigenvalues accounts respectively for the 54%, 21% and 17%

of the total variability), we have h_1*=3m, h_2*=1m and h_3*=8m. If we choose h*=8m we would be driven to think that the whole spatial arrangement of the observed set of plants is given by a rare species.

Once we have the singular value decomposition of Γ we can, as usual, study the MSRF using only few principal components and this allows us to explore the behavior of the SRF using the same tools developed in the univariate case. For example we can use the principal components χ_i to perform ordinary kriging.

2.1 Intrinsic stationarity and large values of *h**

The global range we just defined can be used in several ways, a possible one is to use it to detect intrinsic stationarity. More precisely, analogously to the univariate case we expect that if our MSRF is intrinsic of order zero, the first principal component given by the previous analysis has a variogram with non finite range. However some caution is necessary. Consider a MSRF observed over n locations arranged according to a regular structure, for instance a line transect. By applying our technique we find a value of h_1* close to h_m, for instance in the botanical example h_m=29. In this case two conclusions can be driven: the SRF is second order stationary and the *global* spatial autocorrelation decays for *large* values of the distance between sites, or the SRF is intrinsic stationary and the *global range* is non-finite. A way to choose between these two conclusions is to observe more locations along the transect. In the next section we'll deal again with this problem through several simulated examples.

3. Simulations and Application

We conducted extensive simulation studies of this method (Capobianchi, Jona Lasinio 1997 unpublished manuscript, Capobianchi 1998) for both line transects and grid samples. Here the most representative simulations on line transects are described.

We based our simulations on the following model (Capobianchi, 1998): let first consider $\{\mathbf{X}(s), s\in D\}$a vector valued second order stationary (isotropic) SRF with, say k components and $|D|=n$. It is easy to show that we can write each component of the SRF as a linear combination of totally uncorrelated random variables (r.v.) $Z^{(i)}(s)$, $s\in D$:

$$X^{(i)}(s_j)=\sum_{m=1}^{k}\sum_{r=1}^{n}\sqrt{l_m}\,\sqrt{\xi_r}^{(m)}\,\tau_{im}\,\eta_{rj}^{(m)}\,Z^{(i)}(s_j) \tag{7}$$

if and only if its cross-covariance can be written as

$$C_{ip}(h)=\sum_{m=1}^{k} l_m\,\tau_{im}\,v^{(m)}(h) \tag{8}$$

where the coefficients of the linear combinations (7) and (8) are obtained by the following considerations. We assume that there are two main sources of variation, one is due only to the interaction between the components of the SRF, the other is due to the spatial arrangement of the RF. The two are combined as follows.

- To each component of the multivariate SRF (MSRF) a spatial structure is given as if no correlation between the process components is present. That structure is represented by an n x n matrix $\mathbf{V}^{(i)}$ (i=1,..., k) of spatial auto-covariances. We take its spectral decomposition ($\mathbf{V}^{(i)}=H^{(i)T}\,\Xi^{(i)}\,H^{(i)}$) and we denote by $\xi_r^{(i)}$ and $\eta_{rj}^{(i)}$ its eigenvalues and the elements of its eigenvectors (r,j=1,...,n; ,i=1,...,k) respectively.
- To each site $s_j \in D$ we assign the same correlation structure between the process components (when no spatial correlation is given) and we denote the corresponding covariance matrix by $\mathbf{C}^*(0)$ whose spectral decomposition is $\mathbf{C}^*(0)= \boldsymbol{T}^T\,\boldsymbol{L}\,\boldsymbol{T}$ (and then l_m and τ_{im}, m,i=1,...,k are its eigenvalues and the elements of its eigenvectors).

The MSRF (7) is simulated through the following steps:

a) We fix $\mathbf{C}^*(0)$ and we generate $\mathbf{Z}$ from a k x n Gaussian distribution with zero mean vector and unitary covariance matrix.;

b) We define $\mathbf{Y}_{(i)}^T=\mathbf{Z}_{(i)}^T\mathbf{B}_{(i)}$ where $\mathbf{B}_{(i)}=\Xi_{(i)}^{1/2}\boldsymbol{H}_{(i)}$. Then for each fixed site s_j we have a k-dimensional r.v. with uncorrelated components, and each component $\mathbf{Y}_{(i)}(s_j)$ for j=1,...,n has spatial structure given by $\mathbf{V}^{(i)}$.

c) Finally we compute $\mathbf{X}(s) =\boldsymbol{L}^{1/2}\boldsymbol{T}\mathbf{Y}(s)$.

In order to simulate an intrinsic MSRF we proceed as above, the only difference being in the way we build $\mathbf{V}^{(i)}$ matrices. In this case we use the *generalized covariance* matrix defined in Matheron (1973). Let $\mathbf{K}^{(i)}$ be the generalized covariance matrix associated to the i-th variable, then $\mathbf{V}^{(i)}=\mathbf{P}\mathbf{K}^{(i)}\mathbf{P}$, with $\mathbf{P}=\mathbf{I}-\mathbf{1}_n\mathbf{1}_n^T/n$, $\mathbf{I}$ the n x n identity matrix and $\mathbf{1}_n$ is an n x 1 vector of unitary elements. The generalized covariance matrix has elements $K_{jw}^{(i)}=-\gamma^{(i)}(h)$, ($h=d(s_j,s_w)$, j,w=1,...,n). Through out the simulations we used several variograms models over transects of n=20,30,50 (65 for the intrinsic SRF) sites. The MSRF has 10 components (k=10). In our study we distinguished between "low" (almost diagonal *local* covariance matrix), «medium» and high correlation among the process components. In this paper simulations with «medium» *local* variability structure are shown. We further distinguished between variograms models with *small* and *large* ranges, meaning that in one case the sill is reached for a small value of the distance between sites and in the other it's reached for a large one (*small* and *large* have to be seen w.r.t. the transect's length). We take into account values of the distance h between sites up to $h^0=\lceil h_m-h_m/3\rceil$, as not enough observations are available to estimate the variograms matrices for higher values of h. Then the largest distance values we consider are h^0=42 when n=50, h^0=25 when n=30 and h^0=15 with n=20. In all examples we simulate 500 samples of line transects of n=50 squares. For each simulated sample we compute the matrix Γ and we decompose its eigenvalues according to the proposed method. Then in order to verify the influence of the sample size on the procedure, we repeat the analysis on the same simulated data truncating the transects at n=30 and n=20 squares.

The simulations collected in Table 1 have been performed as follows:

1) Sim.1 In this scheme we fix a Chauchy variogram model (see for instance Cressie 1991, Wackernagel 1995) with parameters a=1, b=5 and α=1 for the components of the SRF with odd index (i=1,3,5,7,9). To the remaining variables we associate a spherical variogram model with parameters a=1 and b=5. The two models reach the sill for small values of the distance between sites (h^*<16 for each component on line transects of 50 locations). The variability between the MSRF components is fixed.
2) Sim.2 We simulate *large* range variograms (h^*>24 for each component on line transects of 50 locations). To 5 components we associate an exponential variogram model with parameters a=1 and b=15 and to the other 5 a spherical model with parameters a=1 and b=35.
3) Sim.3 We simulate from a MSRF whose components have both *small* and *large* range. We «mix» exponential variograms with parameters a=1 and b=3 (*small range*) with spherical variograms with parameters a=1 and b=30 (*large range*).

The *global range* is found (with large confidence) where we expected it. The first two simulation schemes are easily readable being quite homogeneous from both the spatial and «local» point of view. More interesting is to look at Sim.3. here the interaction between the process components has a more relevant role. The different spatial structures influence each other through the existing correlation between the MSRF components. We find the *global range* 54% of the times in the first two classes (n=50), being some of the variables with a *small* range structure and larger variance relatively highly correlated with all the others, the *global* range is «moved» towards small values. The same situation appears when dealing with n=30 and n=20. Analogous results are obtained when we assign to the same variable a *large* range model and to the others a *small* range one.

Table 1. *Second order MSRF: decomposition of the first eigenvalue of Γ and determination of the global range (%).*

Global ranges	Sim.1 n=50	Sim.1 n=30	Sim.1 n=20	Sim.2 n=50	Sim.2 n=30	Sim.2 n=20	Sim.3 n=50	Sim.3 n=30	Sim.3 n=20
1 – 5	40.6	45.4	61.0	0.0	0.0	0.2	20.6	28.2	21.8
6 – 10	32.0	36.6	31.8	0.2	3.4	22.4	20.2	31.8	52.2
11 – 15	16.0	13.4	7.2	2.8	26.2	77.4	14.0	15.0	26.0
16 – 20	6.8	3.8	-	19.6	63.8	-	8.8	11.0	-
21 – 25	2.4	0.8	-	24.0	6.6	-	13.0	14.0	-
26 – 30	2.2	-	-	38.2	-	-	14.6	-	-
31 – 35	0.0	-	-	11.6	-	-	7.4	-	-
≥ 36	0.0	-	-	3.6	-	-	1.4	-	-
Tot.	100.0	100.0	100.0	100.0	100.0	100.0	100.0	100.0	100.0

The simulation of an intrinsic stationary MSRF follows the steps described at the beginning of this section. We build the generalized covariance matrix by choosing an intrinsic (isotropic) variogram model for the components of the process. For all of them we fixed a linear variogram model with parameters a=5 and b=6. Applying our technique we find that the *global range* is found always very close to the maximum admissible value h^0=44. In Table 2 we compare results from simulations of intrinsic MSRF and *large range* second order stationary MSRF at various sample size. The same procedure described above has been performed.

Table 2. *Comparison between intrinsic and large range simulations (%)*

Global range	Intrinsic MSRF n=65	Large range MSRF n=65	Intrinsic MSRF n=50	Large range MSRF n=50	Intrinsic MSRF n=30	Large range MSRF n=30
5 - 9	0.0	0.0	0.0	0.2	0.0	1.2
10 – 14	0.0	1.0	0.2	1.4	11.3	20.6
15 – 19	0.2	4.2	2.6	19.6	72.3	71.4
20 – 24	4.8	26.4	7.8	25.0	16.4	6.8
25 – 29	9.0	23.0	41.2	42.2	-	-
30 – 34	13.2	18.6	39.0	8.6	-	-
35 – 39	36.6	14.8	9.0	3.0	-	-
40 – 44	36.2	12.0	0.2	0.0	-	-
Tot.	100.0	100.0	100.0	100.0	100.0	100.0

From the 65 locations simulations it is clear that over a large enough sample we are able to distinguish the two spatial variability models with large confidence. In table 2 smaller samples analysis are shown too. When n=50, in the *large* range case we find a global range (88%) smaller then h^0 while in the intrinsic case 78% of the simulations find the global range closer to the largest values of the distance . For smaller sample dimensions we cannot discriminate between the two models.

3.1 Application to botanical data

In order to verify the behavior of our technique on real data, we analyze a set of observation previously studied through the technique proposed in Di Bella and Jona Lasinio (1996). The data are measures of abundances of several plants species. The measurements are centimeters covered by each of 8 species and are taken along a line transect of 30 meters length (each site is a square of 1 m^2). Because of the nature of the vegetation we can assume second order stationarity. We applied the method proposed in Di Bella and Jona Lasinio in order to find the "true" scale of the data. At that scale observations of the MSRF can be considered "almost" spatially uncorrelated. The "true" is found so that 9 units are aggregated by a type of spatial moving average. The new data set has been analyzed with our technique and the global range is found at distance $h^*=1$. Then we analyzed the disaggregated data set. The first two eigenvalues accounts for the 77.8% of the total variability. In Table 3 species loadings are reported and in Figure 1 the first two principal components are shown.

Table 3: *Species loadings*

Species	First eigenvector	Second eigenvector
Asparagus acutifolius	0.001313899	0.0402381603
Cistus monspeliensis	-0.077701392	-0.0514799760
Erica multiflora	0.265024609	0.5698000029
Myrtus communis	0.956737575	-0.0879254935
Phillyrea angustifolia	-0.021453125	-0.0423429708
Pistaccia lentiscus	-0.088209856	0.8129921308
Quercus ilex	0.009914978	0.0008337439
Rubia peregrina	0.006034461	-0.0240403769

The spatial behavior of the 8 species is well represented. The first component is mainly characterized by the joint presence of *Myrtus communis* and *Erica multiflora*, describing a primary pattern due to these two dominant species. The second component describe the spatial pattern originated by *Pistaccia lentiscus* a less common plant.

Figure 1: *Data projected on the first and second principal axes.*

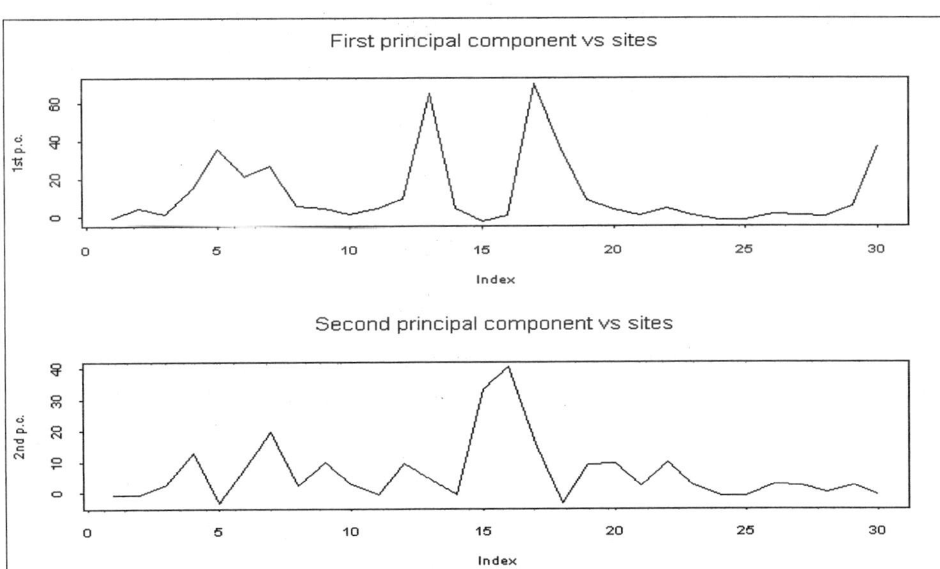

More interesting are applications of the proposed method on grids data (regular and non regular grids). In this contest the authors are working at several theoretical and applied developments

References

Borgman, L.E., Frahme, R.B., (1976). A case study: multivariate properties of bentonite in northeastern Wyoming. In *Advanced Geostatistics in the mining industry*. M. Guarascio eds., D.Reidel, Dordrecht-Holland, 351-390.

Capobianchi, A. (1998). *The Analysis of Multivariate Spatial Data* PhD thesis (in italian). University of Rome «La Sapienza»

Christakos, G. (1992) *Random Field Models in Earth Sciences* Academic Press.

Cressie, N. (1991). *Statistics for Spatial Data* Wiley.

Davis, B.M., Greenes, K.A., (1983). Estimation Using Spatially Distributed Multivariate Data: An Example with Coal Quality. *Mathematical Geology* **15**, N. 2 . 287-300.

Di Bella, G., Jona Lasinio, G., (1996). Including Spatial Contiguity Information in the Analysis of Multispecific Patterns. *Environmental and Ecological Statistics* **3** 269-280.

Journel,A.G., Huijbregts, Ch.J. (1978). *Mining Geostatistics*. Academic Press, London.

Matheron, G., (1973). The Intrinsic Random Functions and their Applications. *Advances in Applied Probability* **5**, 439-468.

Tailiang Xie, Myers D.E., (1995a) Fitting Matrix Valued Models by Simultaneous Diagonalization (Part I: Theory) *Mathematical Geology* **27**, N.7, 867 - 875.

Tailiang Xie, Myers D.E., (1995b) Fitting Matrix Valued Models by Simultaneous Diagonalization (Part II: Application) *Mathematical Geology* **27**, N.7, 877 - 888.

Wackernagel, H. (1995) *Multivariate Geostatistics* Springer Verlag.

On the Assessment of Geographical Survey Units using Constrained Classification

Antonio Giusti - Alessandra Petrucci
Dipartimento di Statistica "G. Parenti", Università di Firenze,
Viale G.B. Morgagni 59, 50134, Firenze, Italy

Abstract: Surveys of spatially distributed phenomena are often conducted using geographical areas as strata. If CATI (computer assisted telephone interviewing) methodology is used to contact the units, it is possible to choose between two different methods of selecting the units to be interviewed: either from a full list of the population units or a selection based on RDD (random digit dialling) technique. On this last case it seems natural to telephone exchange areas as strata. This could be a very interesting solution from many point of views. The aim of this paper is to find a methodology to asses the opportunity of using such a pre-defined geographical stratification in comparison of the usual clustering methods, based on a set of auxiliary variables correlated with the phenomena under study, to define the strata. The choice will depend on the use of some measure of similarity and/or the evaluation of the homogeneity in the strata for the specific phenomenon to be analysed (in our application the evaluation of the loss of homogeneity is verified with respect to a hypothetical set of variables under study).

Keywords: Contiguity Classification, Telephone Surveys, Cluster Validation.

1. Introduction

In statistical surveys the methodologies based on the telephone as a cheap and fast contact tool are growing and they are more and more diffused. In this case it is very interesting to consider the use of computer assisted telephone interviewing (CATI) to survey spatially distributed phenomena for which it is necessary to use geographical areas as strata. Many auxiliary information, that can be very useful in order to analyse a particular phenomenon, are often available at geographical level: for example, a considerable quantity of georeferred census data is available at enumeration district level.

A first CATI selection method is based on the use of a telephone directory as a frame. In this case the units drawn from the directory can be georeferred through the address matching spatial analysis. This is not an easy task, as it needs the processing of character data as the complete address of every telephone subscriber. Moreover, some telephone subscribers are not included in the telephone directory and for this reason they will be excluded from the survey.

An alternative selection method is based on the use of random digit dialling (RDD) technique to choose the units to be called. With this method the telephone directories are not used for the selection and every telephone number has the same probability to bc dialled. To avoid loss of time, the random numbers must be generated keeping into account the code of the telephone lines installed in the telephone exchange of each arca involved in the survey. With RDD methodology a simpler way for georeferring the information is to use the telephone area as geographical unit.
In Italy, the telephone system is based on a hierarchical system of enumeration using the telephone exchange area as the smallest geographical area. The telephone numbers are composed by an area code (telephone district code) and the telephone subscriber number (the first 2, 3, or 4 digits of this number identify the telephone exchange area). Practically, every telephone exchange area is associated with one or more enumeration sequences. Using the enumeration sequences and the coverage of the telephone exchange area it is possible to georefer the telephone number randomly generated.
Through the GIS (geographical information system) functions it is possible to obtain other information on the areas under study (enumeration districts and telephone exchange areas) such as, the centres of gravity, the co-ordinates, and the size of the areas. Furthermore, with a GIS it is possible to conduct any kind of spatial analysis. The geographical organisation of the data would also enable the pattern analysis of the considered phenomenon and the final representation of the results.
In this paper we make a first attempt to define a methodology to verify the opportunity of using the telephone exchange areas instead of a partition of clusters of enumeration districts build through a classification algorithm under a contiguity constraints. In the next section we define the classification problem and the constraints to be imposed. In the following section we make some suggestion on cluster validation based either on some measure of similarity or on the evaluation of the homogeneity of some auxiliary variables that are probably highly correlated with the surveyed variables. Before the conclusions, we report an empirical application using real data on Florence telephone sectors.

2. Constrained classification

The purpose of cluster analysis in this kind of application is the identification of homogeneous areas with respect to the phenomena under study. A geographical area is considered homogeneous if the inhabitants have a similar behaviour about the phenomena under study. The geographical areas are those obtained from the aggregation of smaller areas (as the enumeration districts).
Our problem can be defined as follows: let X be the set of the N spatial objects represented by the enumeration districts: $X = \{X_1, X_2, ..., X_N\}$, and consider a

specific partitioning Y of X in K sets, where the partitioning refers to the enumeration district:

$$Y = \{Y_1, Y_2, \ldots, Y_K\},$$

where each group is a contiguous set of the N objects of X, represented by the telephone exchange areas:

$$Y_k = \{X_{k_1}, X_{k_2}, \ldots, X_{k_{nk}}\}.$$

Our assignment is to produce a new partition of X, i.e. Z, so that:

$$Z = \{Z_1, Z_2, \ldots, Z_L\},$$

where each group is also a contiguous set of the N objects of X:

$$Z_s = \{X_{s_1}, X_{s_2}, \ldots, X_{s_{n_s}}\}.$$

For N sufficiently large this can be accomplished by using a non-hierarchic method. For our application we used the FASTCLUST procedure of SAS imposing K=L (SAS, 1990). But in our classification problem, it is relevant to impose constraints on the set of the allowable solutions.

When a classical clustering technique, such as K-means, is applied to geographically located data, without using the spatial information, the resulting partition has often a sparse appearance over the geographical space (i.e. clusters look dispersed, and reflect only poorly any eventual underlying spatial structure).

Several alternatives have been proposed in order to take into account the geographic location of the data, and produce clusters that are spatially homogeneous. Gordon (1996) presents a complete review on this topic. In this paper we consider only the approaches more suitable for the solution of our problem.

The most common type of constraints is the one specified by contiguity: the objects in a class are required not only to be similar, but also to be spatially contiguous.

Proximity information in two-dimensional space can be incorporated more directly into a classification by making use of the geographical distances separating each pair of objects, or a contiguity graph or matrix (Legendre, 1987, Openshaw, 1977). This implies the definition of a neighbourhood concept.

In this approach, the elements of a contiguity matrix are defined by:

$$C(i,j) = \begin{cases} 1 \text{ if the } i\text{th and } j\text{th objects are contiguous,} \\ 0 \text{ otherwise.} \end{cases}$$

The specification of the contiguity graph of a set of objects is not always straightforward. If the objects are areas of land covering a region, areas sharing a common boundary should clearly be regarded as contiguous, but when the number of these area objects increases some computational problems can arise as in our case: the dimension of the contiguity matrix was very large.
Other approaches to the contiguity-constrain classification consist in the computation of a distance between objects that is a function both of the geographical distance and the distance measured through the values of the variables. In general, the modified distance between the spatial objects i and j can be written as:

$$\begin{cases} \bar{d}_{ij} = f(d_{ij}, \psi_{ij}) \text{ if the } i\text{th and } j\text{th object are contiguous,} \\ \bar{d}_{ij} = d_{ij} \qquad \text{otherwise,} \end{cases}$$

where f is an increasing function of both arguments, d_{ij} is a distance measure between the vectors of the variables on ith and jth objects, and ψ_{ij} is an increasing function of the geographical distance between the ith and jth objects (Zani, 1993). This approach requires the weighting of the geographical distance, ψ_{ij}, and of the distance measured on the variables d_{ij}, imposing $f(d_{ij}, \psi_{ij}) < d_{ij}$; it is not easy to specify the relative weights in an objective manner.
We followed a natural approach that uses the geographical co-ordinates of the centres of gravity of the enumeration districts, more or less heavily weighted, as an additional pairs of variates (Barry, 1966, Jain and Farrokhnia, 1991).

3. Partition assessment

Few constrained classification studies have addressed the problem of cluster validation, but in our case we consider this problem relevant. In our situation the problem is slightly different from a normal cluster validation. As already said, our task is to compare the two partitions of X, where one, Y, was defined to be the telephone exchange areas, and the other, Z, was obtained with a clustering algorithm taking into account spatial relations. In other words, Z is a set of clusters of enumeration districts suitable for data analysis, while, at least for practical reasons, Y would be a clustering suitable for data collection.
The problems to be addressed are now:

1) evaluating the correspondence between the two partitions, i.e. to measure the resemblance between partition Y and partition Z;
2) evaluating the adequacy of partition Y with respect to partition Z for data analysis.

Regarding the first aspect, we have to measure the similarity between the two partitions using an index that compares the partitions. A basic unit of comparison between two partitions is how pairs of objects are clustered. Starting from this concept Rand (1971) proposed an index *c*, varying from 0 to 1, based on the calculation of the individual element-pair placed together in a cluster in each of the two partitions, *Y* and *Y'*. Rand proposed a simple computational form for this index:

$$c(Y,Y') = \frac{\left[\binom{N}{2} - \left[\frac{1}{2}\left\{\sum_i \left(\sum_j n_{ij}\right)^2 + \sum_j \left(\sum_i n_{ij}\right)^2\right\} - \sum\sum n_{ij}^2\right]\right]}{\binom{N}{2}},$$

where N is the number of the objects, and n_{ij} is the number of element simultaneously in the *i*th cluster of *Y* and in the *j*th cluster of *Y'*.

The relative size of the Rand index and of other similar indices is difficult to evaluate and compare, as they are not corrected for chance. So Hubert and Arabie (1985) proposed a correction for chance, and the corrected Rand index assumes this form:

$$c'(Y,Y') = \frac{\sum_{i,j}\binom{n_{ij}}{2} - \sum_i \binom{n_{i.}}{2}\sum_j \binom{n_{.j}}{2} \Big/ \binom{N}{2}}{\frac{1}{2}\left[\sum_i \binom{n_{i.}}{2} + \sum_j \binom{n_{.j}}{2}\right] - \sum_i \binom{n_{i.}}{2}\sum_j \binom{n_{.j}}{2} \Big/ \binom{N}{2}}.$$

The other method to be considered is based on the evaluation of the adequacy of partition *Y* with respect to partition *Z* for data analysis. One methodology can be based on the use of a raw measure of the homogeneity in the groups. For this purpose we assumed the average R-squared weighted by variance. This index must be calculated for each of the two partitions to evaluate the loss in homogeneity implied by the use of the telephone exchange areas.

4. The application

The application was carried out by using the data of the telephone sector of Florence and the census data of the enumeration districts of the same area.

The telephone sector of Florence covers the municipal area of the city of Florence and eight other municipalities of the Florentine Province (Bagno a Ripoli, Calenzano, Impruneta, Fiesole, Rignano, Sesto Fiorentino, Scandicci e Vaglia), which are adjacent to the city of Florence. The map of the geographical coverage of telephone exchange areas was digitised and stored in the GIS ARC/INFO (see Fig. 1). A unique numerical code, that is the key for the linkage of the graphic areas with the enumeration sequence table to be used with the RDD methodology, was assigned to every telephone exchange area. The sequence table allows assessing the potentiality of every telephone exchange device.

Figure 1: *The telephone exchange areas.*

On the other hand, the enumeration districts are often too small geographical areas to be used as strata in a survey; for this reason we grouped the enumeration districts using clustering algorithm with and without the contiguity constrains. The map of enumeration districts of the studied area was already in digital format and stored in the GIS.

The data considered for the cluster analysis were only few of the census variables at the enumeration district level, which the Italian central bureau of statistics (ISTAT) disseminates for the 1991 population census.

To preserve the confidentiality of the data, the variables released by ISTAT are mainly social-demographic variables. Unfortunately, we could not use any economic variables (i.e. income, professional position, etc) that would be more relevant for our study. The social-demographic variables give anyway a first view of some social aspects that can be considered as auxiliary information for the stratification. For this purpose we considered 10 variables: the population density (using the area information supplied by the GIS), the percentage of males, an index for the age population structure, the percentage of self-employed workers, the number of houses per person, and the ratio of occupied houses to total houses.

Some spatial analyses as the polygon overlay was carried on the stored digital maps using the GIS to assign to every enumeration district the corresponding telephone exchange area code.

Figure 2: *The clusters of the enumeration districts*

The telephone exchange areas are a partition of the geographical space, but they are not always an exact aggregation of the enumeration districts. The fact that an enumeration district belongs to different telephone exchange areas can be solved by attributing the district to the telephone area with the highest percentage of overlapping. In this way, we obtain the complete list of the enumeration districts belonging to each telephone exchange area. Of the 5624 enumeration districts enclosed in the above listed municipalities only 3813 were used, the ones overlapping the 49 telephone exchange areas in which the sector of Florence is divided. It is then possible to group at telephone exchange area level all the information at enumeration district level (Mohadjer, 1988).
So the problem was to consider the 3813 enumeration districts overlapping the 49 telephone exchange areas as elements to be grouped before carrying out the analysis. On the other hand, the 49 telephone exchange areas can already be considered a partition of the enumeration districts, which cannot be further divided. The result of the cluster analysis performed is showed in Fig. 2.
At this point, using the Rand index on the two partitions we found a value of c=0.95 while using the corrected Rand index the value of c' is 0.50. That result means that the overlapping between the areas is very high but about half of the value is due to the contiguity condition and to the constrained imposed on the number of cluster (K=L).
We performed also a raw evaluation of the adequacy of partition for data

analysis with the R-squared index. Using the telephone exchange areas as strata, the value of R-squared was 0.09. With a non-hierarchical partition, allowing a number of clusters equal to the number of the telephone exchange areas, the R-squared was of 0.59. To evaluate more precisely the loss of homogeneity using the telephone exchange areas as strata, the same index was computed on groups of enumeration districts verifying the contiguity of the areas that compound each cluster. The R-squared value obtained (0.29) is nearer to that computed on telephone exchange areas, even if the difference is still remarkable.

5. Conclusions

We have proposed the utilisation of the telephone exchange areas to stratify a geographical area. The objection to the ad-hoc clustering can be overcome by some measure of similarity or by some homogeneity indices. We applied some of these indices to assess the stratification in the Florence area and found that the telephone exchange areas were very similar to the constrained clustering of the enumeration districts. In such a situation the utilisation of the telephone exchange area as a frame to conduct an RDD survey would be appropriate. Obviously the operative decisions should to be taken considering the real phenomenon under study for which the data will be collected.
The methodology proposed in this paper is a first attempt to face the problem of comparing a given area partition with a computed clustering. Further studies must be carried out to define a more suitable homogeneity measure.

References

Barry, B. (1966). *Essay on commodity flows and the spatial structure of the Indian economy*, Research paper 111, University of Chicago, Department of Geography.

Gordon, A. D. (1996). A survey of constrained classification, *Computational Statistics & Data Analysis*, 21, 17-29.

Hubert, L. & Arabie, P. (1985). Comparing partitions, *Journal of Classification,* 2, 193-218.

Jain, A. & Farrokhnia, F. (1991). Unsupervised texture segmentation using gabor filters, *Pattern Recognition,* 24, 12, 1167-1186.

Legendre, P. (1987). Constrained clustering, in: *Developments in numerical ecology,* Legendre, P. & Legendre, L. (Eds.), Springer, Berlin.

Mohadjer, L. (1988). Stratification of prefix areas for sampling rare population, in: *Telephone Survey Methodology*, Groves R. M. et al. (Eds.), John Wiley & Sons, New York, 161-173.

Openshaw, S. A. (1977). A geographical solution to scale and aggregation problems in region-building, partitioning and spatial modelling, *Trans. Institute of British Geographers,* NS 2, 459-472.

SAS, (1990). *SAS/STAT User's Guide, Ver. 6, 4th Ed.,* SAS Institute, Cary, NC.

Zani, S. (1993). Classificazione di unità territoriali e spaziali, in: *Metodi Statistici per le Analisi Territoriali*, Zani S. (Ed.), Franco Angeli, Milano, 93-121.

Kalman Filter Applied to Non-Causal Models for Spatial Data

Luca Romagnoli
Dipartimento di Metodi Quantitativi e Teoria Economica
Università "G. d'Annunzio"
Viale Pindaro,42 - 65127 Pescara, Italy - E-mail: romagnol@dmqte.unich.it

Abstract: This paper faces the problem of the application of filtering and smoothing algorithms, in particular Kalman filtering, to spatially dependent data. We take into account the case of first- and second-order homogeneous Gauss-Markov Random Fields (GMRF), and we address the question of parameter estimation for this class of spatial processes; then we consider the possibility of expressing these processes as unilateral ones, so that they can be written in state-space form; and, finally, we present a "classical" Kalman filter algorithm, which is particularly suitable for the case of satellite images contaminated by additive Gaussian noise.

Keywords: Gauss-Markov Random Fields, unilateral representation, Kalman filter.

1. Introduction

In this paper, we shall refer to the case of spatial data, collected on a regular lattice divided into N×M zones; when the random variables (r.v.) taken into account will be followed by a single index, we shall suppose to have numbered the zones from 1 to n (n= N×M), moving from the left to the right, and from the top to the bottom of the lattice.

In general terms, denoting as $S_{ij}^{(p)}$ the p-th order *neighbours set* of the zone (i,j) (chosen on the basis of a certain distance criterion, normally the Euclidean one, and of a given order of spatial lag), we define as *Markov random field* the spatial process for which the following is valid:

$$\Pr(x_{ij} \mid x_{kl}:(k,l)\in D^{ij}) = \Pr(x_{ij} \mid x_{kl}:(k,l)\in S_{ij}^{(p)}) \tag{1}$$

where D^{ij} denotes the parametric space from which the zone (i,j) has been deleted. If we hypothesize that the conditional distribution of the generic r.v. X_i is normal, with parameters:

$$\theta_i = E(X_i|x_j:j\neq i) = \mu_i+\sum_{j=1}^{n} c_{ij}(x_j - \mu_j) \;\; ; \;\; \tau_i^2 = Var(X_i|x_j:j\neq i) \;,$$

we obtain, by applying Brook's factorization theorem (Besag,1974):

$$\mathbf{X}\sim\text{NMV}(\mu, (\mathbf{I}-\mathbf{C})^{-1}\mathbf{M}) \tag{2}$$

where **M** is the diagonal matrix containing the conditional variances, and **I-C** is the NM×NM matrix containing the field *potentials* (that is, the spatial interaction parameters), and is therefore called *potential matrix*.

2. Gauss-Markov random fields: structure and parameters estimation

Under the hypothesis of equal conditional variances, that is, $\tau_i^2=\tau^2$, $\forall i=1,2,\ldots,NM$, the exponent of the density function (2) may be written in compact form:

$$U(\mathbf{x}) = -(2\tau^2)^{-1}\mathbf{x}'\mathbf{A}\mathbf{x}$$

where **A**=**I-C**. The study of GMRFs is, obviously, based on the analysis of this matrix, which completely characterizes the process. In particular, the potential matrix of a *homogeneous* GMRF is always decomposable as: $\mathbf{A} = \mathbf{A}_c+\mathbf{A}_{b.c.}$, where $\mathbf{A}_c$ is the *canonical* potential matrix, independent of boundary conditions (b.c.), while $\mathbf{A}_{b.c.}$ is the *border* potential matrix, i.e., the matrix which contains the interaction parameters among internal and external zones. In the case of a first-order homogeneous GMRF with Dirichlet b.c. (all the off-lattice zones values are put equal to zero), we have that $\mathbf{A}_{c.b.}=\mathbf{0}$, while $\mathbf{A}_c$ may be written as:

$$\mathbf{A}_c = \mathbf{I}_N\otimes\mathbf{B}_1 + \mathbf{H}_N\otimes\mathbf{C}_1 \text{ , where:}$$

$$\mathbf{H}_N = \begin{bmatrix} 0 & 1 & 0 & \cdots & \cdots \\ 1 & 0 & 1 & \cdots & \cdots \\ \cdots & \cdots & \cdots & \cdots & \cdots \\ \cdots & 0 & 1 & 0 & 1 \\ \cdots & \cdots & 0 & 1 & 0 \end{bmatrix} ; \; \mathbf{B}_1 = \begin{bmatrix} 1 & -\beta_{o_1} & 0 & \cdots & \cdots \\ -\beta_{o_1} & 1 & -\beta_{o_1} & 0 & \cdots \\ 0 & -\beta_{o_1} & 1 & -\beta_{o_1} & \cdots \\ \cdots & \cdots & \cdots & \cdots & \cdots \\ \cdots & \cdots & 0 & -\beta_{o_1} & 1 \end{bmatrix} = \mathbf{I}_M-\beta_{o_1}\mathbf{H}_M \tag{3}$$

and β_{o_1} is the only horizontal interaction parameter; and again: $\mathbf{C}_1 = -\beta_{v_1}\mathbf{I}_M$, where β_{v_1} represents the only vertical interaction parameter.

For first-order, and a particular class of second-order GMRF (that for which the diagonal interaction paramers are the same), it is possible to derive exact analytical expressions for the eigenvalues of **A** by applying Kronecker product rules; these eigenvalues are (Jain, 1979; Balram & Moura, 1993):

- first-order fields:

$$\lambda_{ij}(\mathbf{A}_1) = 1-\beta_{v_1}\lambda_i(\mathbf{H}_N) -\beta_{o_1}\lambda_j(\mathbf{H}_M)\text{ , } i=1,\ldots,N\text{ ; }j=1,\ldots,M \tag{4}$$

- second-order fields (with equal diagonal parameters, i.e. $\beta_{dl_{11}}=\beta_{dr_{11}}=\beta_{d_{11}}$):

$$\lambda_{ij}(\mathbf{A}_2)=1-2\beta_{v_1}\cos\frac{i\pi}{N+1}-2\beta_{o_1}\cos\frac{j\pi}{M+1}-4\beta_{d_{11}}\cos\frac{i\pi}{N+1}\cos\frac{j\pi}{M+1}, \quad i=1,\dots,N\ ;j=1,\dots,M \tag{5}$$

We are now able to give some indication for ML estimation of parameters. The negative log-likelihood (multiplied by the positive constant 1/NM) of a first-order, zero-mean GMRF defined on a N×M regular lattice is:

$$L(\mathbf{x};\beta_{o_1},\beta_{v_1},\tau^2) = =(1/2)\ln(\tau^2) -[1/(2NM)]\ln\left|\mathbf{A}_1(\beta)\right| + [1/(2\tau^2 NM)]\mathbf{x}'\mathbf{A}_1(\beta)\mathbf{x}]$$

The quadratic form in L(.), that is $\mathbf{x}'\mathbf{A}_1(\beta)\mathbf{x}$, may be written in a much easier form, remembering of the structure of $\mathbf{A}_1$, and considering the result (4); we obtain:

$$L(\mathbf{x};\beta_{o_1},\beta_{v_1},\tau^2) = =\frac{1}{2}\ln(\tau^2)-\frac{1}{2NM}\sum_{i=1}^{N}\sum_{j=1}^{M}\ln(1-\beta_{v_1}\lambda_i(\mathbf{H}_N)-\beta_{o_1}\lambda_j(\mathbf{H}_M))+ +\frac{1}{2\tau^2}(S_x-2\beta_{o_1}R_x^{o_1}-2\beta_{v_1}R_x^{v_1}) \tag{6}$$

where we put:

$$S_x=\frac{1}{NM}\mathbf{x}'(\mathbf{I}_N\otimes\mathbf{I}_M)\mathbf{x}\ ;\ R_x^{o_1}=\frac{1}{2NM}\mathbf{x}'(\mathbf{I}_N\otimes\mathbf{H}_M)\mathbf{x}\ ;\ R_x^{v_1}=\frac{1}{2NM}\mathbf{x}'(\mathbf{H}_N\otimes\mathbf{I}_M)\mathbf{x}$$

The only closed-form estimate is that of the variance, and, substituting the expression for $\hat{\tau}^2$ in L(.), we derive the so-called *profile log likelihood*:

$$L(.)=\frac{1}{2}\ln(S_x-2\beta_{o_1}R_x^{o_1}-2\beta_{v_1}R_x^{v_1})- -\frac{1}{2NM}\sum_{i=1}^{N}\sum_{j=1}^{M}\ln(1-\beta_{v_1}\lambda_i(\mathbf{H}_N)-\beta_{o_1}\lambda_j(\mathbf{H}_M))+\frac{1}{2} \tag{7}$$

which has to be minimized on the valid parametric space to obtain ML estimates. The same lines are followed for estimating the parameters of a second-order GMRF: the only difference to remark is the definition of the potential matrix, which assumes the form:

$$\mathbf{A}_2 = \mathbf{I}_N \otimes \mathbf{B}_2 + \mathbf{H}_N \otimes \mathbf{C}_2$$

in which $\mathbf{B}_2$ is equal to $\mathbf{B}_1$, and so it has the structure indicated in (3), while $\mathbf{C}_2$ is given by:

$$\mathbf{C}_2 = -\beta_{v_1} \mathbf{I}_M - \beta_{d_{11}} \mathbf{H}_M$$

Under the hypothesis that the sample data, x_{ij}, are contaminated by additive white Gaussian noise, the model becomes: $y_{ij} = x_{ij} + \eta_{ij}$, where the x_{ij}s represent, as it has been told, our (first-order) GMRF, while the η_{ij}s are the noise terms, for which the usual hypotheses are valid:

$$\textbf{i)}\ \eta_{ij} \sim N(0,\sigma^2);\ \ \textbf{ii)}\ E(\eta_{ij}\eta_{kl})=0\ ,\forall\ (i,j)\neq(k,l);\ \ \textbf{iii)}\ E(x_{ij}\eta_{kl})=0\ ,\ \ \forall\ i, j, k, l$$

It is then possible to write (Moura & Balram, 1992):

$$S_y \approx S_x + \sigma^2 \qquad R_y^{o_1} \approx R_x^{o_1} \qquad R_y^{v_1} \approx R_x^{v_1}$$

So, if it is possible to have an estimate of the noise variance, a direct substitution of these terms in (7) will again permit to obtain ML estimates.
We now present some experimental results on parameters estimation for first- and second-order homogeneous GMRF. We have generated, in each case, 30 different realizations of the process (lattice dimension, 64×64; Dirichlet boundary conditions), using the recursive algorithm which will be discussed below; Tables 1 and 2 refer to ML estimation in absence of noise, while Table 3 presents the results in the case of a first-order process corrupted with additive Gaussian noise.

Table 1: *first-order homogeneous GMRF. Parameters:* β_{o_1}=0.25; β_{v_1}=0.15; τ^2=100; *30 simulated samples.*

	Mean	**St. error**
$\hat{\beta}_{o1}$	0.2431	0.0226
$\hat{\beta}_{v1}$	0.1550	0.0265
$\hat{\tau}_2$	100.2280	4.2146

Table 2: *second-order homogeneous GMRF. Parameters:* β_{o_1}=0.05; β_{v_1}=0.05; $\beta_{d_{11}}$=0.15; τ^2=400; *30 simulated samples.*

	Mean	**St. error**
$\hat{\beta}_{o1}$	0.0483	0.0294
$\hat{\beta}_{v1}$	0.0511	0.0264
$\hat{\beta}_{d11}$	0.1477	0.0141
$\hat{\tau}_2$	403.9841	23.5290

Table 3: *first-order homogeneous GMRF. Parameters:* β_{O_1}=0.25; β_{V_1}=0.15; τ^2=100; σ^2=100; *30 simulated samples.*

	Mean	St. error
$\hat{\beta}_{O1}$	0.2370	0.0517
$\hat{\beta}_{V1}$	0.1584	0.0518
$\hat{\tau}_2$	99.1988	9.8835

3. Recursive structure of non-causal GMRF

We will now develop the recursive structure through which it is possible to write a GMRF, maintaining its fundamental characteristic of non-causality. The starting point is the Minimum Mean Square Error (MMSE) representation (Woods, 1972): **Ax**=ε. It is very easy to demonstrate that **Ax**=ε, with ε~MVN(**0**,τ^2**A**), is equivalent to a non-causal autoregressive representation for a GMRF with potential matrix equal to **A**. From this equivalence, it is possible to attain to two equivalent causal (or unidirectional) representations: the first one is defined "backward", the other one "forward". The fundamental consideration, at this purpose, is that, **A** being positive definite, it admits two distinct Cholesky decompositions; the first is written as: **A** = **U'U**, where **U** is upper triangular, while the second decomposition is expressed through the matrix **L**, which is lower triangular: **A**=**L'L**. For example, the "backward" representation becomes:

$$\mathbf{Ux} = \mathbf{w} \tag{8}$$

where: $\mathbf{w} = (\mathbf{U}')^{-1}\varepsilon$, with: $\Sigma_{\mathbf{w}} = E(\mathbf{ww}') = \tau^2\mathbf{I}$, and: $\Sigma_{\mathbf{xw}} = E(\mathbf{xw}') = \tau^2\mathbf{U}^{-1}$

It is essential to point out that the very regular structure of **A** reflects on the structure of **U**. So, as **A** is block tridiagonal, **U** has only two block diagonals different from **0**. Then, **U** can be written as follows:

$$\mathbf{U} = \begin{bmatrix} \mathbf{U}_1 & \Theta_1 & \mathbf{0} & \cdots & \cdots \\ \mathbf{0} & \mathbf{U}_2 & \Theta_2 & \mathbf{0} & \cdots \\ \cdots & \cdots & \cdots & \cdots & \cdots \\ \cdots & \cdots & \mathbf{0} & \mathbf{U}_{N-1} & \Theta_{N-1} \\ \cdots & \cdots & \cdots & \mathbf{0} & \mathbf{U}_N \end{bmatrix}$$

where $\mathbf{U}_i$ and Θ_i are M×M blocks (as always, we are presenting the case of a first-order process, but the extension to higher order GMRFs is straightforward). The unilateral representation of a GMRF leads to a state-space form for the process; in fact, making use of the structure of the matrix **U**, we obtain the following formulas:

$$\mathbf{x}_N = \mathbf{G}^b_N \mathbf{w}_N;$$
$$\mathbf{x}_i = \mathbf{F}^b_i \mathbf{x}_{i+1} + \mathbf{G}^b_i \mathbf{w}_i \quad \text{for } 1 \le i \le N-1 \tag{9}$$

where $\mathbf{G}^b_i = \mathbf{U}_i^{-1}$, and $\mathbf{F}^b_i = -\mathbf{U}_i^{-1}\Theta_i$; the problem of the calculation of the blocks $\mathbf{U}_i$ and Θ_i is solved addressing to a *Riccati-type iteration*. Let us define: $\mathbf{S}_i = \mathbf{U}_i^{'}\mathbf{U}_i$, then, basing upon the Cholesky decomposition of the potential matrix and the block structure of **A**, we obtain the following recursive formulas:

$$\mathbf{S}_1 = \mathbf{B};$$
$$\mathbf{S}_2 = \mathbf{B} - \mathbf{C}^{'}\mathbf{B}^{-1}\mathbf{C};$$
$$\mathbf{S}_i = \mathbf{B} - \mathbf{C}\,\mathbf{S}_{i-1}^{-1}\mathbf{C}; \quad \text{for } 3 \le i \le N-1 \tag{10}$$
$$\mathbf{S}_N = \mathbf{JBJ} - \mathbf{C}^{'}\mathbf{S}_{N-1}^{-1}\mathbf{C}$$

To obtain the blocks Θ_i, $1 \le i \le N-1$, it is sufficient to observe that the decomposition of **A** makes it possible to write:

$$\mathbf{U}_1^{'}\Theta_1 = \mathbf{C}_1;$$
$$\mathbf{U}_i^{'}\Theta_i = \mathbf{C}; \tag{11}$$
$$\mathbf{U}_{N-1}^{'}\Theta_{N-1} = \mathbf{C}_{N-1}$$

Through (10)-(11) it is possible to implement a very efficient simulation algorithm. Further simplifications are reached by considering that, under general conditions, the succession $\{\mathbf{S}_i\}$ rapidly converges to a fixed matrix, $\mathbf{S}_\infty$.

At the end of this section, we present the neighbourhood system for the unilateral processes we are discussing [see (8)], and the correspondent neighbourhood system for the equivalent non-causal models (2), this latter obtained making use of Euclidean distance.

Figure 1: *Neighbourhood system of the zone indicated with ◊ for GMRFs of order 1-4 in* (a) *unilateral representation, and* (b) *non-causal representation.*

o	o	o	o	o	o	o
o	o	o	o	o	o	o
o	o	o	◊	1	1	1
1	1	1	1	2	3	3
3	3	3	3	4	o	o
o	o	o	o	o	o	o
o	o	o	o	o	o	o

(a)

o	o	o	o	o	o	o
o	o	4	3	4	o	o
o	4	2	1	2	4	o
o	3	1	◊	1	3	o
o	4	2	1	2	4	o
o	o	4	3	4	o	o
o	o	o	o	o	o	o

(b)

4. A Kalman-type algorithm applied to image analysis

Let us formally define the problem: we have at our disposal an image, defined as a lattice of N×M regularly spaced pixels, whose values vary in the range of the integers between 0 and L-1, where L is the number of gray levels (usually, L=256); the image being received from a satellite, it is very likely that it is corrupted by errors of various sorts (observational errors, measurement errors, and so on); we suppose that the model employed to describe the "true" image under analysis is a GMRF: $\mathbf{Ax}=\varepsilon$; we hypothesize, moreover, that the observations are altered by Gausian white noise; the model for the observations, in other terms, is:

$$\mathbf{y}=\mathbf{x}+\eta \tag{12}$$

where: $\eta \sim NMV(0,\sigma^2\mathbf{I})$. The recourse to formulas (9)-(11), clearly, leads in a direct way to the use of Kalman filter; the algorithm (Rauch, Tung, Striebel, 1965) consists of two "passages" over the image, to obtain better results: the first one is the filter passage (forward), the second one is the smoothing passage (backward).

1st phase (forward)

Initialization: $\begin{cases} \hat{\mathbf{x}}_{1|0} = \mathbf{0} \\ \mathbf{P}_{1|0} = \tau^2 \mathbf{I} \end{cases}$

1) Kalman gain: $\mathbf{K}_i = \mathbf{P}_{i|i-1}(\mathbf{P}_{i|i-1} + \sigma^2\mathbf{I})^{-1}$

2) Filter update: $\hat{\mathbf{x}}_{i|i} = \hat{\mathbf{x}}_{i|i-1} + \mathbf{K}_i(\mathbf{y}_i - \hat{\mathbf{x}}_{i|i-1})$

3) Filter covariance matrix update: $\mathbf{P}_{i|i} = (\mathbf{I} - \mathbf{K}_i)\mathbf{P}_{i|i-1}$

4) Forecast update: $\hat{\mathbf{x}}_{i|i-1} = \mathbf{F}^f_i \hat{\mathbf{x}}_{i-1|i-1}$

5) Forecast covariance matrix update: $\mathbf{P}_{i|i-1}=\mathbf{F}^f_i \mathbf{P}_{i-1|i-1}(\mathbf{F}^f_i)' + \sigma^2 \mathbf{G}^f_i (\mathbf{G}^f_i)'$

2nd phase (backward)

Initialization: $\begin{cases} \hat{\mathbf{x}}_N = \hat{\mathbf{x}}_{N|N} \\ \mathbf{P}_N = \mathbf{P}_{N|N} \end{cases}$

1) Smoother gain matrix: $\mathbf{Z}_i = \mathbf{P}_{i|i}(\mathbf{F}^f_{i+1})' \mathbf{P}^{-1}_{i+1|i}$

2) Smoother update: $\hat{\mathbf{x}}_i = \hat{\mathbf{x}}_{i|i} + \mathbf{Z}_i(\hat{\mathbf{x}}_{i+1} - \hat{\mathbf{x}}_{i+1|i})$

3) Smoother covariance matrix update: $\mathbf{P}_i = \mathbf{P}_{i|i} + \mathbf{Z}_i(\mathbf{P}_{i+1} - \mathbf{P}_{i+1|i})\mathbf{Z}_i'$

As an application, we present here the results obtained with the use of the Kalman filter algorithm on an image of dimension 100×100 pixels. Fig.2(a) is the original image; Fig 2(b) is the noisy image; Fig.2(c) and 2(d) are, respectively, the restorations on the basis of a first- and second-order GMRF.

Figure 2: *Application of Kalman filtering and smoothing algorithm.*

(a) Original image (100×100)

(b) Image corrupted with additive Gaussian noise (σ^2=514; SNR=8)

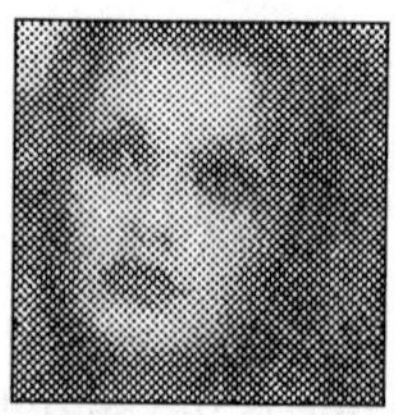

(c) Restoration on the basis of a homogeneous first-order GMRF; estimated parameters: $\hat{\beta}_{o1}$=0.288, $\hat{\beta}_{v1}$=0.212, $\hat{\tau}_2$=368.377

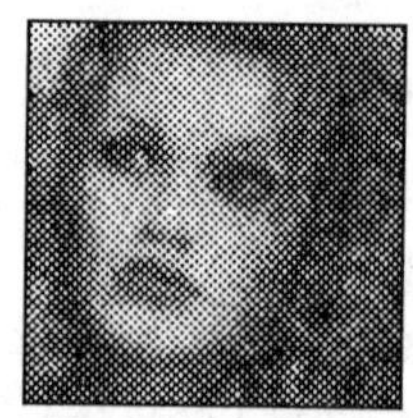

(d) Restoration on the basis of a homogeneous second-order GMRF; estimated parameters: $\hat{\beta}_{o1}$=0.054, $\hat{\beta}_{v1}$=0.053, $\hat{\beta}_{d11}$=0.197, $\hat{\tau}_2$=506.542

As it can be noted, though we have used very simple models (at maximum 3 different spatial interaction parameters), the results are visually satisfactory, especially in the case of second-order GMRF. In order to have a statistical measure of the "goodness of reconstruction", we calculated the Mean Squared Error (MSE) between the original image and each of the other ones; the results are the following: Fig.2(b), MSE=526.086; Fig.2(c), MSE=467.876; Fig.2(d), MSE=308.420.

References

Balram, N. & Moura, J.M.F. (1993). Noncausal Gauss Markov random fields: parameter structure and estimation, *IEEE Trans. Inform. Theory*,39,1333-55.

Besag, J. (1974). Spatial interaction and the statistical analysis of lattice systems, *Journal of the Royal Statistical Society*, B, 36, 192-236.

Jain, A.K. (1979). A sinusoidal family of unitary transforms, *IEEE Trans. Pattern Anal. Machine Intell.*, PAMI-1, 356-365.

Moura, J.M.F. & Balram N. (1992). Recursive structure of noncausal Gauss-Markov random fields, *IEEE Trans. Inform. Theory*, 38, 334-354.

Woods, J.W. (1972). Two-dimensional discrete Markovian fields, *IEEE Trans. Inform. Theory*, 18, 232-240.

A Paradigmatic Path for Statistical Content Analysis Using an Integrated Package of Textual Data Treatment

Sergio Bolasco
Univ. of Rome "La Sapienza"
bolasco@scec.eco.uniroma1.it

Adolfo Morrone
ISTAT - DCPT/1
morrone@istat.it

Francesco Baiocchi
ISTAT - CEN/A
baiocchi@istat.it

Abstract: In this paper different phases of the treatment of text are sketched, in order to link them both with some lexical characteristics of the analised corpus and with multidimensional techniques useful for the statistical content analysis of the latter. Our proposal is directed towards maintaining intact the system of meanings present in the corpus and to bettering the degree of monosemy of words. In this way a corpus vocabulary of mixed units of analysis is realised.

Keywords: Pre-processing of textual data, Textual variables, Mixed units of analysis, Disambiguation, Locutions and Polyrhematics.

1. Domain of study and aims

The principal changes in the evolution of statistical-quantitative studies of natural languages could be synthesized through many adjectives: from the linguistic (Zipf, Waring-Mandelbrot, Yule: see Herdan 1964) to the lexical (Guiraud, Quemada, Imbs, Brunet: see Muller 1977), from the textual (Lebart and Salem 1994) to the lexico-textual (Reinert, Labbé, Silberztein, Elia: see Bolasco *et al.* 1995). Often in the course of this evolution, the theoretical disputes - for instance, the one relative to the choice of the unit of analysis (headword or graphic form, words or segments) - draw our attention away from our empirical evidence, in this case the criterion of goodness of fit to the semantics of text.

Throughout these years, multidimensional methods for the analysis of natural languages, borrowed from the analysis of numerical data (qualitative and quantitative) have been proposed, with their corresponding statistical packages, but without any serious attention to "text care" nor has software for this very time-consuming phase been developed. This care should be understood as the automatic lexico-textual treatment of the linguistic information with the aim of the content analysis of a corpus which is the object of study. In this way, we deal with the problem of the transformation of text into statistical data. Text has, in fact, a sequential type structure. Therefore, each operation of segmentation into units of analysis causes a loss of information.

The transformation from the linguistic information to textual data requires, apart from the logic of the study, an accurate work of "pre-processing" in order to minimise this loss. This is an aspect completely neglected by the statistical techniques of textual analysis; these latter, indeed, expect the text to be analysed without any influence on the starting information (Lebart, Salem 1994).

On the contrary, it is possible to improve the statistical quality of the linguistic data and to guarantee, on the one hand, monosemy (intended as stability of the meaning of the word in spite of context changes) and, on the other, coherence with the content of the text. The latter in particular is satisfied by choosing, as elementary units of analysis, mixed forms (for instance *lexias* or textual forms, Bolasco 1997).

2. Some paradigms in order to perform text segmentation

If we want to identify the least meaningful units in their discourse, we should consider both single forms and polyforms. With this aim in mind, it is fundamental to conform to sequencing when considering the steps to follow. Here, some general principles for the segmentation and the disambiguation of the text are defined, in the form of 5 paradigms.

A) The graphic forms (GF), by the means of which a text is presented, are carriers of the specific contents of the discourse (subjects, times, modality, intentions). In order to preserve intact the system of the meanings present in the corpus, we need to observe the following.

1- To maintain the graphic form as a textual variable of the first level.

2- To resolve ambiguity, be it semantic or grammatical, in the graphic forms where the ambiguity is not irrelevant in terms of occurrences. A grammatical disambiguation could lead to a semantic disambiguation: the single form <abito>, if it is a noun, means "an article of clothing"; if it is a verb, it is the first person singular present indicative of "abitare", and so <abito> means "I live". A semantic disambiguation, starting from an analysis of concordance, could lead to a grammatical one; for example <un dato di fatto> is an idiomatic expression (meaning "in point of fact") in which we can clearly identify <dato_noun> and <fatto_noun>, once their classification as verbs ("given" and "made"), has been excluded (which becomes possible if the analysis is conducted at the level of single words as: datum of fact, or datum of made, or given of fact, or given of made).

3- The lack of isofrequency between two unlemmatised forms derived from the same lexeme is a sign of an imbalance of use, which is a function of meanings. In this case, the disambiguation of the most frequent form is the most opportune thing to do (Bolasco 1998).

B) Every discourse is dense with locutions, or rather with fixed-meaning sequences (idiomatic expressions). These locutions are called polyrhematics, or rather, expressions "whose total meaning is not calculable from the component lexemes" (De Mauro *et al.*, 1993: 153), that is to say their whole sense is different from the sum of the meanings of their component terms; for example, <materie prime> ("raw materials") or <capo dello stato> ("head of state").

4- The preliminary and automatic recognition of the more common locutions eliminates at a stroke the ambiguity of the homographic types which make up the locutions (see for example the aforementioned <dato di fatto> meanings "in point of fact"). The availability of frequency dictionaries of polyforms facilitates the recognition of these locutions (Bolasco, Morrone 1998).

5- The lexicalisation of polyrhematics should be carried out according to the following criteria:

i) The need to capture the sequences from the most expanded to the most shortened, < fino in fondo> ("to the furthermost point"), <in fondo> ("basically"). If two sequences are of equal length, the one with the highest number of full words should be lexicalised first: <più presto possibile> ("at the earliest time possible") before <al più presto> ("as soon as possible").

ii) The need to recognize the grammatical locutions (adverbial, adjectival, prepositional and conjunctive) and the nominal polyrhematics at the graphic form level <al fine di> "with the purpose of" <alla fine di> "at the end of"; on the contrary, phrasal verbs, given the high number of inflectional forms from a verb, should be captured at the headword level, otherwise they would not be easily describable in terms of frequency.

iii) The need to lexicalize sequences not only when they are very common, but particularly when they are strongly absorbent with respect to the single forms of which they are composed; for example <lavori agricoli> ("agricultural works") 28 occurrences, of which "lavori" (58) and "agricoli" (30). In order to select the nominal polyrhematics, reference could be made to a particular index of relevance (Morrone 1993).

3. An environment for the treatment of textual data

In order to perform these paradigms, we propose an integrated package of tools (hereafter called TALTAC: italian acronym of Automatic Lexico-Textual Treatment for Content Analysis) that allows us to check and select the important linguistic information in order to make it consistent with the statistical analysis.

TALTAC gathers together in a single "environment" both several base linguistic resources, available from outside the corpus which is the object of study (frequency dictionaries, electronic lexicons), these being useful in the lemmatisation phase, together with a series of tools for "text pre-processing". TALTAC is independent of any specific statistical package for the analysis of textual data, but it allows a series of interventions on the corpus which are directly linked to the different phases of an applied strategy of statistical analysis.

4. Phases of a strategy of statistical analysis on textual data

For our purposes, we can break down every strategy of statistical analysis on textual data oriented to a content analysis, into at least some of the following steps:
a) Definition of the *type of corpus* and storage of the text: possible identification of the fragments (propositions, phrases) and "numeration" of the words; possible matching of the subtexts with categorical variables; b) Creation of various *indexes*: dictionary of graphic forms and/or inventory of repeated segments; c) *Analysis of concordance*: study of the local contexts for the purpose of producing semantic and grammatical fusions/disambiguations; d) *Categorisations* of the words with meta-textual information: recoding by means of equivalencies for grammatical tagging, lexematic reductions, formation of families of words and reconstruction of thematic fields; e) *Selection* of lexical units of a set of forms as object of study: extraction of the specific language, thresholds of frequency; f) *Partition* of the corpus in subtexts: to obtain statistics on use and specificity or to construct data matrices in order to apply statistical techniques; g) Analysis of simple (Lafon 1980) or chronological (Salem 1988) *specificity*; h) Application of *multidimensional methods* of content analysis: correspondence analysis, discriminant analysis (Lebart 1995), cluster analysis (Reinert 1995), semantic networks (Scott 1997); i) Links with *external information*: positioning of qualitative categorical variables on the results of the analysis.

5. Phases of the automatic lexico-textual treatment of text

The TALTAC treatment is characterised by the adoption of a strategy of a recursive nature that does not precede the statistical analysis, but integrates with it. The package of textual analysis provides the raw information that, processed by TALTAC, allows us to make the text more consistent with further statistical analysis. The TALTAC environment is made up of a number of phases each consisting of different steps that are in interaction with the statistical programs of textual analysis. The flow-chart in fig. 1 reproduces the typical path of text treatment, which is determined by the combination of some modules of "textual pre-processing" on the corpus with others of "lexico-textual analysis" on the linguistic data. This path is only indicative because, even if some modules are to a certain degree preparatory to others, it is possible to personalise the path, by repeating the execution of modules already inserted or by adding steps of

Fig. 1 - Typical Path of an *Automatic Lexico-Textual Treatment for Content Analysis*

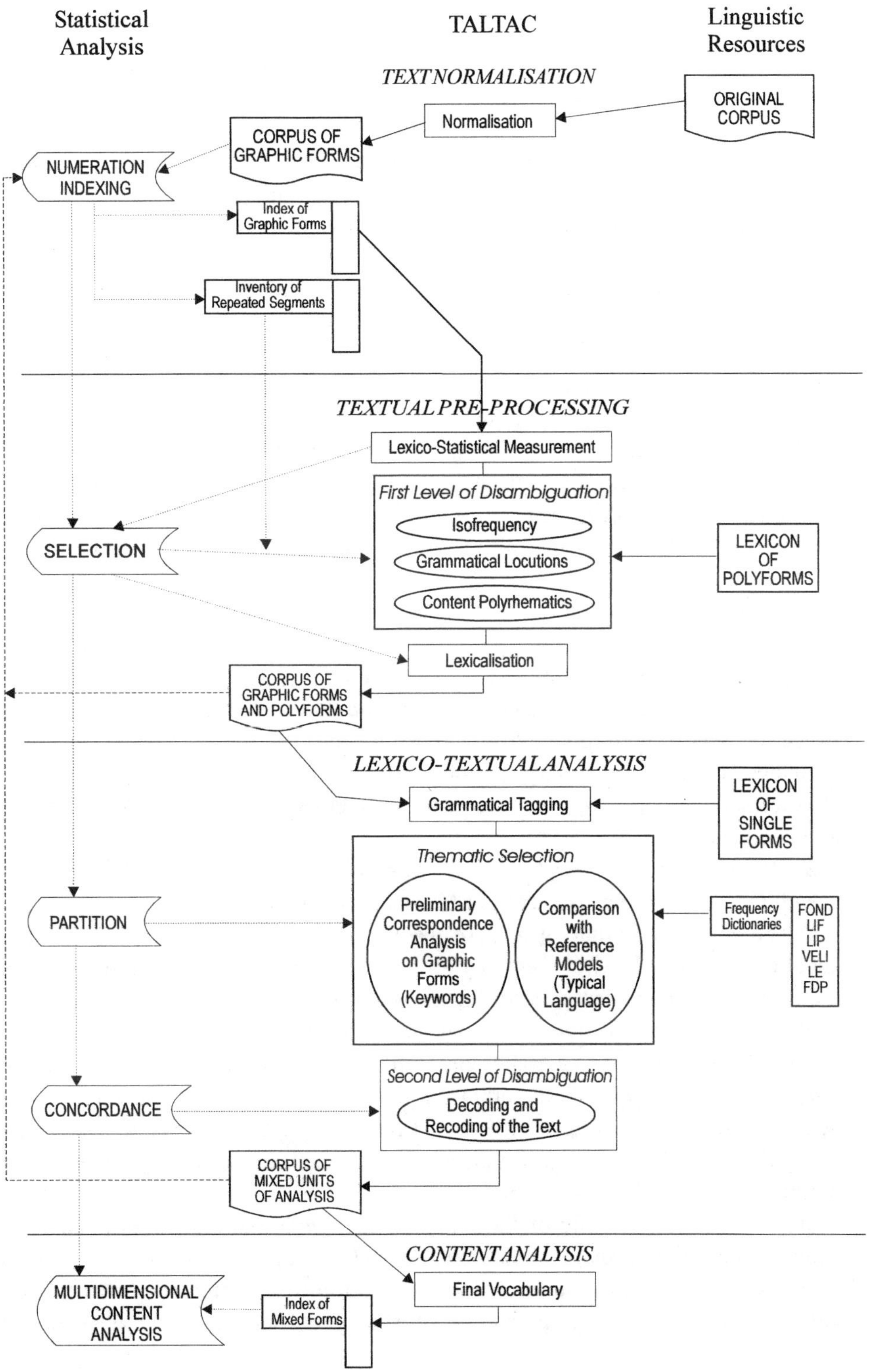

analysis according to each individual's specific demands. Our path is articulated in four phases, for a total of eight steps, each with one or more modules.

A) ***Text Normalisation***. This first phase is the zero step of any automatic treatment of text. Its objective is to guarantee transparency and to assure exhaustivity in the automatic reading of the text. This step allows us to minimize the number of original graphic forms in the first vocabulary of the text and also to produce corpora available for comparison with variation in elaboration type.

In textual statistics, units of analysis are defined as successions (chains) of characters included between delimiters (Lafon 1984, Lebart, Salem 1994), where the valid characters and the delimiters define the reference alphabet.

The normalisation module implements the transformation from an automatically identified succession of characters to the graphic form, as type or unlemmatised form (Silberztein 1993), which represents the most elementary level of textual data, eliminating possible sources of data splitting.

This step includes the following points: definition of an alphabet (valid characters + delimiters); normalisation of dates, numbers, names, acronyms and of accents, apostrophes, spaces, capital letters; association of categorical attributes to the different parts of the corpus (fragments and/or subtexts).

B) Textual Pre-processing. In this phase the objective is to find, in the corpus of textual data, the information which is crucial for the preparing of the best unit of analysis. This phase is composed of three steps.

1. The first step (*Lexico-Statistical Measurement* in fig. 1), useful for the choice of level of frequency threshold in the following content analysis, calculates, on the vocabulary of the normalised corpus, some basic lexico-statistics such as: rank, range of frequency, coverage of the text, lexical richness of vocabulary.

2. The second step (*First Level of Disambiguation* in fig. 1) makes a primary disambiguation of types by means of three distinct modules:

a) The selection module of non *Iso-frequency* situations (Bolasco 1994, 1998) that allows us to highlight potentially ambiguous words on which an intervention of disambiguation is necessary for the purposes of analysis.

b) The identification module of the most important basic structures of the discourse, through the comparison with the fundamental dictionary of *Grammatical Polyforms* as adverbial, adjectival, prepositional and conjunctive locutions (Bolasco, Morrone 1998b). Their recognition allows a systematic disambiguation of 25% of occurrences of very usual ambiguous types or words (homographs).

c) The selection module of the repeated segments that allows us to identify automatically the principal *Content Polyrhematics* or nominal locutions (Morrone 1995). Due to the operational definition of repeated segment (Lebart, Salem 1994), the original inventory of segments, selected by the statistical package, is difficult to consult, since it presents many elements which are not meaningful for the purpose of the analysis. In order to obviate this problem, the module identifies, by means of an index of relevance (Morrone 1993), the grammatically complete segments that are specific expansions of a catalyzer form.

3. The third step (*Lexicalisation* in fig. 1) builds the complex units (lexias) by means of lexicalisation, that is composing sequences of words (idiomatic expressions) to be treated as single forms. This module attends to the recognition of such structures (grammatical locutions, phrasal verbs and nominal groups), which are semantically very specific (for instance as polyrhematics).

At the end of this second phase, a first corpus is made available, segmented into units of analysis of mixed type (single forms and polyforms) in order to submit it to statistic analysis.

C) Lexico-Textual Analysis. The third phase (composed of the fourth and fifth steps) realize a lexico-textual analysis for which it is fundamental to interact with linguistic external resources like dictionaries of single or complex forms, local grammars and frequency dictionaries. This phase has the objective of exploiting these bases of linguistics knowledge in order to select meaningful textual variables for content analysis.

4. The fourth step consists of the module of *Grammatical Tagging* (fig. 1), available for the Italian or the French languages (Morrone 1995), which is a necessary condition for text lemmatisation. The module call up some lexicons, which can be upgraded and also integrated with sectorial languages, and attributes the grammatical category and the headword to each word. In the case of homography, the module associates it with all the possible couples category/headword that are pertinent. This operation allows us to identify ambiguous forms, to convert the text to graphic forms with grammatical categories or to headwords (dictionary address), and to get for each headword the list of the inflectional forms really present in the text.

This step should be performed *a latere* of the statistical analysis, in that the lemmatised corpus doesn't always produce a gain in information. In fact, whatever the method used - markovian processes (Grigolli et al. 1991), or local grammars (Silberztein 1993) -, automatic lemmatisation is still subject to a rate of specific error superior to 15%. For the purpose of reducing errors, it is preferable to lemmatise the corpus after the recognition of polyforms and fixed phrases. In general, for the goals of content analysis, an almost systematic lemmatisation of verbs and of adjectives is useful, while the lemmatisation of nouns is not always opportune.

5. The fifth step (*Thematic Selection* in fig. 1), permit the selection of relevant textual variables. This step could develop in two different perspectives: A) lexical or B) textual. The first allows us to individualize the specific language with the help of linguistic external resources (language models of reference); the

second, by means of factorial preliminary analysis on graphic raw forms, allows us to individualize the "key" words for further content analysis.

A- Comparison with reference models. When the corpus is of a large dimension and the vocabulary is composed of many thousands of headwords, it is necessary to limit "textual variables" only to the typical language (original words + Positive/Negative specific words). Frequency dictionaries (VELI, Lip, Lif, LE, Tpg, FdP) permit us to select such words, and at the same time to indicate intrinsic specificity.

To select the intrinsic specificity or the typical language it is necessary to compare the vocabulary of the corpus with the model (frequency dictionary of reference). This involves the calculus of the index of use and some measures of lexical correlation. The index of use is based on the measure of dispersion (Muller 1977) that presupposes the construction of the contingency table, words by texts (parts of the corpus), available directly from the statistical package.

The measures of lexical correlation, based on the comparison between rank of words in the vocabulary and the correspondant rank in the frequency dictionary, allows us to delineate the fundamental core of vocabulary. The technique of parallel coordinates (Wegman 1990) represents graphically the differences between ranks and allow us to select, for instance for each part of speech, the more specific forms. This step is useful for the selection of verbs (Bolasco 1998).

B- Preliminary correspondence analysis on graphic forms. If the comparison with a frequency dictionary of reference is not possible (when the corpus is small or a pertinent lexicon of reference doesn't exist) we may move on to an exploratory analysis of the factorial type on the graphic forms with high frequency, to characterise the structural words (key words) on which the interventions of second level of disambiguation (semantic meaning) will be focused.

6. The sixth step (*Decoding and Recoding of Text* in fig. 1) is useful in deciding which corrections should be made directly on the text, those which have to be made virtually via software and those which should be abandoned since they would reduce the quality of the textual data. Beginning from an index of the "thematic forms" (typical language or key words) choices of intervention are made, usually involving less than 10% of the selected forms.

With this aim in mind, in the first place the analysis of concordance is made by means of calling up the integral text in order to obtain a reading of local contexts. In this way, for instance, it is possible to identify the different meanings of a single form. Afterwards some hypotheses of both grammatical and semantic disambiguations or fusions are examined. For some doubtful or critical cases, it is possible to undertake validation processes with a bootstrap method on a factorial plane (Balbi 1995). The application of resampling techniques allows us to build confidence regions (as convex hulls) of single inflections, in order to verify the impact of the choices which could be made (Bolasco 1998).

On the basis of the information collected during the previous phase, the list of the recodings is compiled whose realisation, effected inside the package of statistical analysis, does not modify the original text.

D) Content Analysis by Multidimensional Methods.

7. In this last step (*Construction of the Final Vocabulary* in fig. 1), the vocabulary of mixed forms with a frequency higher than the select threshold for the content analysis is set. Such units of analysis are forms of lexico-textual type (headwords, graphic forms, lexias as locutions and nominal groups) with high monosemic content. A final treatment balance is also produced in terms of the disambiguated or lemmatised forms, as well as of the singled out lexicalised sequences. Finally, the coverage rate is calculated before and after the treatment. A rate higher than 80% is already obtainable with less than 12% of graphic forms of the vocabulary, in correspondence with the rank of the first decile of the low frequencies. For corpus of about 50,000 occurrences, the absolute frequency of such a threshold is not inferior to 10.

This vocabulary, in which significant variability has not been reduced and in which, at the same time, monosemia and variables robustness increases, becomes the point of departure for the study of the text in order to perform the further content analysis.

References

Balbi, S. (1995). Non symmetrical correspondence analysis of textual data and confidence regions for graphical forms, in: *JADT 1995 Analisi statistica dei dati testuali,* Bolasco, S., Lebart, L., Salem, A. (Eds.), CISU, Roma, II, 5-12.

Bolasco, S. (1994). L'individuazione di forme testuali per lo studio statistico dei testi con tecniche di analisi multidimensionale, in: *Proceedings of XXXVII Riunione Scientifica della SIS - Sanremo*, CISU, Roma, 95-103.

Bolasco, S. (1997). L'analisi informatica dei testi in: *La ricerca qualitativa*, Ricolfi, L. (Ed.), Nuova Italia Scientifica, Roma, 165-203.

Bolasco, S. (1998). Meta-data and Strategies of Textual Data Analysis: Problems and Instruments, in: *Data Science, Classification and Related Methods*, Hayashi, C., Ohsumi, N., Yajima, K., Tanaka, Y., Bock, H.-H., Baba, Y. (Eds.), Springer, Tokio, 468-479.

Bolasco, S., Lebart, L., Salem, A. (Eds.), (1995). *JADT 1995 - Analisi statistica dei dati testuali*, CISU, Roma, 2 tomes.

Bolasco, S. & Morrone, A. (1998). La construction d'un lexique fondamental de polyformes selon leur usage, in: *Proceedings of JADT 1998*, Mellet, S. (Ed.) Univ. Sophie Antipolis de Nice, 155-166.

Bolasco, S. & Morrone, A. (1998b). A frequency dictionary of polyforms as a linguistic database for text disambiguation in TALTAC, in: *Proceedings of VI Conference of the IFCS*, Univ. of Rome, Short Papers Volume, 32-35.

De Mauro, T., Mancini, F., Vedovelli, M., Voghera, M. (1993). *Lessico di frequenza dell'italiano parlato,* EtasLibri, Milano.

Elia, A. (1995). Per una disambiguazione semi-automatica di sintagmi composti: i dizionari elettronici lessico-grammaticali, in: *Ricerca qualitativa e computer* Cipriani R. & Bolasco S., (Eds.), Franco Angeli, Milano, 112-141.

Grigolli, S., Maltese, G., Mancini, F. (1991). Un sistema per la lemmatizzazione automatica di testo libero in: *Atti del convegno AICA Text Processing*, Milano.

Herdan, G. (1964). *Quantitative Linguistics*, Butterworths, London (tr. it. 1971, *Linguistica quantitativa*. Il Mulino, Bologna).

Lafon, P., (1984), *Dépouillements et statistiques en lexicométrie,* Slatkine, Genève-Paris.

Lebart, L. & Salem, A., (1994). *Statistique textuelle,* Paris, Dunod.

Lebart, L. (1995). Discriminazione in base a dati testuali in: *Ricerca qualitativa e computer*, Cipriani R. & Bolasco S. (Eds), Franco Angeli, Milano, 184-202.

Morrone, A., (1993). Alcuni criteri di valutazione della significatività dei segmenti ripetuti, in: *Jadt 1993. Secondes Journées internationales d'Analyse statistique de Données Textuelles*, Anastex, S. J. (Eds.), TELECOM, Paris, 299-309.

Morrone, A., (1995). Una strategia di trattamento del testo per l'individuazione di variabili testuali rilevanti, in: *JADT 1995 - Analisi statistica dei dati testuali* Bolasco, S., Lebart, L., Salem, A. (Eds.), CISU, Roma, 135-142.

Muller, Ch. (1977). *Principes et méthodes de statistique lexicale*, Hachette, Paris.

Reinert, M. (1995). I mondi lessicali di un corpus di 304 racconti di incubi attraverso il metodo Alceste in: *Ricerca qualitativa e computer*, Cipriani R. & Bolasco S. (Eds), Franco Angeli, Milano, 203-223.

Salem, A. (1988). Approches du temps lexical. Statistique textuelle et series chronologiques. *Mots* , 17, 105-143.

Scott, J. (1997). *L'analisi delle reti sociali.* NIS, Roma.

Silberztein, M., (1993). *Dictionnaires électroniques et analyse automatique des textes,* Collection informatique linguistique, Masson, Paris.

Wegman, E. J. (1990). Hyperdimensional Data Analysis Using Parallel Coordinates. *J.A.S.A.*, 85, 411, 664-675.

The Analysis of Auxological Data by Means of Nonlinear Multivariate Growth Curves

Marcello Chiodi*, Angelo M. Mineo**

* Institute of Statistics - Faculty of Economics - University of Palermo
Viale delle Scienze - 90128 - Palermo, Italy, e-mail: chiodi@unipa.it

** Department of Mechanical Technology & Production - University of Palermo
Viale delle Scienze - 90128 - Palermo, Italy, e-mail: amineo@unipa.it

Abstract: In this paper we treat the problem to analyse a data set constituted by multivariate growth curves for different subjects; thus in this context we deal with 3-way data tables. Nevertheless, it is not possible using factorial techniques proposed to deal with 3-way data matrices, because the observations are generally not equally spaced; moreover a multilevel approach founded on polynomial models is not suitable to deal with intrinsic nonlinear models. We propose a non-factorial technique to analyse auxological data sets using an intrinsic nonlinear multivariate growth model with autocorrelated errors. The application to a real data set of growing children gave easily interpretable results.

Keywords: Longitudinal studies, multivariate growth models, nonlinear regression, serial correlation, MLE, three-way data.

1. Introduction[1]

The analysis of data sets constituted by multivariate observations depending on time for different subjects is a widely studied topic; it depends on many conditions concerning the kind of data, their quality, the purpose of the analysis, and so on. In this paper, we are concerned mainly with the analysis of real data constituted by multivariate growth measures of a set of children, surveyed on different times. Therefore, at least formally, we have a 3-way data table and so we could think to use one of the specific techniques proposed to deal with 3-way data matrices, based mainly on different types of factorial decompositions. Three common methods proposed to deal with 3-way matrices "individuals x variables x occasions" are:

a) STATIS (Escoufier, 1987), that can be seen as a principal component analysis where different statistical studies with many variables are compared, by obtaining a graphical representation where the points are the studies and the proximity of the points gives a similarity among the studies;
b) the Tucker3 model (Tucker, 1966), and

[1] This research has been supported by MURST grants.

c) the PARAFAC (PARAllel FACtor) model (Harshman, 1970).
Both models b) and c) try to decompose the initial 3-way matrix by considering sets of virtual units, variables and occasions according to a minimisation function (Rizzi, Vichi, 1995). Many other methods have been proposed but few means are available to decide which method is better than the others when we have real data disposed in a 3-way matrix (Kroonenberg, 1992). Furthermore, these methods deal with 3-way matrices in which the occasions are always the same for all subjects and give generally linear decompositions. Therefore, these methods are entirely unusable for data sets constituted by individual observations with different survey times, as in our case. It is also difficult with these methods to deal properly with a serial correlation structure that could be present in the individual data.
In the following sections, we first present a data set and describe the multilevel models that could be used with growth data. Then, in section 3 we present an explicit nonlinear multivariate growth model with autocorrelated errors and in section 4 we treat the problem of the estimation of the involved parameters. In section 5 we present the results obtained by the analysis of our data set.

2. Analysis of multivariate growth curves

A longitudinal data set is constituted by k variables observed on n subjects in different occasions; in particular, our data set is a sample data set in the framework of an auxological study in order to assess growth standards: we have the weight and the height (k=2) of babies (n=64) observed in different occasions, starting mainly in the first three months from the birth and ending at an age between 3 and 5 years old. For the i-th baby and for each variable the relevant information is the observed growth curve with m_i different occasion t_{ij} ($i=1,2,...,n$; $j=1,2,...,m_i$). Lags of successive surveys are in general very different among subjects and within the same subject so that the t_{ij} are unequally spaced; also, the number of occasions m_i varies for each subject. Typical growth curves are reported in figure 1 and figure 2.

Figure 1: *Growth curves of height and weight for a single subject*

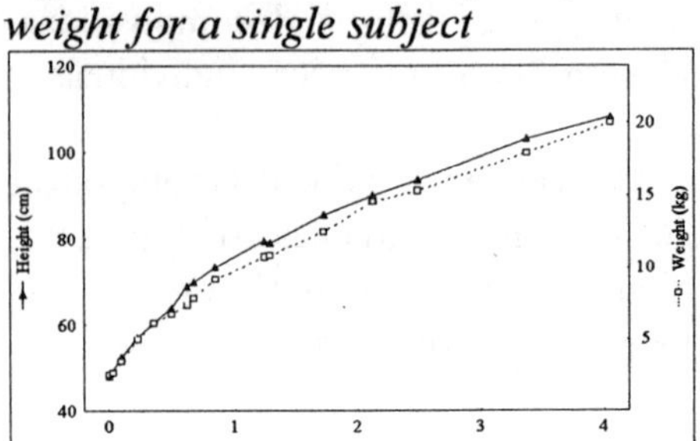

Figure 2: *Growth curves of height for five different subjects.*

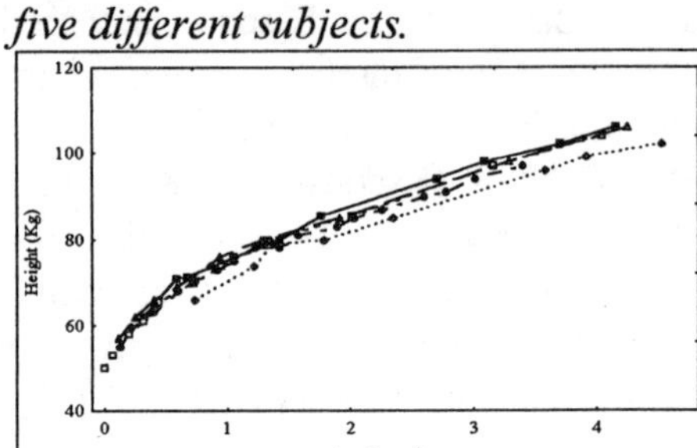

Classification approaches founded on cluster analysis techniques (Mineo, 1987; Chiodi, 1989) are not useful in this context because the data cannot be seen as T matrices of equal dimensions n×k; indeed *sections* of the 3-way data set are possible only along each subject.
Among useful approaches to the analysis of longitudinal data set, with different survey times for different subjects and with non constant time lags for each subject, the multilevel models can be taken particularly into account; in these models variability components of 1st level (different measurements of one subject) and of 2nd level (the different subjects) are considered. Besides multilevel factorial approaches (Borra, Di Ciaccio, 1996), it is interesting the 2-level growth model, proposed by Goldstein and others (1994):

$$y_{ij}=\sum_{u=0}^{p}\gamma_{iu}t_{ij}^{u}+\sum_{v=1}^{q}\alpha_{v}z_{ijv}+e_{ij} \tag{1}$$

where y_{ij} is the j–th measurement on the i–th subject, γ_{iu} are the polynomial coefficients for the level 1 (the successive measurements), t_{ij} is the time of the j-th measurement on the i-th subject, z_{ijv} are the covariates, α_v are the coefficient for the covariates z's (level 2) and e_{ij} are the level 1 random terms that usually are assumed to be distributed independently with zero mean and constant variance. So these models can be considered an extension of the polynomial models for growth curves (Rao, 1965).
However, we did not use this model basically for two reasons: we have been not interested, at least in this paper, in examining random coefficient models (in the multilevel model introduced above the γ_{iu} coefficients are random at level 2 with coefficient values varying and covarying between individuals); moreover the level 1 systematic components, i.e. the time dependence of the individual measurements, have to be in our case expressly nonlinear: so we can not consider a polynomial model, even though of high degree, to obtain individual parameter estimates with a well defined biological meaning. Another opportunity is to consider a 2-level model with nonlinear systematic components on the parameters, but linear by using Taylor approximations of the first order (Milani, Bossi, 1988). In the next section, we analyse the presented data set using an explicit multivariate nonlinear growth curve model with fixed parameters.

3. Nonlinear multivariate growth model with autocorrelated errors

The main purpose of the present paper is an *exploratory* analysis of an auxological data set, to understand if the children have a similar growth with respect to the observed variables, and to understand which model can be adopted to describe the dynamic of the growth. Given very short time series, with variable time lags, we found very hard, or even impossible, to deal with this data with proper dynamic models, so that we preferred an approach based on a

nonlinear growth model, which has the advantage of summarising the behaviour of each individual with a small set of parameters easily interpreted.
For sake of simplicity, and only to look for simple descriptive quantities which can summarise such complex data, we tried to fit, for each individual and for each variable, a general nonlinear growth curve of the family of Von Bertalanffy curves (Von Bertalanffy, 1957), that is a three parameter exponential curve:

$$E(y_t)= \gamma + \alpha\,(1-e^{-\beta t}) \tag{2}$$

Of course, the whole human growth can not be well described by only 3 parameters (Tanner, 1981): many curves have been proposed for the description of human growth even with seven parameters (Jolicoeur, Pernin, and Pontier, 1988); however, our data set concerns only the first years of human life, when growth speed decreases and this aspect is satisfactorily described by simple models. The model (2) has an easy interpretation since γ is the value of y at birth, α is a scale parameter related to the whole growth and β depends on the logarithmic growth speed: the individual fits have resulted generally better than those obtained by Gompertz or logistic curves.
The whole model is:

$$y_{ijh}=\gamma_{ih}+\alpha_{ih}[1-\exp(-\beta_{ih}t_{ij})]+\varepsilon_{ijh} \qquad i=1,2,...,n;\ j=1,2,...,m_i;\ h=1,...,k \tag{3}$$

where y_{ijh} is the value of the h-th variable observed at the j-th occasion t_{ij} of the i-th individual, γ_{ih}, α_{ih}, β_{ih} are the parameters of the i-th individual and h-th variable, ε_{ijh} is the random error.
A peculiarity of growth curves is the possible presence of serial correlation between the measurements of an individual (Palmer, Phillips and Smith, 1991); so we assumed that random errors ε_{ijh} are normally distributed and the generic random vector ε_{ih}, (constituted by the m_i errors of the h-th variable and the i-th individual) has covariance matrix:

$$E(\varepsilon_{ih}\,\varepsilon'_{ih})= \sigma^2_{ih}\,\mathbf{R}_{ih}, \tag{4}$$

where σ^2_{ih} is the common variance and $\mathbf{R}_{ih}$ is a correlation matrix with generic (j,s) element $\rho_{ih_{js}}$ representing the correlation between elements of ε_{ih} at times t_j and t_s. Of course, we need a model for the autocorrelations, in order to employ a limited number of parameters. Since the times t_{ij} are not equally spaced, we could not employ ordinary discrete time ARMA models, so that we modelled the autocorrelations according to an exponential decay (Diggle, 1988):

$$\rho_{ih_{js}} = E(\varepsilon_{ijh}\,\varepsilon_{ish})/\sigma^2_{ih} = \rho_{ih}^{|t_{ij}-t_{is}|} \qquad \rho_{ih}\geq 0. \tag{5}$$

This is the autocorrelation function of a continuous AR(1) process (Jones, Ackerson, 1990), which allows only non negative serial correlation. At the first stage, the autocorrelations ρ_{ih} have been supposed different for each individual and each variable. Finally, we supposed that random errors ε_{ih} are not correlated among different individuals and different variables. Individual correlations among variables are taken into account in the systematic component of the model (3).

4. Estimation of the parameters of the model

With the assumptions of the previous section, the log-likelihood function l_{ih} for the m_i data of the i-th individual and the h-th variable is given by:

$$l_{ih}(\alpha_{ih},\gamma_{ih},\beta_{ih},\rho_{ih},\sigma^2_{ih}|\mathbf{y}_{ih})=$$
$$=-n\log(\sigma^2_{ih})/2-\log(|\mathbf{R}_{ih}|)/2-(\mathbf{y}_{ih}-\mathbf{f}_{ih})'\mathbf{R}_{ih}^{-1}(\mathbf{y}_{ih}-\mathbf{f}_{ih})/(2\sigma^2_{ih})$$
$$(i=1,2,\ldots,n;\ h=1,\ldots,k) \qquad (6)$$

being $\mathbf{y}_{ih}$ the vector of observed data and $\mathbf{f}_{ih}$ the vector of fitted data, depending on the unknown parameters α_{ih}, γ_{ih}, β_{ih}, according to the model defined by (3), and $\mathbf{R}_{ih}$ is defined through the relations (4) and (5).

In order to estimate the whole set of parameters, we have to maximise the above quantities; for sake of brevity we do not report in this paper the explicit expressions of the inverse and the determinant of $\mathbf{R}_{ih}$, since simple expressions are given by Núñez-Antón and Woodworth (1994): in fact, as in the usual case of equally spaced times and discrete time AR(1) process, the inverse of $\mathbf{R}_{ih}$ is a tridiagonal or Jacobi matrix depending only on ρ_{ih} and the set of t_{ij}, while its determinant is given by a simple factorisation.

As usual in ML estimation in regression models, we can estimate σ^2_{ih} as an explicit function of the other parameters α_{ih}, γ_{ih}, β_{ih}, ρ_{ih} and then maximise the likelihood concentrated on the latter set of parameters. In fact, the MLE s^2_{ih} of the variance component σ^2_{ih} is:

$$s^2_{ih}(\alpha_{ih},\gamma_{ih},\beta_{ih},\rho_{ih})=(\mathbf{y}_{ih}-\mathbf{f}_{ih})'\mathbf{R}_{ih}^{-1}(\mathbf{y}_{ih}-\mathbf{f}_{ih})/n, \qquad (7)$$

so that by substitution in $l_{ih}(.)$ we have the concentrated log-likelihood:

$$l_{ih}(\alpha_{ih},\gamma_{ih},\beta_{ih},\rho_{ih},s^2_{ih}(.))=-n\log((\mathbf{y}_{ih}-\mathbf{f}_{ih})'\mathbf{R}_{ih}^{-1}(\mathbf{y}_{ih}-\mathbf{f}_{ih})/n)/2-\log(|\mathbf{R}_{ih}|)/2-n/2 \qquad (8)$$

which is maximised with respect to α_{ih}, γ_{ih}, β_{ih}, ρ_{ih}, with ordinary optimisation methods. When we deal with models with reduced sets of parameters or however with constrains on the parameters, overall sample likelihood has to be used: of course the whole log-likelihood is obtained adding $l_{ih}(.)$ for all values of i and h, since we supposed the independence of the random errors among individuals and variables. Specific values of the parameters could be tested by comparing the unconstrained maximum with the maximum obtained imposing ν constraints to the parameters and then using the LR (Likelihood Ratio) test; asymptotically -2log(LR) follows a χ^2 distribution with ν degrees of freedom, but unfortunately the number m_i of observations for each individual is generally too small in our data set, so that the χ^2 approximation to LR could be used only to give a rough judgement on the reliability of specific hypothesis.

5. Application to a real data set

The main aim of the proposed parameterisation for our data set is to deal with a 2-way data set, because the parameters of the systematic part of the model, γ_{ih}, α_{ih}, β_{ih}, summarise the third way, i.e. time. In a first stage we applied the above parameterisation to our data set, obtaining a set of 3×n×k parameter estimates: in fact we have 3 estimated parameters for each of the n=64 individuals and for each of the k=2 variables (height and weight). The analysis of the relationships between the estimates suggested some reductions in the number of parameters; for the i-th individual we put:

$\rho_{i1}=\rho_{i2}=\rho_i$ (autocorrelations are equal for the two variables but generally different among individuals);

$\beta_{i1}=\beta_{i2}=\beta_i$ (equal individual growth speeds for the two variables but generally different among individuals). A similar simplification is used by Lundbye-Christensen (1991).

Indeed the last simplification is also strongly suggested by the data, as well as the need of using all the information at disposal to estimate individual growth speeds; in fact the height has a 12% average percentage of missing data. Furthermore, the strong internal (infra-individual) linear correlation between the height and weight suggested us this simplification. The decrease of likelihood of this simplified model was not significant, so that we summarised the data set by means of the estimates of the five parameters of the systematic component: $\hat{\alpha}_{i1}$, $\hat{\gamma}_{i1}$, $\hat{\alpha}_{i2}$, $\hat{\gamma}_{i2}$ and the common slope $\hat{\beta}_i$. This estimated common individual slope $\hat{\beta}_i$ resulted to be highly correlated (R=0.95) with the individual slopes $\hat{\beta}_{i1}$ and $\hat{\beta}_{i2}$ estimated separately for the two variables.

The data did not present any evidence of difference between male and female parameters. Two individual parameter estimates appeared to be very far from the bulk of the data, so that we eliminated them from subsequent stages: they belong to children for which the above assumptions lead to unrealistic parameter estimates, since their observed growth curves are almost linear. In table 1 we report the mean and standard deviation of the individual estimates of the parameters, computed on the remaining 62 subjects.

Table 1: *Mean and standard deviation of the individual parameter estimates, computed on 62 subjects*

Estimate	$\hat{\gamma}_{i1}$	$\hat{\alpha}_{i1}$	$\hat{\gamma}_{i2}$	$\hat{\alpha}_{i2}$	$\hat{\rho}_i$	$\hat{\beta}_i$
Mean	4.48	16.69	54.58	64.78	0.05	0.42
Std. Dev.	1.19	5.48	4.99	13.97	0.06	0.20

An interesting aspect is the strong non-normality of the joint distribution of the estimates, as can be seen from figure 3, where we plotted the pairs of values of the estimates of α_{i1} and β_i. We see some evidence against the joint normality of the sampling distribution of the estimates, as it can be expected given the intrinsic nonlinearity of the model (Seber, Wild, 1989) also from figure 4, where we reported the likelihood contour plot of the 22nd individual with respect to same pair of parameters (α_{i1} and β_i, with i=22).

Figure 3: *Plot of 62 pairs of estimates of α_{i1} (x-axis) and β_i (y-axis)*

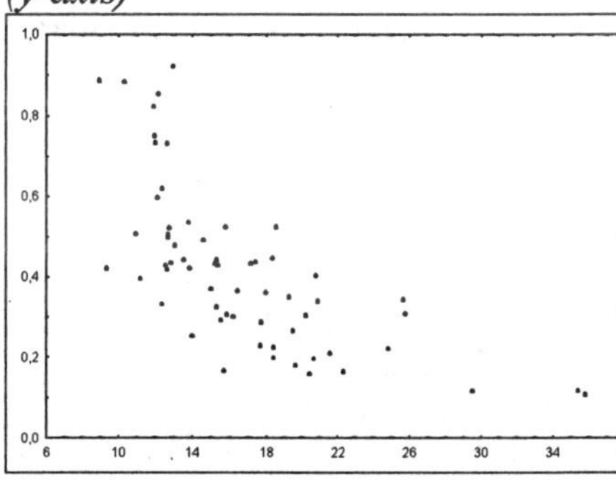

Figure 4: *Likelihood contour plot of the 22nd individual for the parameters α_{i1} and β_i*

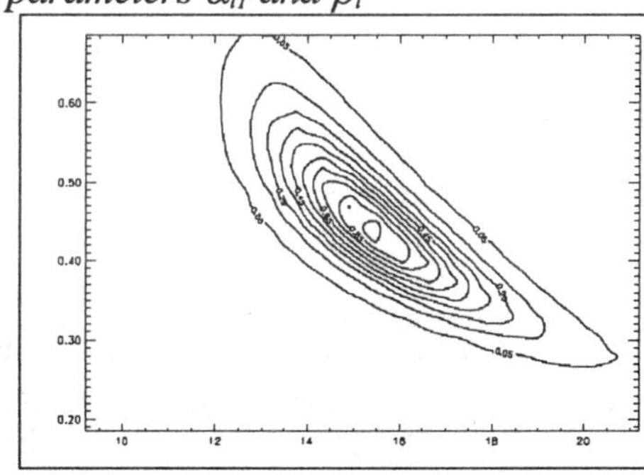

6. Conclusion

The above analysis shows that the model (3), together with the assumptions made on serial correlations, is suitable to analyse the growth of children. The data have suggested some reduction on the number of parameters: in particular we estimated a common individual slope $\hat{\beta}_i$ and a common individual serial correlation $\hat{\rho}_i$ for both variables; even if there is still a strong collinearity among the estimates of the remaining parameters, in the present paper we do not mention any further reduction of parameters.

A strong non normality of the sampling distribution of the parameter estimates is suspected, as usual in intrinsic nonlinear models.

The obtained promising results induced us to deal, in a forthcoming paper, with random coefficient nonlinear models, in order to better deepen the study of the variability among individual growth parameters.

References

Borra S., Di Ciaccio A. (1996). Analisi fattoriale multilevel:potenzialità di analisi nell'ambito della valutazione scolastica. In *Nuove metodologie per l'analisi di dati a tre indici*; Workshop held on November, 19th, 1996, in Dipartimento di Statistica, Probabilità e Statistiche Applicate; Roma, 20-21.

Chiodi, M. (1989). The clustering of longitudinal multivariate data when time series are short. In: *Multiway data analysis*. Editors: Coppi,R. and Bolasco,S. Elsevier Science Publisher B.V. (North-Holland).

Diggle, P.J. (1988). An Approach to the Analysis of Repeated Measurements. *Biometrics*, 44, 959-971.

Escoufier, Y. (1987). Three-mode data analysis: the STATIS method, *Methods for Multidimansional Data Analysis*, ECAS, 259-272.

Goldstein H., Healy M.J.R., Rasbash J. (1994). Multilevel time series models with applications to repeated measures data. *Statistics in Medicine*, 13, 1643-1655.

Harshman R.A. (1970). Foundations of the PARAFAC procedure: Models and methods for an "explanatory" multi-mode factor analysis, *UCLA Working Papers in Phonetics*, 16, 1-84.

Jolicoeur, P., Pernin, M.O, Pontier, J. (1988). A Lifetime Asympotic Growth Curve for Human Height. *Biometrics*, 44, 995-1003.

Jones, R.H. and Ackerson, L.M. (1990). Serial correlation in unequally spaced longitudinal data. *Biometrika*, 77, 4, 721-731.

Kroonenberg, P.M. (1992). Three-mode component model: a survey of the literature, *Statistica Applicata*, 4, 4, 619-633.

Lundbye-Christensen, S. (1991). A Multivariate Growth Curve Model for Pregnancy. *Biometrics*, 47, 637-657.

Milani S., Bossi A. (1988). Relazione tra modelli lineari classici per lo studio dell'accrescimento somatico, *Atti della XXXIV Riunione Scientifica della Società Italiana di Statistica*, Siena 27-30 Aprile 1988, 2, 2, 77-84.

Mineo, A. (1987). Solution using clustering method. In *Data analysis:The ins and outs of solving real problems*. Editors: Janssen,J., Marcotorchino,F. and Proth,J.M.; Plenum Press, New York.

Núñez-Antón, V., Woodworth, G.G. (1994). Analysis of longitudinal data with unequally spaced observations and time-dependent correlated errors. *Biometrics*, 50, 445-456.

Palmer, M.J., Phillips, B.F., Smith, G.T. (1991). Application of Nonlinear Models with Random Coefficients to Growth Data. *Biometrics*, 47, 623-635.

Rao, C.R. (1965). The theory of least squares when the parameters are stochastic and its application to the analysis of growth curves. *Biometrika*, 52, 447-458.

Rizzi, A., Vichi, M. (1995). Three-way data set analysis. In: *Some Relations Between Matrices and Structures of Multidimensional Data Analysis*, Editor: Rizzi, A., Consiglio Nazionale delle Ricerche; Giardini editori, Pisa, 93-166.

Seber, G.A.F., Wild, C.J. (1989). *Nonlinear regression*. John Wiley, New York.

Tanner, J. M. (1981). *Auxologia dal feto all'uomo: la crescita fisica dal concepimento alla maturità*. Ed. italiana a cura di L.Benso, UTET, Torino.

Tucker, L.R. (1966). Some mathematical notes on three-mode factor analysis, *Psychometrika*, 31, 279-311.

Von Bertalanffy, R. (1957). Quantitative laws in metabolism and growth. *Quarterly Review of Biology*, 32, 217-231.

The Kalman Filter on Three Way Data Matrix for Missing Data: A Case Study on Sea Water Pollution*[1]

Mauro Coli, Luigi Ippoliti, Eugenia Nissi
Dipartimento di Metodi Quantitativi e Teoria Economica
Università G. d'Annunzio
Viale Pindaro, 42, 65127 Pescara

Abstract: This paper proposes a method for the reconstruction of missing data in a three-way data array, based on six modified procedures of the optimum Kalman filter in relation to the structural data analysis. The case study regards environmental data on sea water pollution observed in the Adriatic sea.

Key Words: Kalman filter, state-space model, missing data, three way environmental data matrix.

1. Introduction

The aim of this paper is to propose a methodology for the reconstruction of missing data in a three way data array, supposing that the environmental observations $y_{(i,j,t)}$, characterised by three co-ordinates variables, space and time, are represented through a state-space dynamic system.
The above mentioned representation, based on a particular Markovian stochastic processes, allows the change of state from past to future information through an optimum filtering proposed by R. E. Kalman, to study the system theory (Kalman and Bucy, 1960). Such a dynamic system results completely specified if it is structured in terms of two equations, where the first, namely of transition or state equation, shows the functional composition of the state parameter, while the second equation, namely of observation, tries to forecast the future data or, as in our case, to reconstruct the missing information. As far as we know, the treatment of missing data on three-way arrays has not been fully studied and general and suitable techniques have not been proposed. The most frequently adopted techniques in time series analysis, use methods linked to ARIMA class models, with stationary and isotropic restrictions as well as equally spaced units of measurements of variables.

[1] * The paper has been supported by a grant MURST 40% titled "Analisi dei dati spaziali", national coordinator Prof. Mauro Coli.

Fig.1 An Exemplification of three way data matrix object of study.

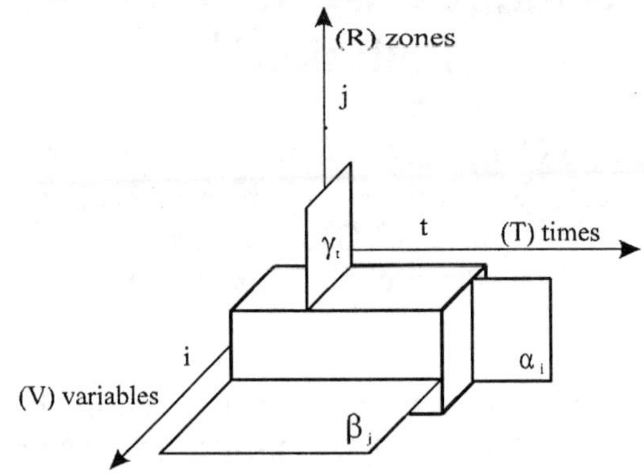

$\mathbf{Y}=\{\ y_{(i,j,t)}$ where $i = 1,2,..,V$ variables; $j = 1,2,.....,R$ zones ; $t = 1,2,..,T$ times;$\}$

2. The model

Any interpretative dynamic model of three-dimensional data, should be able to reproduce, with its own structure, the variables interrelationship in any of the three directions of the data set. In particular to analyse a phenomenon characterised by a three-way array $\mathbf{Y}=\{y_{(i,j,t)}$: $i=1,2,..,V$ variables; $j=1,2,.....,R$ zones; $t=1,2,..,T$ times;$\}$ the missing part can be completely reconstructed through a procedure that requires the collapse of one or two dimensions (fig. 1). Choosing, for example the i-th variable, the corresponding slice $\alpha_i(T,R)$, parallel to axes T and R, is a space-time matrix, where the bidimensional optimum Kalman filter may be applied, to reconstruct the missing data. Similarly, choosing the j-th zone, the corresponding slice $\beta_j(T,V)$ parallel to axes V and T, identifies a matrix containing a multiple time series, on which the Kalman filter for autoregressive vector models (VAR) may be applied.
In our paper, we intend propose six procedures: four are based on a bidimensional Kalman filter, and two are based on the ordinary Kalman filter, in which, the first is an ARIMA model and the second is an autoregressive vector model. Moreover, the procedures, whose names include the letter S, we have applied a smoothing algorithm.
From the above considerations, a flexible and general approach for the reconstruction of missing data in a three-way array can be carried out considering a linear model of two equations expressed in vectorial terms. The first, state equation, is $\mathbf{X}_{s+1} = \mathbf{\Phi}\mathbf{X}_s + \mathbf{W}_{s+1}$,while the second, observation equation, is $\mathbf{Y}_s = \mathbf{A}^T\mathbf{\Psi} + \mathbf{H}_s\mathbf{X}_s + \mathbf{V}_s$, where $\mathbf{\Phi}$ $\mathbf{H}$ and $\mathbf{A}^T$ are parameter matrices; $\mathbf{W}_s$ e $\mathbf{V}_s$ are observations and state Gaussian zero mean noises with covariances equal to $\mathbf{Q}$w and $\mathbf{Q}$v; $\mathbf{\Psi}$ is a vector of exogenus or predetermined

variables. In such terms, the Kalman filter can be expressed through two phases: forecasting and updating, and determines the optimal estimate of the state vector X_s, whenever new information becomes available. The optimal estimate of $\mathbf{X}_{s+1}$ is given by: $\hat{\mathbf{X}}_{s+1/s} = \mathbf{\Phi}\hat{\mathbf{X}}_{s/s}$, while the covariance matrix of the forecast error is $\mathbf{P}_{s+1/s} = \mathbf{\Phi}\mathbf{P}_{s/s}\mathbf{\Phi}' + \mathbf{Q}_w$. These two equations are known as forecasting equations. Once the new information $\mathbf{Y}_s$, becomes available the estimate $\hat{\mathbf{X}}_{s/s}$, can be updated. The updating equations are: $\hat{\mathbf{X}}_{s/s} = \hat{\mathbf{X}}_{s/s-1} + \mathbf{K}_s\left[\mathbf{Y}_s - \mathbf{H}\hat{\mathbf{X}}_{s/s-1}\right]$ and $\mathbf{P}_{s/s} = \left[\mathbf{I} - \mathbf{K}_s\mathbf{H}_s\right]\mathbf{P}_{s/s-1}$, where $\mathbf{K}_s$ is the Kalman gain matrix. It should be pointed out that, when we have missing data the updating equations become: $\hat{\mathbf{X}}_{s/s} = \hat{\mathbf{X}}_{s/s-1}$ and $\mathbf{P}_{s/s} = \mathbf{P}_{s/s-1}$.

In some cases the state vector can be interpreted in structural terms, so it is more appropriate to estimate its value at a particular point, using all the information and not just a part of it. Such an inference is called the smoothed estimate, while the corresponding estimator is called smoother. Since the smoother is based on more information than the filtered estimator, it will have a mean squared error which, generally is smaller than the filtered estimator. In statistical literature several smoothing algorithms in linear models have been proposed, and in our paper we will use fixed-interval and fixed-lag smoothing algorithms. The first computes the full set of smoothed estimates for a fixed span of data and implies a backward recursion of the Kalman filter. The latter algorithm computes smoothing estimates for a fixed delay and runs in parallel with the Kalman filter (Anderson and Moore, 1979). It is clear that when we apply the bidimensional Kalman filter, the smoothing procedure is characterized by a considerable computational effort, infact in each iteration of the backward procedure both state vector and covariance matrix of the forecast error must be stored.

As we can see in fig. 1, the slice $\alpha_i(T,R)$ is referred to a space-time matrix associated to the i-th variable and we can apply the ordinary unidimensional or bidimensional Kalman filter method. A key feature in two-dimensional application is that there is great freedom in deciding which data are to be considered available for processing. As in the 2-D signal processing literature the area of the lattice (slice $\alpha_i(T,R)$) which is used to process the data is called "the support region". For example, in the whole slice $\alpha_i(T,R)$ the smoothing may be performed on the basis of a support which in principle involves all the data. In practice, considering that the filtering is optimum only for a minimum dimension of the state vector, only particular supports are meaningful in the prediction and filtering tasks. To by-pass the high dimension of the state vector problem, the model for the reconstruction of the missing data that we propose, regards the Ordinary Reduced Kalman Filter-Smoothing (ORKFS) (Fig. 2). Next, the fixed lag smoothing will be applied. In this case, the updating procedure is based on a limited number of information near to $y_{\alpha i}(j,t)$ regarding slice α of the i-th variable at the t-th instant and in the j-th zone. Thus we choose to update only those elements of the state vector within a fixed distance from $y_{\alpha i}(j,t)$. We expect this procedure to give a good approximation because significant update will be confined to a region around the observation $y_{\alpha i}(j,t)$. Therefore, omitting the update of distant elements should only minimally impact the performance. Since in our case the filter needs a direction, we will assume that the updating is done starting with the upper left axis, moving along observations from left to right and row after row (Woods and Radewan 1977) (fig. 2.b).

Fig. 2 Restriction dependence for ORKFS and OMKFS

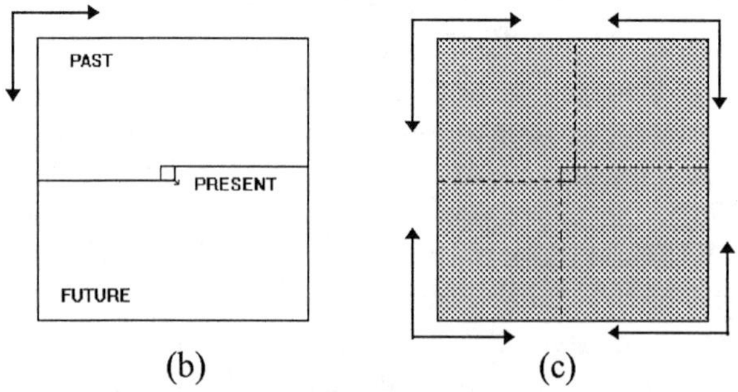

The support region R_M related to the ORKFS is defined by:

$$R_M(j,t) = \left[(t-g, j-h) \middle| (1 \le h \le M; 0 \le g \le M) \cup (-M \le h \le 0; 1 \le g \le M)\right]$$

where M =1,2 defines the order of the recursive model.

As an explanation, Figure 3 shows the support region R_M for a first order model (M=1).

Fig. 3 Support Region R_M for a first order model

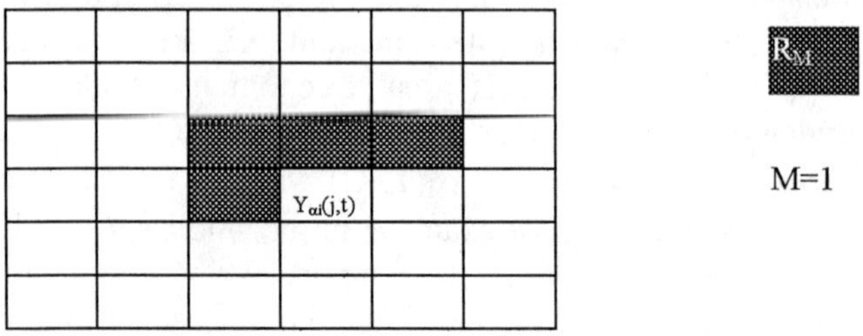

This allows us to define the generic elements $y_{\alpha_i}(j,t)$ through the following state equation: $x_{\alpha_i}(j,t) = \sum_{(j-h;t-g)\in \mathbf{R_M}} \phi_{hg} x_{\alpha i}(j-h, t-g) + w(j,t)$ where ϕ_{hg} represent the coefficients that regulate the relation between $x_{\alpha i}(j,t)$ and $x_{\alpha i}(j-h,t-g)$ and w(j,t) is the realisation of a Gaussian stochastic field.

The resulting ORKFS equations can be written in a scalar form as given below.

In these equations the subscript "a" and "b" indicate "after" and "before" updating, respectively; the argument represents the position of the data on slice $\alpha_i(T,R)$. State prediction and update are as follows:

$$\hat{x}^{b}_{\alpha i}(j,t) = \sum_{(j-h,t-g)\in R_M} \phi_{hg}\hat{x}^{b}_{\alpha i}(j-h,t-g)\,, \tag{1}$$

$$x^{a}_{ai}(m,n) = \hat{x}^{b}_{ai}(m,n) + k(j-m;t-n)\left[y_{\alpha i}(j,t) - \hat{x}^{b}_{ai}(j,t)\right], \tag{2}$$

with: $(m,n) \in R_{\oplus+} = \{m \geq 0,\ n \geq 0\} \cup \{m < 0, n > 0\}$. (3)

It should be noted that the estimate defined in the last equation can be a filtered estimate (if j=m, t=n) or a fixed lag smoothed estimate (if (m,n) occurs prior to (j,t)). Since the implementation of the ORKFS requires the coupling coefficient vectors ϕ and the process noise variance Q_w, the general and Bias-Compensated Least Squares is used to identify these quantities directly from the data (as suggested by H.Kaufman, J.Woods et al. 1983).

In order to estimate Q_w, consider the following expression:

$$P = E\left|y_{\alpha i}(j,t) - \phi^T \mathbf{y}^{R_M}_{\alpha i}(j-1,t)\right|^2\,, \tag{4}$$

where:

$$\mathbf{y}^{R_M}_{\alpha i}(j-1,t) = \mathbf{x}^{R_M}_{\alpha i}(j-1,t) + v_{\alpha i}(j-1,t)\,, \tag{5}$$

and $\mathbf{x}^{R_M}_{\alpha i}(j-1,t)$ is a vector consisting of those elements in R_M.

Expansion of the above expression, taking into account the fact that $v_{\alpha i}(j,t)$ is an uncorrelated zero mean sequence, gives

$$P = E\left|x_{\alpha i}(j,t) - \phi^T \mathbf{x}^{R_M}_{\alpha i}(j-1,t) + v(j,t) - \phi^T v(j-1,t)\right|^2 = E\left[w(j,t) + v(j,t) - \phi^T v(j-1,t)\right]^2 = Q_w + Q_v + \phi^T Q_v \phi\,, \tag{6}$$

thus

$$Q_w = E\left[y_{\alpha i}(j,t) - \phi^T \mathbf{y}^{R_M}_{\alpha i}(j-1,t)\right]^2 - Q_v(1+\phi^T\phi)$$

Consequently, this result suggests the following procedure for identifying Q_w:

1) Collect a representative set W of data $\{\, y_{\alpha i} = (j,t) \colon (j,t) \in W \,\}$, and compute an estimate of ϕ with the procedure to be discussed subsequently. Next approximate P with an estimated value $\hat{P}$, namely:

$$\hat{P} = \frac{1}{\mathbf{N}_w} \sum_{(j,t)\in W} \left((\mathbf{y}_{\alpha i}(j,t) - \phi^T y_{\alpha i}^{R_M}(j-1,t)) \right)^3 \tag{7}$$

where N_w denotes the number of data in W.

2) Compute an estimate of Q_w as follows:

$$\hat{Q}_W = \hat{P} - Qv(1+\phi^T\phi). \tag{8}$$

The most straightforward procedure for identifying the coefficient vector ϕ is to perform a least-squares fit over a representative block W of data using the observations $y_{\alpha i}(j,t)$ in place of the true densities $x_{\alpha i}(j,t)$.

That is, the estimate of ϕ would be determined so as to minimize $\hat{P} = \sum_{(j,t)\in W} \left((y_{\alpha i}(j,t) - \phi^T y_{\alpha i}^{R_M}(j-1,t)) \right)^2$.

Setting to zero the gradient of P with respect to ϕ results in:

$$\hat{\phi} = \left[\sum_{(j,t)\in W} \mathbf{y}_{\alpha i}^{R_M}(j-1,t)\, \mathbf{y}_{\alpha i}^{R_M\,T}(j-1,t) \right]^{-1} \left[\sum_{(j,t)\in W} \mathbf{y}_{\alpha i}^{R_M}(j-1,t)\, y_{\alpha i}(j-1,t) \right] \tag{9}$$

It should be noted that, since $\mathbf{y}_{\alpha i}^{R_M}(j,t)$, the term that multiplies ϕ, contains noise, the estimate will be significantly biased.
In order to reduce this bias, the following correction is suggested:

$$\hat{\phi} = \left[\sum_{(j,t)\in W} y_{\alpha i}(j,t) y_{\alpha i}^{R_M\,T}(j-1,t) - \mathbf{I}N_w Q_w \right]^{-1} \left[\sum_{(j,t)\in W} y_{\alpha i}(j,t) y_{\alpha i}(j-1,t) \right] \tag{10}$$

where **I** is the identity matrix.

By changing the dependence restrictions, the Ordinary Reduced Kalman Filter (ORKFS) procedure (fig. 2.b.), can be repeated starting from each vertex of the slice. The final estimate of the missing data is an average of the obtained results. This new procedure is a modification of the (ORKFS) and has been called the Ordinary Modified Kalman Filter-Smoothing (OMKFS) (fig. 2.b). Keeping in mind slice $\alpha_i(T,R)$, an alternative procedure called RWKF (Reduced Weighted Kalman Filter), consists in highlighting an eventual recurring data structure. In such way, the matrices $\mathbf{A}^T$, $\mathbf{H}$ and Ψ, can be useful for weighting the estimates obtained with the ORKFS procedure with seasonal index, temporal and/or spatial mean. Again, like before, by changing the dependence restrictions, the procedure can be repeated starting from each corner of the slice (procedure MRWKF). The last two proposed procedures require an univariate and multivariate time series analysis. Particularly, on slice $\alpha_i(T,R)$, we can identify a time series for each zone and utilise both ordinary Kalman Filter and fixed-interval smoothing for ARIMA class models (FKARMAS). On the other hand, on slice $\beta_t(V,T)$ we can filter with an autoregressive vectorial model (VAR), and one time too, we are able to apply the Kalman filter and smoothing algorithm (procedure FKVARS).

4. The case study

The data of this study, are related to the results of a monitoring project and concerns the presence of some polluting substances supposed responsible for the Eutrophications phenomenon. The three-way array proves to be defined by 10 variables, 17 zones and 49 times. From the above mentioned matrix, several observations were eliminated in such a way as to internally obtain a "cloud" of missing data. This cloud was then reconstructed using the six adopted procedures and the goodness of fit was evaluated using the R squared measure. The following tables show the measure and ranks of the R squared, calculated for our first five procedures and for each of the ten observed variables. In our case study, looking at both tables, the best method results the MRWKF procedure. This confirm that the previous study of the data structure composition is fundamental to obtain the best reconstruction of the missing data.

Tab 1. Values of R squared calculate for the first five procedures for each of ten variables

	Variables									
Procedures	**1**	**2**	**3**	**4**	**5**	**6**	**7**	**8**	**9**	**10**
ORKFS	0.90	0.95	0.90	0.93	0.96	0.92	0.90	0.96	0.97	0.90
OMKFS	0.95	0.89	0.95	0.92	0.95	0.95	0.91	0.90	0.94	0.86
RWKF	0.91	0.78	0.92	0.91	0.72	0.97	0.97	0.94	0.90	0.92
MRWKF	0.87	0.84	0.91	0.89	0.90	0.99	0.95	0.96	0.98	0.97
FKARMAS	0.92	0.93	0.87	0.85	0.68	0.90	0.86	0.83	0.88	0.84

Tab 2. Ranks of the R squared obtained with each of the five procedures

	Variables											
Procedures	**1**	**2**	**3**	**4**	**5**	**6**	**7**	**8**	**9**	**10**	**Average**	**Ranks**
ORKFS	4°	1°	4°	1°	1°	4°	4°	1°	2°	3°	2.5	2°
OMKFS	1°	3°	1°	2°	2°	3°	3°	4°	3°	4°	2.6	3°
RWKF	3°	5°	2°	3°	4°	2°	1°	3°	4°	2°	2.9	4°
MRWKF	5°	4°	3°	4°	3°	1°	2°	1°	1°	1°	2.4	1°
FKARMAS	2°	2°	5°	5°	5°	5°	5°	5°	5°	5°	4.2	5°

As demonstrated in the table 3, even the last procedure (FKVARS) based on the

VAR models, shows for a multiple time series, a satisfying reconstruction capacity of missing data.

Tab. 3 R squared obtained using the FKVARS procedure

	Zones with missing data							
	7	8	9	10	11	12	13	14
Rsquare	0.89	0.91	0.93	0.94	0.79	0.97	0.91	0.86

References

Akaike H. (1974), "Markovian Representation of Stochastic Processes and its Application to the Analysis of Autoregressive Moving Average Processes", *Annals of the Institute of Statistical Mathemathics*, 26, pp. 363-387.

Anderson T.W. (1984), *An Introduction to Multivariate Analysis*, John Wiley and Sons.

Anderson B.D.O., Moore J.B. (1979), *Optimal Filtering*, Englewood Cliffs, Prentice Hall.

Arabie P., Carroll J.D., De Sarbo W.S., (1987), "*Three way scaling and clustering*", Sage London

Harvey A.C. (1989), *Forecasting structural time series models and the Kalman filter*, Cambridge University Press.

Kalman R.E., (1960), "New Approach to Linear Filtering and Prediction Problems", *Transaction of ASME, Journal of Basic Engineering* D, 82, 35-45.

Kalman R.E., Bucy R. S: (1961) "New Results in Linear Filtering and Prediction Theory" *Transaction of ASME, Journal of Basic Engineering* D, 83, 95-108.

Kaufman H, Woods J.W., Dravida S, Tekalp, A.M. (1983), Estimation and identification of two dimensional Images, *IEEE Transaction on Automatic Control*, 28, 7, 745- 756.

Rizzi A., Vichi M. (1995), "Three way data sets analysis", in *Some relations between matrices and structures of multidimensional data*, Giardini Editore.

Woods J.W., Radewan C.H. (1977), "Kalman Filtering in Two Dimension", *IEEE Transaction Information Theory*, IT-23, 473-482.

Three-Way Data Arrays with Double Neighbourhood Relations as a Tool to Analyze a Contiguity Structure

Pierre-André Cornillon†, Pietro Amenta*, Robert Sabatier‡
† Unité de biométrie, INRA - ENSA.M - UM II, Place Paul Viala, 34060 Montpellier, France, e-mail: pac@helios.ensam.inra.fr
*Dipartimento di Matematica e Statistica, Università "Federico II" di Napoli, Via Cinthia, 80126 Napoli, Italia, e-mail: amenta@unina.it
‡ Laboratoire de Physique Moléculaire et Structurale, 15 av. Ch. Flahaut 34060 Montpellier, France, e-mail: sabatier@pharma.univ-montp1.fr.

Abstract: In this paper we present two methods to analyze three-way data arrays with double neighbourhood relations. The first procedure use Kronecker product between graph matrices to construct a neighbourhood operator. Some of the most significant eigenvectors of this operator allows modelization of the underlying phenomena. The second methods make Kronecker product between neighbourhood operators of each graph matrices and is equivalent to a particular STATIS. A comparison between these two procedures on ecological data set is then performed.

Keywords: Graph, Geary coefficient, Kronecker product, Three-ways data, STATIS method, Neighbourhood operator.

1 Introduction

In the analysis of three-way data arrays, and also two-way matrices, relations between observations are unfortunately often ignored. Following the works of Lebart (1969), Cliff and Ord (1981), Le Foll (1982) and more recently Méot *et al.* (1993), proximity between observations can be used in a multivariate data analysis framework. Extension to three-way data with double neighbourhood can be done in a natural way (Cornillon *et al.*,1993). In order to understand what is double neighbourhood, let us consider some variables collected in different sites at different times. To take into account the relationship between two sites (for instance the distance) it is useful to consider a neighbourhood matrix. Another one is obviously needed to modelize time relations between observations. The ecological data set we will study presents a similar structure as described in Figure 1. In section 2 we first recall a definition of neighbourhood matrix and local variance (Lebart, 1969). From these definitions we introduce the neighbourhood operator and give a brief review of methods using it, with a special emphasis on Geary's index (1954) and local variance (Lebart, 1969). This method, after a diagonalization of a particular operator, produces an orthogonal base with optimal autocorrelation properties. The extension of this methodology in a three-way data context lead us to introduce two different methods. Both approaches take into account the two neighbourhood matrix, by either the direct Kronecker product of the graph matrices or Kronecker product of the neighbourhood operators. The first analysis generalizes the Upton and Fingleton approach (1985) while the second one is directly connected to the STATIS method (Lavit, 1988).

2 Some neighbourhood analysis in the two-way case

To define the neighbourhood relation between p statistical units, we can use a symmetric neighbourhood valued graph (e.g. Lebart 1969, Méot *et al.* 1993). In the sequel we will restrict ourselves to unvalued graph. We can associate to these graph a boolean symmetric matrix $\mathbf{M}$ of order p, which general term is m_{ij} $(1 \leq i \leq p)$ such that $m_{ij} = 1$ if vertices i and j are linked, zero otherwise. Let x be the vector of the p observations of the variable x.

Figure 1: *Frame of the ecological data set*

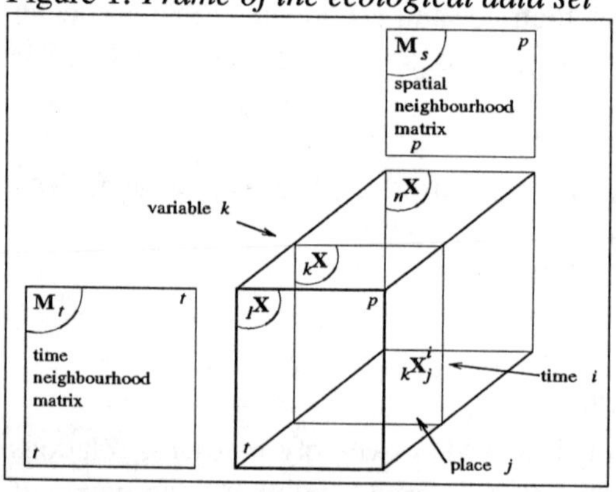

This contiguity information can be introduced into empirical variance of variable x, $S^2(\mathrm{x})$, by a suitable decomposition of $S^2(\mathrm{x})$ in two terms: the first one depends on the graph and is called the local variance while the second one depends on the complementary graph. Let define a weight matrix of statistical units $\mathbf{D} = diag(d_1, d_2, \cdots, d_p)$, the empirical variance can be written as :

$$\begin{aligned} S^2(\mathrm{x}) &= \frac{1}{2}\sum_{i,j} d_i d_j \,(x_i - x_j)^2 \\ &= \frac{1}{2}\sum_{i,j} m_{ij} d_i d_j \,(x_i - x_j)^2 + \frac{1}{2}\sum_{i,j} (1 - m_{ij}) d_i d_j \,(x_i - x_j)^2. \end{aligned}$$

Putting $\mathbf{D}^* = diag(d_1^*, d_2^*, \cdots, d_p^*)$, where $d_i^* = \sum_{j=1}^{p} m_{ij} d_j$, which is the sum of weights of vertix i's neighbours, we can write the last expression as:

$$S^2(\mathbf{x}) \quad = \quad \mathbf{x}'\mathbf{D}\mathbf{Q}_0\mathbf{x} = \mathbf{x}'\mathbf{D}\mathbf{E}\mathbf{x} + \mathbf{x}'\mathbf{D}(\mathbf{Q}_0 - \mathbf{E})\mathbf{x},$$

where $\mathbf{E} = \mathbf{D}^* - \mathbf{MD}$ is called the neighbourhood operator associated with the matrix $\mathbf{M}$, and $\mathbf{Q}_0$ is the projector on the orthogonal complement of the subspace spanned by $\mathbf{1} \in \mathbb{R}^p$. The matrix $\mathbf{Q}_0 - \mathbf{E}$ is also an neighbourhood operator, that is a $\mathbf{D}$-symmetric positive matrix $\mathbf{A}$ which verify $\sup_{\|\mathbf{x}\|_\mathbf{D} \leq 1} (\mathbf{x} \,|\mathbf{A}\mathbf{x})_\mathbf{D}$.
Diagonalization of $\mathbf{X}'\mathbf{DAXQ}$ where $\mathbf{Q}$ is the metric of the space of variables lead us to consider a Principal Components Analysis (PCA) of proximities and give principal components which maximizes the following criterion: $\sup_{\|\mathbf{a}\|_\mathbf{D}=1} (\mathbf{AXQ}(\mathbf{a}) \,|\mathbf{XQ}(\mathbf{a}))_\mathbf{D}$ (Méot *et al.* 1993). This can be interpreted as a linear combination of the variables of $\mathbf{X}$, namely $\mathbf{XQ}(\mathbf{a})$, which has a local variance maximum. If $\mathbf{D} = 1/p\mathbf{I}_\mathbf{p}$ then the eigenvectors of $\mathbf{E}$ maximize the generalized autocorrelation Geary's index. Finally, Principal Component Analysis with respect to Instrumental Variables (PCAIV) on the eigenvectors of $\mathbf{E}$ can be performed. This analysis allows to explain the data matrix by the space spanned by the vectors of generalized autocorrelation Geary's index maximum (see Méot *et al.* 1993 or Escoufier, 1987).

3 Extension in the three-way data arrays case

Let $\mathbf{X}_{t \times p \times n}$, a three-way data array. We can associate to $\mathbf{X}$ two contiguity matrices $\mathbf{M}_\mathbf{s}$ and $\mathbf{M}_\mathbf{t}$, one for each of the two first dimensions. We note $\mathbf{E}_\mathbf{s}$ and $\mathbf{E}_\mathbf{t}$ their respective neighbourhood operator. We can vectorialize the two-way ${}_\mathbf{k}\mathbf{X}$ matrices of $\mathbf{X}$ along the third dimension: ${}_\mathbf{k}\mathbf{X}^\mathbf{c} = Vec({}_\mathbf{k}\mathbf{X})$ and place this vector, by columns, in the super-matrix $\mathbf{Y} = [{}_\mathbf{1}\mathbf{X}^\mathbf{c} \cdots {}_\mathbf{n}\mathbf{X}^\mathbf{c}]$.
Hence we can apply the local framework to the matrix $\mathbf{Y}$ of dimension $tp \times n$. In order to take into account the time and spatial contiguity relations into a variance decomposition, we can consider the Kronecker product of the graphs. This approach generalize Méot *et al.* (1993), using the matrix $\mathbf{M}$ of the graph given by Kronecker product of $\mathbf{M}_\mathbf{s}$ and $\mathbf{M}_\mathbf{t}$: $\mathbf{M} = \mathbf{M}_\mathbf{s} \otimes \mathbf{M}_\mathbf{t}$. The resulting operator $\mathbf{E}$ verifies:

$$\mathbf{E} = \mathbf{D}^* - \mathbf{MD} = \mathbf{D}^*_\mathbf{s} \otimes \mathbf{D}^*_\mathbf{t} - (\mathbf{M}_\mathbf{s} \otimes \mathbf{M}_\mathbf{t})\,(\mathbf{D}_\mathbf{s} \otimes \mathbf{D}_\mathbf{t})$$

If $\mathbf{Y}$ is univariate this approach is similar to the one proposed by Upton and Fingleton (1985). As mentioned before, we can consider a PCAIV on the most significant eigenvectors of $\mathbf{E}$, that is a regression of each variable on the space spanned by the chosen eigenvector of $\mathbf{E}$. Hence we try to fit $\mathbf{Y}$ with the most autocorrelated variables in the sense of Geary's index for the given matrix $\mathbf{M}$. As in principal component regression we only keep $m < n$ axis and the model could be written as:

$$\widehat{\mathbf{Y}}_{\mathbf{k}} = \alpha_1 \mathbf{c}^{(1)} + \cdots \alpha_{\mathbf{n}} \mathbf{c}^{(\mathbf{m})},$$

where $\mathbf{c}^{(1)} \cdots \mathbf{c}^{(\mathbf{m})}$ are the m first axis of the PCAIV, and $\widehat{\mathbf{Y}}_{\mathbf{k}}$ is the fitted model for the kth variable.

As a particular case, if we consider two complete graphs ($m_{ij} = 1, \ \forall(i,j)$), this approach yields to a PCA of the centered matrix $\mathbf{Y}$.

We can consider a different approach: the Kronecker product of the operators $\mathbf{E_s} \otimes \mathbf{E_t}$:

$$\mathbf{E} = \mathbf{D_s^*} \otimes \mathbf{D_t^*} - \mathbf{D_s^*} \otimes \mathbf{M_t D_t} - \mathbf{M_s D_s} \otimes \mathbf{D_t^*} + \mathbf{M_s D_s} \otimes \mathbf{M_t D_t}.$$

By analogy with the local variance, which in the two-ways case could be written as: $S^2_{loc}(\mathbf{y}) = \mathbf{y}'\mathbf{D}\mathbf{E}\mathbf{y}$, we can consider another type of local variance:

$$\gamma_{kk} = {}_{\mathbf{k}}\mathbf{X}^{\mathbf{c}\prime} \left(\mathbf{D_s E_s} \otimes \mathbf{D_t E_t} \right) {}_{\mathbf{k}}\mathbf{X}^{\mathbf{c}} = tr \left({}_{\mathbf{k}}\mathbf{X}' \mathbf{D_s E_s} \, {}_{\mathbf{k}}\mathbf{X} \, \mathbf{D_t E_t} \right).$$

This value can be interpreted as the product between matrices in ${}_{\mathbf{k}}\mathbf{X}$, or, as the scalar product between columns of $\mathbf{Y}$. Obviously, this scalar product take into account the double neighbourhood structure. These remarks lead us to see this method as a STATIS procedure with particular weight metrics $\mathbf{D_s E_s}$ and $\mathbf{D_t E_t}$. Hence the matrix $\mathbf{\Gamma}$ of general term γ_{kl} represents the variance matrix of the interstructure stage. The others steps of the STATIS method remain unchanged (see for instance Lavit, 1988). We can notice that, if we have two complete graphs and if every ${}_{\mathbf{k}}\mathbf{X}$ is both centered by column and centered by row, then the coefficient γ_{kl} can be written as $\gamma_{kl} = Cov({}_{\mathbf{k}}\mathbf{X}^{\mathbf{c}}, {}_{\mathbf{l}}\mathbf{X}^{\mathbf{c}})$.

4 Adriatic sea pollution

In order to analyze the Adriatic sea pollution we have considered a set of 10 physico-chemical variables (Temperature [Temp], Salinity [Sal], Transparency [Tran], Chlo-rophyll [ClorA], Ph [PH], Ammonia [Ammo], Nitric Nitrogen [ANITRC], Nitrous Nitrogen [ANITRO], Ortho Phosphorus [FosOrt], Global Phosphorus [FosTot]) sampled in 17 different stations along the Abruzzo coast at 49 different dates. The three-way data set dimension was $t = 49$, $p = 17$, $n = 10$. Moreover these data set was sampled at two different distances from the coast (500 and 3000 meters). To take into account the time and spatial contiguity relations we used two different linear neighbourhood graph matrices. Every sampling time is placed along a line and it is linked to another one if the segment of the line between them do not include another sampled time point. A more complex graph is constructed for the spatial contiguity: we consider the mutual distances between stations; two stations will be connected if their distance is less than the greatest distance between two following stations. A more graphical explanation is in Figure 2. We have developed two different analysis, each for different distance from the coast: the Kronecker product of the graphs and the Kronecker product of the operators. For the first approach, starting from the contiguity matrices we compute the Kronecker product between them, obtaining the neighbourhood operator **E**. We diagonalize this operator and the eigenvectors associated will be utilized for an PCAIV with respect to these eigenvectors. That is we try to explain the variability of **Y** by the most autocorrelated variables in the sense of Geary's index. In order to choose the most significative eigenvectors we have performed a simple regression of the 10 variables on each eigenvector.

Figure 2: *Neighbourhood matrix construction*

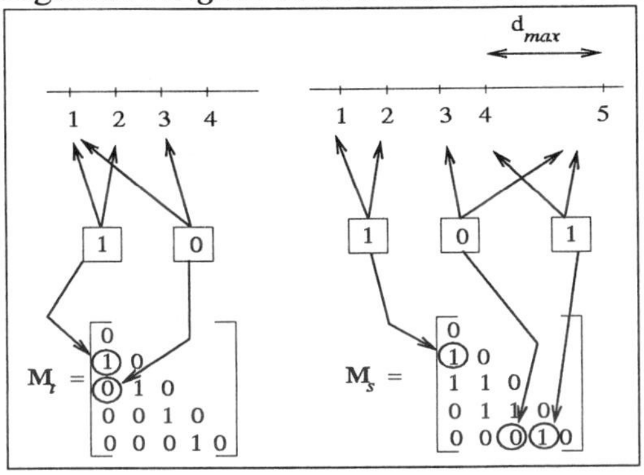

We call "Periodograph" the graphical representation of the squared correlation coefficients of this regression. The eigenvectors are chosen such that their squared correlation coefficients are greater than a fixed value. Figure 3 indicates that eigenvectors with a great contribution are those associated to the first sampling stations, whatever the distance considered. This follows from the fact that the Adriatic sea main stream is in direction of Venice. Because stations are located from Venice to Pescara in a progressive order, the first sampling stations are those with a greater contribution in explanation of the Adriatic sea pollution.

Figure 3: *Squared correlation coefficients of each variables with the eigenvectors of the neighbourhood operator (periodograph)*

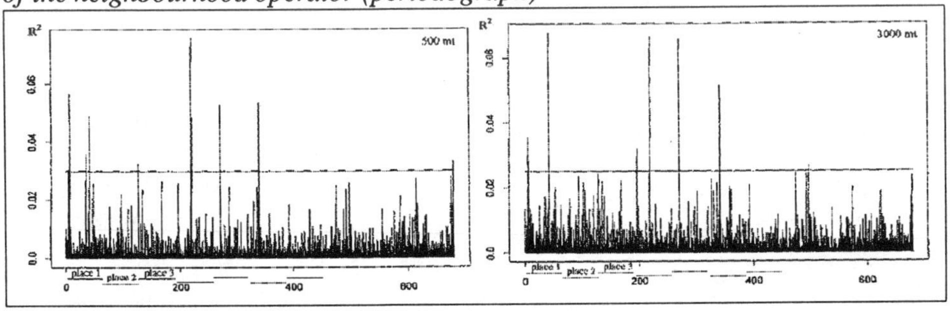

Figure 4 displays the plot of the variables FosOrt and Temp at 3000 m, and thc model fitted for the first component of the PCAIV ($m = 1$). As we could see, on one side the model selected has good ability to explain the variation of FosOrt, but on the other side some of the variability is not captured for the temperature (Temp). The one dimension model achieve a good explanation of the global variation and allow to detect unusual comportment.

Figure 4: *Data points and fitted data using one-dimensional model (solid lines)*

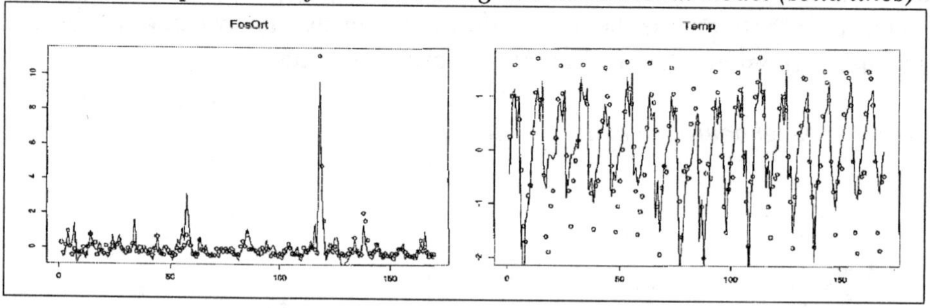

The second approach (Kronecker product of the operators) leads us to a STATIS on the array $\mathbf{X}$'s with semi-metrics $\mathbf{D}_s\mathbf{E}_s$ and $\mathbf{D}_t\mathbf{E}_t$. Obviously the interstructure step is equivalent to a PCA of the statistical triplet $(\mathbf{Y}, \boldsymbol{\Delta}, \mathbf{N})$ where $\boldsymbol{\Delta} = \mathbf{D}_s\mathbf{E}_s \otimes \mathbf{D}_t\mathbf{E}_t$ is a semi-metric for $\mathbb{R}^{pt}$ and $\mathbf{N}$ define a metric in $\mathbb{R}^n$, (see Lavit, 1988).

Figure 5: *Interstructure with and without double neighbourhood.*

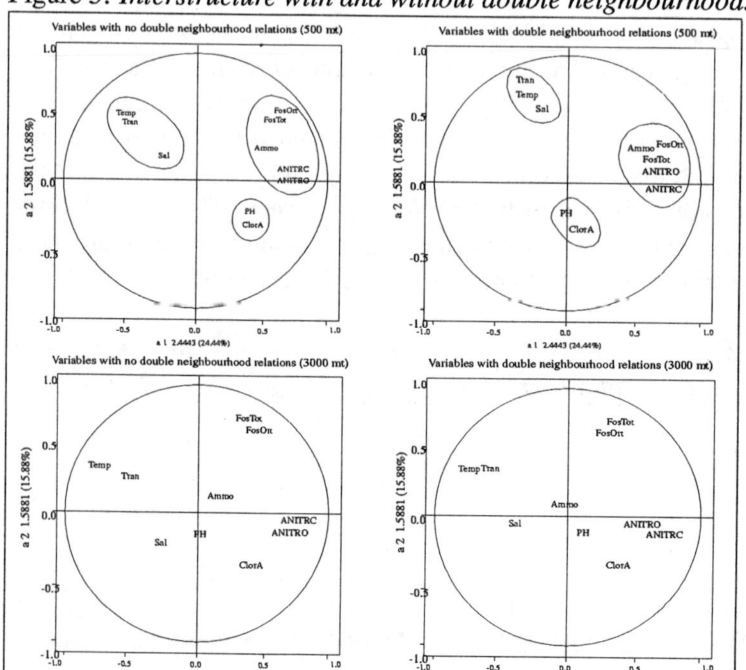

The plot of the $n = 10$ variables from interstructure step allows to detect a common structure between variables at different distances and different times. We can notice on figure 5 that variables' clusters at distance 500 m, are far more distinct when we use the neighbourhood constraints. This effect disappear when we consider the distance 3000 m. This is the direct consequence of a main ecological difference between 500 and 3000 m: at 3000m away from the coast this is open sea and the dilution of the pollution is far more efficient than at 500m.
A "mean" (called compromis) of all the n arrays is computed at the compromise step. From this array we can extract principal components and axis which are a new basis for their respective space (i.e. $\mathbb{R}^t$ and $\mathbb{R}^p$). Using this basis, a plot of each array (one per variable) can be done - these plots are called intrastructure step. For instance we can notice that for Nitric Nitrogen (figure 6) all the stations have different correlation between them for 500m and 3000m. Fine tuning this interpretation yields to note that stations 1 and 8 have both a good projection in the 500m and 3000m cases. That suggest they have a global behavior (in the time sense) close to the "mean" (i.e. compromis) behavior for this variable. Of course similar plot can be done with the other variable.

Figure 6: *Intrastructure for Nitric Nitrogen with neighbourhood relations.*

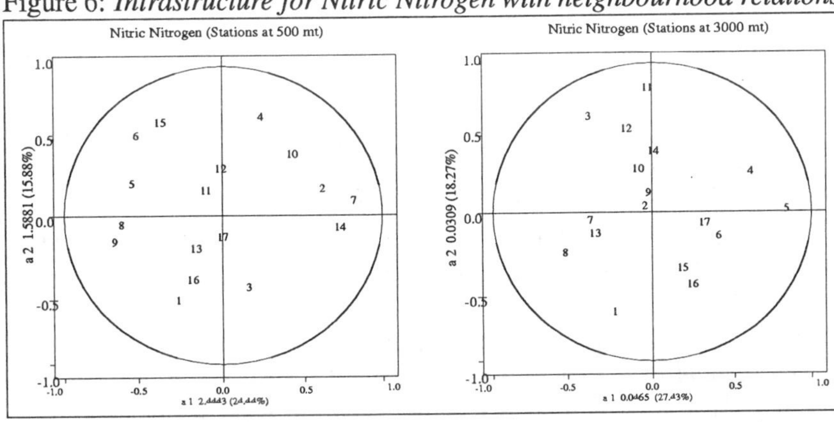

5 Concluding remarks

In order to analyze the Adriatic sea pollution we have developed two different analysis: the Kronecker product of the graphs and the Kronecker product of the operators. We have noticed that for the first method difficulties arise from the choice of components used in the model: obviously different components yield to different results. A better criterion than the so-called "periodograph" is needed, to improve the ability of this method to fit the data set. This second approach seems to be a good exploratory tool in the double neighbourhood framework: it give better results than standard STATIS and, moreover, is easy to implement on a computer. This study was realized on standard S-Plus software and functions are available from the second author.

References

Cliff, A. D. and Ord, J. K. (1981). *Spatial processes : models and applications.* Pion, London.

Cornillon, P. A., Sabatier, R., and Chessel, D. (1993). Analyse d'un cube sous double contrainte de voisinage. In *Bulletin of the international statistical institute*, volume 1, pages 285–286.

Escoufier, Y. (1987). Principal components analysis with respect to instrumental variables. In *European courses in advanced statistics*, pages 27–28. University of Napoli.

Geary, R. (1954). The contiguity ratio and statistical mapping. *The incorporated statistician*, 5:115–145.

Lavit, C. (1988). *Analyse conjointe de tableaux quantitatifs*. Masson, Paris.

Le Foll, Y. (1982). Pondération des distances en analyse factorielle. *Statistique et Analyse des Données*, 7:13–31.

Lebart, L. (1969). Analyse statistique de la contiguïté. *Publication de l'institut de statistiques de l'université de Paris*, 28:81–112.

Méot, A., Chessel, D., and Sabatier, R. (1993). Opérateur de voisinage et analyse de données spatio-temporelles. In Lebreton, J. and Asselain, B., editors, *Biométrie et Environnement*, pages 45–71. Masson, Paris.

Upton, G. J. G. and Fingleton, B. (1985). *Spatial analysis by example*, volume 1. John Wiley and Sons, New-York, 2 edition.

Acknowledgments: We thank Professor M. Coli for providing the Adriatic sea pollution data set and Professor D'Ambra for helpful remarks. This paper has been supported with a grant from C.N.R. 1996, (Prof. Luigi D'Ambra).

Firm Performance Analysis with Panel Data

Achille Lemmi
Department of Quantitative Methods
University of Siena

Duccio Stefano Gazzei
PHD Doctor
Statistical Department "G.Parenti", University of Florence

Abstract: This paper deals with an "ecumenical" approach to the productive processes analysis. In particular we suggest a research strategy based on five known estimation methods for the production frontier function one of each characterised by a proper type of flexibility in results evaluation. In such a way, we obtain through a bootstrap methodology, an interval estimation of the efficiency score of any firm.

Key words: Productive performance, Efficiency, Frontier Function, Bootstrap, Panel Data.

1. Introduction

In production analysis performance measures have the task to measure the "capacity" of a economic unit to transform input into output and to point out the achievement degree of a given production standard considered as the optimal one. Such a standard is normally expressed by a function representing both a production or a cost function.

Three different approaches are generally used for frontier estimation in international econometric and statistical literature.

- Deterministic approach[1]: all the observations are supposed lying below the frontier and the error term (one sided) capturing inefficiency only;
- Stochastic approach: observations are affected by an error term made up of two independent components one representing the statistical noise and the other inefficiency (Aigner, Lovell e Schmidt, 1977);
- Non parametric approach: no distributive hypothesis is advanced for the frontier shape. In some applications only frontier convexity is assumed. On the contrary Free Disposal Hull (FDH) technique, we will consider in this paper, ignores convexity as well.

---

[1] In the sense of Cornwell and Schmidt (1996): "*...technical efficiency for each firm can be calculated essentially as in the COLS procedure for the deterministic frontier*".

This paper deals with an "ecumenical" approach to the productive processes analysis. In particular we suggest a research strategy based on five known estimation methods for the production frontier function one of each characterised by a proper type of flexibility in results evaluation.

This is particularly relevant when a performance analysis of several firms is conducted using panel data; in this case we have to deal with productive processes dishomogeneity. Using methods characterised by different degree of flexibility in terms of efficiency, it is possible to control such dishomogeneous behaviours avoiding heavy estimation biases.

2. Frontier Estimation with Panel Data

Consider a production function

$$Y_{it} = \alpha + x_{it}\beta + E_{it}\,, \tag{1}$$

where Y_{it} indicates the proper production function for the i-th sample firm ($i=1,2,...,N$) in the t-th period ($t=1,2,...,T$); x_{it} is a vector $(1 \times k)$ of proper input function associated to the i-th sample firm considered at time t (the first element should be one); β is the vector $(k \times 1)$ of the regressor coefficients, α is a constant.

Two different forms can be used to represent the error term: (i) in the deterministic approach only the simple condition $E_{it} \leq 0$ is used since it represents inefficiency only; (ii) in the stochastic approach, proposed independently by Aigner, Lovell e Schmidt (1977) e Meeusen e Van Den Broeck (1977), the error term is divided into two components $E_{it} = V_{it} - U_i$: V_{it} are assumed as $N(0,\sigma_v^2)$ independent random variables, identically distributed (IID) and independent from U_i random variables; U_i are also still IID and non negative and defined by the truncation (at zero) of the $N(\mu,\sigma_v^2)$ distribution. In addition, it assumed that the V_{it} e U_i random variables are independent of the input variables in the model.

Coming back in the deterministic approach, if we have more than one observation for each unit and we suppose that each unit has its own characteristics different from the other units, it is possible to add a t suffix to every observation and to introduce an individual effect (α_i) constant in time but different for each unit. With the following model

$$Y_{it} = \alpha_i + x_{it}\beta + V_{it} \tag{2}$$

where $\alpha_i = \alpha - U_i$. As it is well known the efficiency score (Technical Efficiency Degree-TED) of every unit (firm) under the hypothesis of time constancy can be obtained by two estimations strategies referred to model (2): i) within estimator and ii) least squares dummy variables, LSDV (Baltagi, 1995).

Among parametrical deterministic methods, we mention Aigner e Chu (1968) proposal. Their model can be written as (1) and the vector of parameters can be estimates via linear or quadratic programming. In other words minimising the sum of the residuals absolute value under the constraint that every residual is non positive.

TED can be directly estimate by the residual vector. With panel data it is possible to obtain a global efficiency score as an average of annual TEDs.

For the stochastic approach still assuming a time constant TED we referred to a the classic Battese e Coelli (1988) method.

They consider an error term is divided into two components $E_{it} = V_{it} - U_i$: V_{it} are assumed as $N(0,\sigma_v^2)$ independent random variables, identically distributed (IID) and independent from U_i random variables; U_i are also still IID and non negative and defined by the truncation (at zero) of the $N(\mu,\sigma_v^2)$ distribution. Given all the distributive assumptions, joined density function of the vector $[V_{i1} - U_i, V_{i2} - U_i, \ldots, V_{iT} - U_i]$ can be derived, and it is possible to obtain maximum likelihood estimator (MLE) of parameters (α, β and the parameters of distributions of V and U).

Finally TEDs are calculated as $U_i = E(U_i | V_{i1} - U_i, V_{i2} - U_i, \ldots, V_{iT} - U_i)$ where $V_{it} - U_i$ are residuals $Y_{it} - \hat{\alpha} - x'_{it}\hat{\beta}$.

Non parametric analysis is conducted as Free Disposal Hull approach (FDH - Deprins, Simar e Tulkens, 1984), modified by Tulkens (1993) in a panel framework.

The last estimation method considered is the so-called semi-parametric approach proposed by Viviani (1996); it's a two-stage estimation procedure: in the first step undominated (efficient) firms are determined via FDH filter. Then from the FDH data set just identified, is possible to estimate a function specification using OLS. If the FDH filter has satisfactory results, the error term, originally divided into two components $E_{it} = V_{it} - U_i$, loses U. Only V_{it} remain and they are $N(0,\sigma_v^2)$.

3. A Bootstrap Methodology for Constructing Confidence Intervals for Estimated Efficiency Scores

Define a set of panel data efficiency scores obtained from several estimation methods:

$$TED = \{ted_{im} | i = 1,\ldots,N; m = 1,\ldots,M\} \tag{3}$$

i indicates the sample firm and *m* the estimation procedures for calculating TED: in this case M=5, 1=*within* estimator (*wit*), 2=stochastic approach (*sto*), 3=non parametric approach (*npa*), 4=parametric-deterministic approach (*pda*) e semi-parametric approach (*spa*).

In the discussion that follows, we consider the problem of computing a confidence interval for a set of firm means $\overline{ted_{i\bullet}} = \frac{1}{M}\sum_{m=1}^{M} ted_{im}, \forall i = 1,\dots,N$.

Assume that for each $i=1,\dots,N$ TEDs represent a set of IID random variables $\left\{ted_{im}\right\}_{m=1}^{M}$, with constant mean, $\mu_{get_i} = E\left(ted_{im}\right)$, and constant finite variance, $\sigma^2_{get_i}$ [2], the Lindberg-Levy theorem (LLT - Atkinson, Wilson, 1995) indicates that the sample firm mean $\overline{ted_{i\bullet}}$ is asymptotically normally distributed with μ_{get_i} and variance $\sigma^2_{get_i} / M$, regardless of the distributions of the $\left\{ted_{im}\right\}_{m=1}^{M}$.

An unbiased, consistent estimator of $\sigma^2_{get_i}$ is given by $\hat{\sigma}^2_{get_i} = (M-1)^{-1}\sum_{m=1}^{M}\left(ted_{im} - \overline{ted_{i\bullet}}\right)^2$. Thus, the variance of the sample mean $\overline{ted_{i\bullet}}$ can be consistently estimated by $\hat{\sigma}^2_{\overline{ted_{i\bullet}}} = \hat{\sigma}^2_{ted_i} / M$.

Given large M and the other assumptions of the LLT, $Z_M = \left(\overline{ted_{i\bullet}} - \mu_{ted_i}\right)\sigma^{-1}_{\overline{ted_{i\bullet}}}$ is asymptotically standard normal. Unfortunately in our case, sample of TEDs for each firm $\left\{ted_{im}\right\}_{m=1}^{M}$ is small so that LLT does not guarantee asymptotic normality for $\overline{ted_{i\bullet}}$. Fortunately, when we have small samples that are not normally distributed, we can use the bootstrap to obtain approximate confidence intervals for μ_{get_i}. For a given $i=1,\dots,N$ we have a random sample $\left\{ted_{im}\right\}_{m=1}^{M}$. The following steps lead to a bootstrap estimate of the confidence interval for μ_{get_i} (Atkinson, Wilson, 1995):

1. Compute the sample time mean $\overline{ted_{i\bullet}} = \frac{1}{M}\sum_{m=1}^{M} ted_{im}, \forall i = 1,\dots,N$;
2. Compute $\widetilde{ted}_{im} = ted_{im}\sqrt{M/(M-1)} + \overline{ted_{i\bullet}}\left(1-\sqrt{M/(M-1)}\right)$;
3. Independently draw N times from the set $\left\{\widetilde{get}_{im}\right\}_{m=1}^{M}$ with replacement, such that each observation has equal probability of selection, to obtain $\left\{get^*_{im}\right\}_{m=1}^{M}$;
4. Compute $\overline{ted_{i\bullet}}^*(j) = \frac{1}{M}\sum_{m=1}^{M} ted^*_{im}$;

[2] The same hypothesis is made by Atkinson e Wilson (1995) to computing a confidence interval for a set of annual firm TEDs means.

5. Repeat steps (3)-(4) J times to obtain $\left\{\overline{get_{i\bullet}}*(j)\right\}_{j=1}^{J}$; where J is appropriately large in magnitude.

The correction in (2) is necessary, however, to avoid type-I errors in small samples as proved by Atkinson e Wilson (1995).

The bootstrap values $\left\{\overline{ted_{i\bullet}}*(j)\right\}_{j=1}^{J}$ approximate the exact small-sample distribution of $\overline{ted_{i\bullet}}$. Thus, the values in $\left\{\overline{ted_{i\bullet}}*(j)\right\}_{j=1}^{J}$ can be sorted by algebraic value to construct confidence intervals for μ_{get_i} via the bootstrap percentile method described by Efron (1982). Letting $\overline{ted_{i\bullet}*}^{(\alpha)}$ denote $(100\times\alpha)th$ percentile of the J bootstrap replications $\left\{\overline{ted_{i\bullet}}*(j)\right\}_{j=1}^{J}$, the percentile method gives the bounds $\left(\overline{ted_{i\bullet}*}^{(\alpha)}, \overline{ted_{i\bullet}*}^{(1-\alpha)}\right)$ for the $\left[(1-2\alpha)\times 100\right]$ percent confidence interval for $\overline{ted_{i\bullet}}$. In other words, the $\left[(1-2\alpha)\times 100\right]$ percent confidence interval for $\overline{get_{i\bullet}}$ is obtained by deleting αJ values from both ends of the sorted array of J bootstrapped values and taking the endpoints of the newly truncated, sorted array as the boundaries of the confidence interval (Atkinson e Wilson, 1995).

4. Empirical Analyses

Such theoretical results have been applied to a sample of firms operating in Tuscany in 1993/94 (Lemmi and others, 1996) mostly diffused and of relevant innovative interest.

Such firms have been selected from official databases (called in Italy CERVED and held by Chambers of Commerce). They belong to the sector of manufacturing firms (textile, food, wood, mechanics, steel, chemical, etc.) and they have more than 6 employees to avoid fragmentation and recording problems.

Models variables (Giusti 1994, Griliches, Ringstad 1971) are represented by:

- Total Production Value (Y);
- Labour Hours (L);
- Capital Value (K);
- Material Input (MP)

Production function (Cobb-Douglas type) is represented by:

$$LnY_{it} = \alpha_i + \beta_1 LnL_{it} + \beta_2 LnMP_{it} + \beta_3 LnK_{it} + V_{it} \quad \text{[deterministic approach]} \quad (4)$$

$$LnY_{it} = \alpha + \beta_1 LnL_{it} + \beta_2 LnMP_{it} + \beta_3 LnK_{it} + V_{it} - U_i \quad \text{[stochastic approach]} \quad (5)$$

Estimation results are contained in tables 1,2,3:

Table 1: *Parameters of production function.*

	WITHIN ESTIMATOR ($R^2 = 0{,}996$)			ML ESTIMATES (Battese and Coelli) (*Log-Likelihood=286,803*)		
Variable	**Coefficient**	**St.Error**	**t-Statistic**	**Coefficient**	**St.Error**	**t-Statistic**
Intercept	-	-	-	0,950	0,419	2,267
LnL	0,539	0,103	5,225	0,480	0,095	5,031
LnMP	0,240	0,041	5,854	0,242	0,048	5,042
LnK	0,191	0,048	3,978	0,220	0,029	7,586
$\mu / \sigma u$				2,141	0,004	535,250
$\sigma^2 u / \sigma^2 v$				23,582	4,506	5,233
$\sigma^2 (v)$				3,13E-02	0,003	10,433

Table 2: *Two-stage estimation procedure (Viviani ,1996)*

	OLS ON UNDOMINATED FDH '93 ($R^2 = 0{,}963$)			OLS ON UNDOMINATED FDH '94 ($R^2 = 0{,}935$)		
Variable	**Coefficient**	**St.Error**	**t-Statistic**	**Coefficient**	**St.Error**	**t-Statistic**
Intercept	2,131	0,411	5,176	2,094	0,559	3,745
LnL	0,202	0,049	4,055	0,261	0,066	3,901
LnMP	0,414	0,040	10,320	0,321	0,046	6,839
LnK	0,294	0,048	6,074	0,338	0,064	5,249

Table 3: *Parametric and deterministic approach (Aigner and Chu, 1968)*

	Intercept	***LnL***	***LnMP***	***LnK***
YEAR 1993	2,210	0,316	0,297	0,378
YEAR 1994	4,178	0,205	0,383	0,130

N.W.:Estimates on a sample of 99 firms which provided their own budget

Several authors consider FDH as the method leading to results very closed to a possible competitive measure in the market. Therefore with FDH is possible to compare market shares and thus product power. The non parametric procedure results can give an indication of the competitiveness of the considered firms; such indication is usually combined with a measure of the merely technical efficiency.

Tab.4 shows some emblematic results. For privacy respect, individual data and complete results (available from the authors if interested) are not shown.

Fig.1 contains the TEDs averages and the confidence intervals for the firms in Tab.4.

Interpretation of the figure is immediate. All the firms with an high efficiency score and a reduced confidence interval show substantial agreement of the five procedures. Therefore in this case the firm is not only efficient from technical point of view but is winner in direct comparison with the other local units. Therefore is also competitive.

The contrary happens when a narrow confidence interval combines with a very low average efficient score. In case of firms with a wide confidence interval we have the following interesting situations: in fact the measures obtained using the five chosen procedures do not agree. This can mean either that firms are technically more efficient than competitive or, and it is the most frequent case, the exact contrary.

Table 4: *Results of five estimation procedures.*

FIRMS	TED_{wit}	TED_{pad}	TED_{sto}	TED_{npa}	TED_{spa}	TED_{med}	TED_{ll}	TED_{ul}
9	0,14	0,18	0,16	0,21	0,43	0,23	0,16	0,33
10	0,15	0,20	0,16	0,22	0,49	0,24	0,16	0,37
13	0,25	0,30	0,13	1,00	0,79	0,49	0,22	0,82
14	0,20	0,31	0,16	1,00	0,75	0,48	0,20	0,79
17	0,17	0,23	0,34	0,46	0,52	0,34	0,24	0,46
39	0,09	0,11	0,11	0,24	0,25	0,16	0,10	0,22
40	0,28	0,45	0,13	1,00	1,00	0,57	0,28	0,89
41	0,32	0,37	0,11	1,00	0,90	0,54	0,21	0,84
45	0,31	0,33	0,12	1,00	0,81	0,51	0,20	0,82
47	0,13	0,25	0,20	1,00	0,75	0,47	0,18	0,80
54	0,15	0,20	0,22	0,25	0,46	0,25	0,19	0,36
68	0,79	0,94	0,76	1,00	1,00	0,90	0,81	0,99
82	0,10	0,27	0,25	1,00	0,84	0,49	0,20	0,79
89	0,14	0,19	0,20	0,41	0,46	0,28	0,18	0,39
92	1,00	1,00	0,82	1,00	1,00	0,96	0,89	1,00

TED_{mean} = TED mean
TED_{ll} = lower limit (95%).
TED_{ul} = upper limit (95%).

Figure 1: *TED_{mean} and confidence interval*

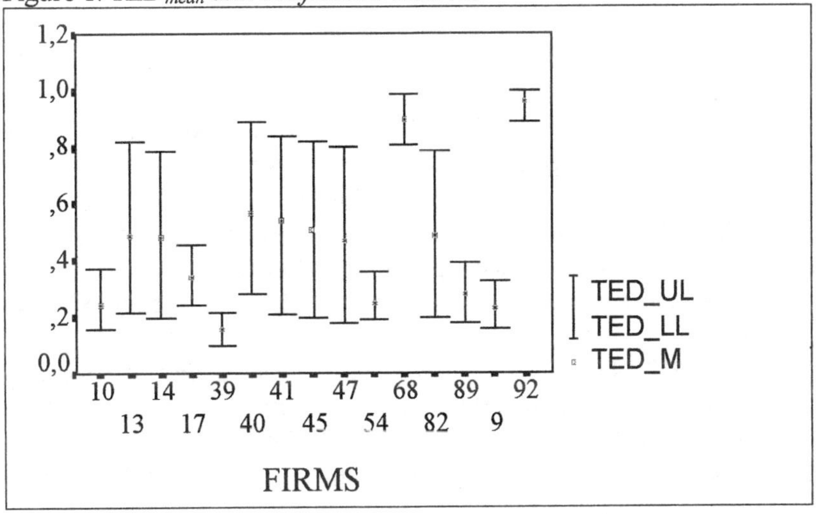

5. Finally Remarks

One of the most frequent problem connected with the analysis of firm productive processes and with the operative standard definition is the choice of a general method to be used in any situation.

Nevertheless we know that the choice of a particular approach has often appreciable effects on the efficient scores of the firms under control.

In this paper we have proposed an "ecumenical" approach to the productive processes analysis which uses five different estimation methods for the production frontier function. In such a way we obtain also using a bootstrap technique an interval estimation of technical efficiency scores for any firm.

As shown above in the empirical analysis such a methodological choice is particularly useful when firms under observation are characterised by high dishomogeneity in their productive processes.

REFERENCES

AIGNER D.,LOVELL C.A.K., SCHMIDT P. (1977), "Formulation and Estimation of Stochastic Frontier Production Function Models", *Journal of Econometrics*, Vol.6, pp.21-37

AIGNER D., CHU S. (1968), "On Estimating the Industry Production Function", *American Economic Review*, Vol.58, pp.826-839.

ATKINSON S.E., WILSON P.W. (1995), "Comparing Mean Efficiency and Productivity Scores from Small Samples: A Bootstrap Methodology", *The Journal of Productivity Analysis*, 6, pp.137-152

BALTAGI B.H. (1995), *Econometric Analysis of Panel Data*, J.Wiley & Sons

BATTESE G.E., COELLI T.J. (1988), "Prediction of Firm-Level Technical Efficiencies with a Generalized Frontier Production Function and Panel Data", *Journal of Econometrics*, 38, pp.387-399

CORNWELL C., SCHMIDT P. (1996), "Production Frontiers and Efficiency Measurement", *The Econometrics of Panel Data*, Matyas L. and Sevestre P. ed., Kluwer Academic

DEPRINS D., SIMAR L., TULKENS H. (1984), "Measuring Labor-Efficiency in Post Offices", *The Performance of Public Enterprises: Concepts and Measurement*, Amsterdam, North-Holland, pp.243-267

EFRON, G. (1982), *The Jackknife, the Bootstrap, and Other Resampling Plans*, Philadelphia: Society for Industrial and Applied Mathematics.

GIUSTI F. (1994), *Modelli neoclassici di produzione*, Università degli Studi di Roma "La Sapienza".

GRILICHES Z., RINGSTAD V. (1971*), Economics of Scale and the Form of Production Function*, North Holland Publishing Company.

LEMMI A., GAZZEI D.S., GHELLINI G., PANNUZI N. (1996), "La performance del settore manifatturiero toscano a metà degli anni '90", Impresa Toscana, 2, Supplemento

MEEUSEN W., VAN DEN BROECK J. (1977), "Efficiency Estimation from Cobb-Douglas Production Functions with Composed Error", *International Economic Review*, Vol.18, n°2 pp.433-444.

TULKENS H. (1993), "On FDH Efficiency Analysis: Some Methodological Issues and Applications to Retail Banking, Courts, and Urban Transit", *The Journal of Productivity Analysis*, 4, pp.183-210

VIVIANI A. (1996), "*Tecnologie di produzione ed efficienza nella distribuzione commerciale*", Atti della XXXVIII Riunione della SIS

Detection of Multivariate Outliers by Convex Hulls

Maria Rosaria D'Esposito*
Dipartimento di Sociologia e Scienza della Politica, Università di Salerno
Via ponte don Melillo, 84084 Fisciano, Salerno

Giancarlo Ragozini*
Dipartimento di Matematica e Statistica, Università di Napoli "Federico II"
Via Cinthia, Monte S. Angelo, 80126, Napoli

Abstract: This paper deals with the problem of identifying multiple outliers in multivariate data. Detection of anomalous values is achieved by looking at the variations in the convex hull of the data set as block of observations are deleted.

Key words: Outliers, Convex Hull.

1. Introduction

Outliers are defined as the observations which are surprisingly extreme with respect to the remainder of that set of data, and in univariate data sets they can be easily spotted by numerical or graphical inspection. On the contrary, detection of outliers in multivariate data clouds is a difficult task. The observed data, in fact, are points in a p-dimensional space and, for $p>3$, cannot be inspected directly. Neither the analysis of usual 2D or 3D projections is useful, because outliers do not necessarily stick out in some of the coordinate directions. Many exploratory and graphical methods therefore have been put forward; the most commonly used declare, as in the univariate case, outlying any extreme observation which is too far from the center of the data cloud. To judge extremeness, classical methods use the Mahalanobis distance based on the sample mean and the sample covariance matrix (Gnanadesikan and Kettering, 1972) but they fail to correctly detect the anomalous observations when there are groups of outliers, due to the masking effect. The most popular robust alternative procedures (Rousseeuw and van Zomeren, 1990; Hadi, 1992; Aktinson, 1994) have two main drawbacks: their computational complexity is high and they, like the Mahalanobis distance, evaluate extremeness with respect

* Work supported by ex-40% MURST Research Project "Nuovi Metodi di Classificazione e Analisi dei Dati". M. R. D'Esposito wrote sections 1, 2 and 6. G. Ragozini sections 3, 4 and 5. Computations are due to G. Ragozini and were made by S-Plus code.

to an ellipsoidal hull, hence work well in a neighborhood of normal distribution and give preference to linear association among the data.

In this paper we propose, for the detection of multiple outliers, an entirely data oriented procedure, which does not depend on initial estimates of center and variability, and is instead based on the analysis of the convex hull (*CH*) of the data sample. In fact, we suggest judging outlyingness by looking at the modifications of the *CH* volume (*CHV*) when groups of observations are omitted in turn. We declare outlying those observations whose omission mostly decreases the convex hull volume.

The proposed procedure is presented in section 2 and 3, and is illustrated by simulated and real data sets in section 4. Related computational problems are in section 5. Some conclusions and directions of future work are in section 6.

2. Proposed procedure

Outliers in a given data set can a result of recording or transmission errors, misplaced decimal points, exceptional phenomena, observations on different population slipping in the sample, etc. In any case, they are observations which differ very much from the others. By a geometric point of view, outlying data could be seen as points that lie on the periphery of the points cloud, very far from the others.

To detect them both explorative analysis and indexes of outlyingness have been proposed. In the class of explorative methods graphical representations can be very useful. Unfortunately, to visualize multivariate data we need very complex plots, such as parallel coordinate plot, coneplot, dynamic plot, grand tour and so on, all of which need an expert user; furthermore identification of outliers cannot be performed through an algorithm.

This is why many procedures have been proposed based on synthetic indexes of outlyingness. However:

a) the indexes often impose a prechosen form to the data cloud (usually ellipsoidal) to declare as outliers the observations which have the highest distance from the center of the cloud.

b) most of the indexes are designed to detect a single outlier. Their application to the case of multiple outliers often is affected by the danger of masking (some outliers go unnoticed) or swamping (spurious outliers are detected) (Barnett and Lewis, 1994; pag. 109).

To overcome all these problems, we propose to look at ***i)*** the convex hull of the data to identify the periphery and the form of cloud and at ***ii)*** the convex hull volume to measure the dispersion of observations. Outliers will coincide with the vertices of the most outer convex hulls in the sequence of nested convex hulls, and their presence will inflate the convex hull volume.

To be more specific, let *S* be a finite set of *n* observations $\boldsymbol{x}_i$ in $\Re^p$ ($i = 1,...,n$). The convex hull of *S* is defined as the set of all convex combinations of *S*.

$$CH(S)=\left\{x \,\middle|\, x=a_1x_1+\ldots+a_nx_n,\ 0\le a_i\le 1\,,\ \sum_{i=1}^{n}a_i=1.\right\} \tag{1}$$

By its definition, the convex hull is the best fit for the periphery of any data cloud. A point $\boldsymbol{x}_i$ is an *extreme point* or *vertex* of *S* if $CH(S)\neq CH\left(S-\{x_i\}\right)$. Given the vertices, the set of *edges* and the set of *facets* are identified.

A sequence of nested convex hulls can be constructed by considering the *CH* of the entire sample and in turn the *CH* of the remaining sample after the deletion of the vertices. For any convex hull, the volume *CHV* can be obtained by partitioning the hull in *m* simplexes with *p*+1 vertices (for example, in $\Re^2$ a convex hull can be partitioned in triangles). The *CHV* is then obtained as the sum of volumes of the simplexes (Grunbaum, 1967) and does not depend on the particular partition to individual simplexes.

The omission of one or more observations decreases the *CHV*. Clearly, when outliers are omitted, the *CHV* falls down.

In our proposal, the variations in the *CHV* when a subset ${}_kI$ of size *k* is omitted are then measured by the index:

$$C({}_kI)=\frac{CHV({}_k\bar{I})}{CHV(S)}, \tag{2}$$

where $CHV({}_k\bar{I})$ is the convex hull volume of the subset $\{S-{}_kI\}={}_k\bar{I}$ (the original data set without the ${}_kI$ observations). Observations are deleted in block to avoid the masking effect. Hence, the index is constructed starting from a quantity (the *CHV*) which is unambiguously defined and computed, and no estimation of center is needed. For the univariate case (*p*=1), the *CH* of a given set of points is the line segment with $x_{(1)}$ and $x_{(n)}$ as end points, and the index in (2) reduces to a Dixon type statistic (Barnett and Lewis, 1994; pag. 90): $T=\left(x_{(s)}-x_{(r)}\right)/\left(x_{(n)}-x_{(1)}\right)$.

3. Algorithm

The building blocks of the overall detection procedure are so designed to avoid any swamping effect and to lessen the computational complexity.

The procedure for identifying multiple outlier is as follow:

Step 1. Given the set *S* of *n* points x_i, $CH(S)$ is constructed and $CVH(S)$ is evaluated.

Step 2. To lessen the computational burden the index in (2) is computed only for some specific ${}_kI_i$. Precisely, let V_1 be the set of vertices of $CH(S)$ and consider the set S without the points in V_1. Let this set be $\{S-V_1\}$. Given $CH(\{S-V_1\})$, let V_2 be its set of vertices (for example, in *Fig.* 1, V_1 is the set of points $\{20,27,17,14,7,15,25,6\}$ and V_2 is the set $\{19,10,24,12,16\}$). This process is repeated until a percentage $[\alpha n]$ of the original data points are taken as vertices, with α chosen so that $[\alpha n]$ is the maximum number of outliers which the sample is assumed to contain. An upper limit for $[\alpha n]$ is $[n/2]$, typically $0.1 \le \alpha \le 0.3$. Let $V_\alpha = V_1 \cup V_2 \cup \ldots$; from now on only the points in V_α will be considered for the analysis.

Step 3. For each possible subset ${}_kI_i$ of size k in V_α, for $k = 1,\ldots,[\alpha n]$, the vertices of $CH(S-{}_kI_i)$ are computed.

Step 4. For each ${}_kI_i$ in step 3, $CHV({}_k\bar{I}_i)$ and the ratio $C({}_kI_i) = CHV({}_k\bar{I}_i)/CHV(S)$ are computed, for $k = 1,\ldots,[\alpha n]$.

Step 5. *Decision rule.* For each k there are $\binom{N}{k}$ $C({}_kI_i)$, with N the number of elements in V_α. The indexes $C({}_kI_i)$ show lower values when there are outlying points in ${}_kI_i$. Hence, we can look at the minimum of $C({}_kI_i)$ over i. When one more observation is deleted and it is an outlier, $\min_i C({}_kI_i)$ is much lower than $\min_i C({}_{k-1}I_i)$, and $D_k = \min_i C({}_{k-1}I_i) - \min_i C({}_kI_i)$ is large. Higher values of D_k point to outliers among the ${}_kI_i$. If there are at most h outliers, $C({}_hI^*)$ is the minimum value among all $C({}_hI_i)$, with ${}_hI^*$ the set of the h outliers, and $\min_i C({}_kI_i)$ for $k = h+1, h+2,\ldots$ stabilizes at $C({}_hI^*)$ and the difference D_h reaches a maximum. When subgroups of outliers are deleted, the index D_k could show local maxima. Note that $C({}_0I)$ is equal to 1 by position. A typical pattern for D_k is in *Fig.* 2.

Stopping rule. Since $\min_i C({}_kI_i)$ stabilizes at $C({}_hI^*)$, from the analysis of the index plot $\{k, D_k\}$ for $k = 1,2,\ldots$ we can decide to stop the procedure when the D_k values decrease less then a fixed small constant ε.

4. Some illustrative example

In order to verify the behaviour of the proposed diagnostic procedure and to show how the stopping rule on the D_k's works, we applied both on three different data sets.

The first data set refers to 28 observations on the body and brain weights of different animal species (in Rousseauw and Leroy, 1987) (Fig. 1). The first two nested convex hulls contain 13 observations and the three observations $\{6,16,25\}$ appear as outliers.

Fig. 2 and Table 1 have respectively the index plot and the values for D_k differences. D_k reaches its maximum for k=3 and falls down afterwards. The three outliers are, hence, correctly identified.

Fig. 1: Brain and Body weights sets. Two nested convex hulls are plotted

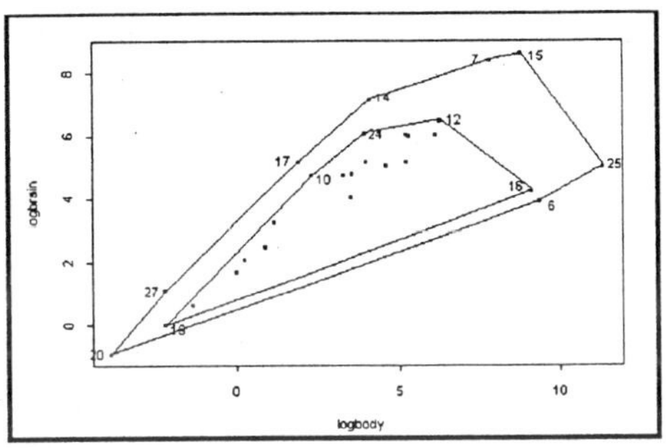

Fig. 2: Brain and Body Weights. Index plot of $D_k = \min_i C({}_{k-1}I_i) - \min_i C({}_k I_i)$

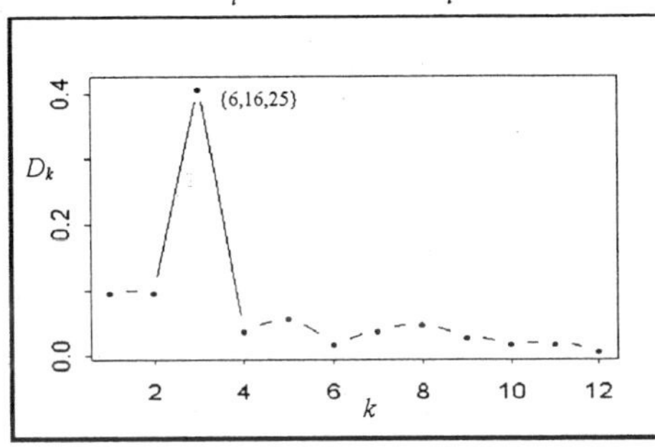

Table 1: Brain and Body. $D_k = \min_i C({}_{k-1}I_i) - \min_i C({}_k I_i)$ *values for k=1,...,12*

k	1	2	**3**	4	5	6	7	8	9	10	11	12
D_k	0.10	0.10	**0.41**	0.04	0.06	0.02	0.04	0.05	0.03	0.02	0.02	0.01

The second example is given by an artificial data set. 33 observations were generated by a mixture of two normal distributions: $(1-\alpha)\Phi(x|\mu_1,\Sigma)+\alpha\Phi(x|\mu_2,\Sigma)$, $\mu_1=\begin{bmatrix}0\\0\end{bmatrix}$, $\mu_2=\begin{bmatrix}1\\1.5\end{bmatrix}$, $\Sigma=\begin{bmatrix}1 & 0.9\\0.9 & 1\end{bmatrix}$, α=0.1 (i.e. there are at most four outliers). The data set, along with the two outer nested convex hulls, is portrayed in Fig. 3 The total number of vertices is 18 and by graphical inspection four outliers are easily spotted (observations 30, 31, 32 and 33.) .

Fig. 3: Artificial Data Set. Scatter plot *and the two outer nested convex hulls.*

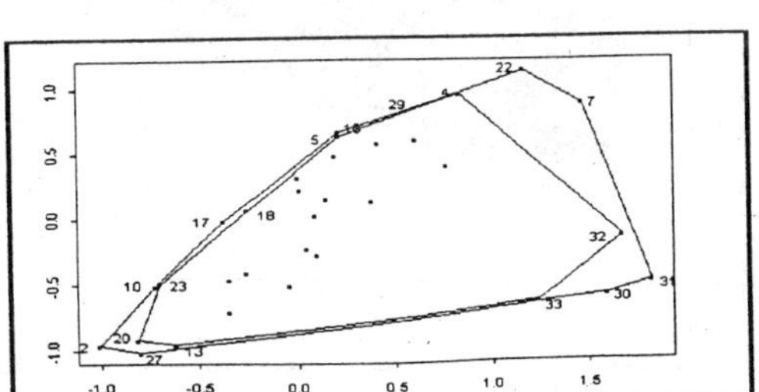

Fig. 4: Artificial Data Set. Index *plot of* $D_k = \min_i C({}_{k-1}I_i) - \min_i C({}_k I_i)$

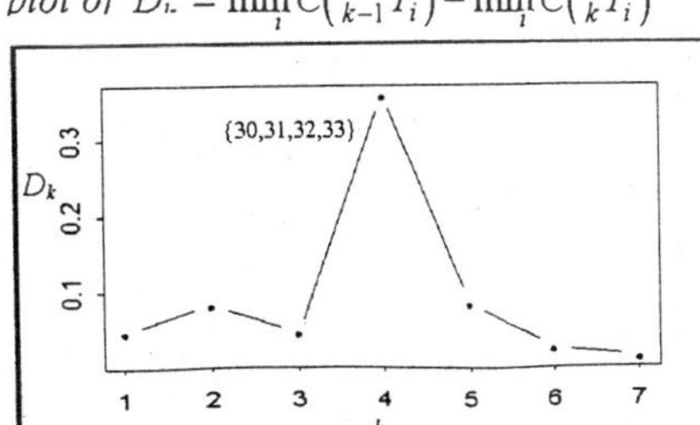

Table 2: Artificial Data Set. Some sorted values of $C({}_kI_i)$, for k= 1,...,4.

${}_1I_i$	$C({}_1I_i)$	${}_2I_i$	$C({}_2I_i)$	${}_3I_i$	$C({}_3I_i)$	${}_4I_i$	$C({}_4I_i)$
32	1	32,33	1	9,32,33	1	7,22,30,31	0.82
22	0.98	2,27	0.98	16,17,10	0.98	7,22,31,32	0.78
31	0.97	30,31	0.94	7,22,31	0.85	4,7,22,29	0.76
7	0.95	7,22	0.87	4,7,22	0.82	**30,31,32,33**	**0.46**

Table 2 reports some $C({}_kI_i)$ values when up to four observations are omitted in turn. When one, two or three observations are omitted no points appear as outlier. Only when the four observations {30,31,32,33} are omitted in block the index falls down to 0.46. When blocks of five observations are omitted, low values of the index appear in correspondence of the 5-tuples containing the four outliers. To decide on the number of outliers (4 or more) the decision rule in step 5 was applied.

In Table 3 the differences D_k are shown and plotted in Fig. 4.

Table 3: Artificial data set. $D_k = \min_i C({}_{k-1}I_i) - \min_i C({}_kI_i)$

k	1	2	3	**4**	5	6	7
D_k	0.0495	0.0828	0.0464	**0.3570**	0.0852	0.0278	0.0152

The procedure clearly detects the four outliers: D_k reaches its maximum for k=4 and stays constant at lower values afterwards. Therefore the iterations can be stopped at k=7.

Finally, the p=3 case is here exemplified by using the set of the explanatory variables in the stack loss data set, analyzed by Hadi (1992). The data describe the operation plant for oxidation of ammonia to nitric acid. The three predictors are the Air flow, the Cooling water temperature and Acid concentration and are shown in Fig. 5. Four outliers appear in the upper part of the 3D plot and correspond to observations {1,2,3,21}.

Fig. 5: Stack Loss data set. 3D plot

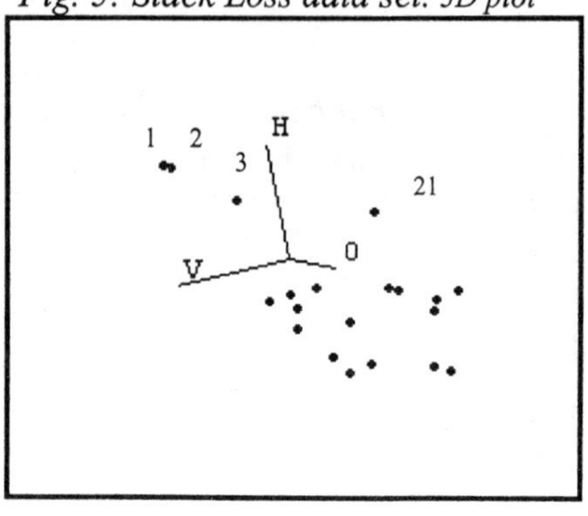

Fig. 6: Stack Loss data: Convex Hull

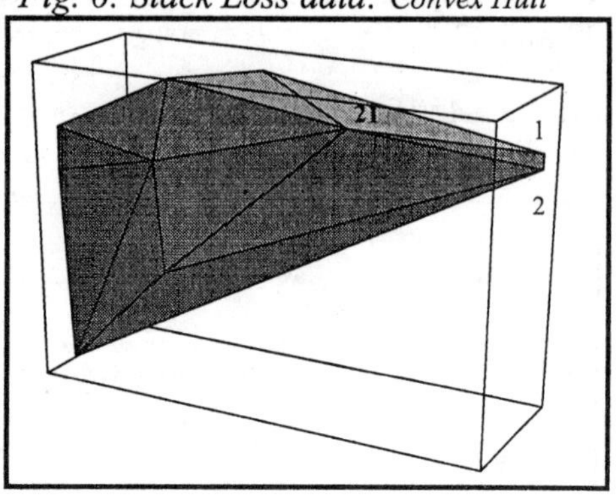

Fig. 7: Stack Loss Data. Index plot of $D_k = \min_i C({}_{k-1}I_i) - \min_i C({}_kI_i)$

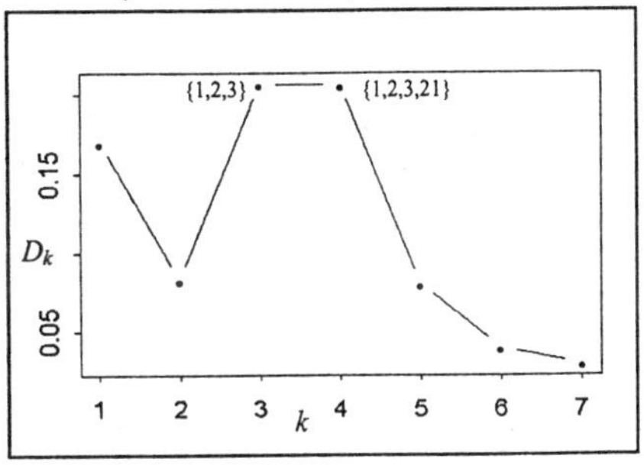

Table 4: Stack Loss Data set: $D_k = \min_i C({}_kI_i) - \min_i C({}_{k-1}I_i)$

k	D_k
1	0.17
2	0.0083
3	**0.2065**
4	**0.2062**
5	0.080
6	0.040
7	0.030

The convex hull and the index plot for D_k are respectively, in Fig. 6 and Fig. 7. The four outliers are correctly detected. D_k has two maximum points because the outliers come in two distinct groups $\{1,2,3\}$ and $\{21\}$. This example shows the reliability of the procedure when outliers appear on opposite sides of the data cloud.

5. Computational issues

For the computation of convex hull vertices several algorithms have been proposed. The first computational efficient algorithm is due to Chand and Kapur (1970) based on the so called gift-wrapping principle, that requires $O(n^2)$ operations. A more efficient algorithm that requires $O(n \log n)$ operations, even if up to three dimensions, was proposed by Preparata and Hong (1977). Later on, Edelsbrunner (1987) provided a version in the primal space of the Seidel's algorithm (1981) in the dual space. The Edelsbrunner's algorithm is based on the beneath-beyond method and requires $O(n \log n)$ in two dimensions and $O\left(n \log n + n^{[(p+1)/2]}\right)$ in more dimensions. The algorithm by Clarkson and Shor (1989), based on the random sampling approach, requires instead $O(n \log C)$ expected number of operations, where C is the number of vertices. It still works, indeed, up to three dimensions.

Recently, Barber, Dobkin and Huhdanpaa (1996) have proposed the Qhull algorithm used in this paper. It combines the beneath-beyond method, the Eddy's quickhull algorithm in two dimensions (1977) and the Clarkson and Shor's derandomized algorithm. Its computational cost is output depending and it requires $O(n\log C)$ for $p \leq 3$ and $O(nf_C/C)$ operations for $p \geq 4$, where f_C is the maximum number of facets for C vertices. In practice, it results to be a very fast algorithm. It evaluates contextually the convex hull volume without additional cost.

To evaluate the index $C({}_kI_i)$ is, therefore, computationally feasible. In sample of size n the index $C({}_kI_i)$ should be computed $\binom{[\alpha n]}{k}$ times, for k=1,..., $[\alpha n]$ (i.e. $2^{[\alpha n]}$ times), which can be quite a large number. The stopping rule introduced in step 5 of the procedure can help in lessen the number of iterations required.

6. Conclusions

The procedure proposed appears very promising. It is simple and suitable for automation. A modification based on a clustering around the vertices of the first convex hull is under study to further lessen the computational cost. The effectiveness and the power in dealing with masking and swamping problems have to be further investigated, and the method's power should be compared with other available proposed procedures.

References

Atkinson, A.C. (1994),"Fast Very Robust Methods for the Detection of Multiple Outliers", *Journal of the American Statistical Association*, 89, pp.1329-1339.

Barber, C. B., Dobkin, D. P. and Huhdanpaa, H. (1996), "The Quickhull Algorithm for Convex Hulls", *ACM Trans. on Math. Softw.*, 22, pp. 469-483

Barnett, V. and Lewis, T. (1994), *Outliers in Statistical Data*, John Wiley, New York

Chand, D. and Kapur, S. (1970), "An Algorithm for Convex Polytopes", *Journal of ACM*, 7, 78-86

Clarkson, K. and Shor, P. (1989), "Applications of Random Sampling in Computational Geometry", ii. *Discrete Computational Geom.*, 4, pp. 387-421

Eddy, W. (1977), "A New Convex Hull Algorithm for Planar Sets", *ACM Transaction on Mathematical Software*, 3, pp. 398-403

Edelsbrunner, H. (1987) *Algorithms in Combinatorial Geometry*, Springer-Verlag.

Gnanadesikan, R. and Kettering, J.R. (1972), "Robust Estimates, Residuals, and Outlier Detection with Multiresponse Data", *Biometrics*, 28, pp.81-124.

Grunbaum, B. (1967). *Convex Polytopes*, John Wiley, New York.

Hadi, Ali S. (1992), "Identifying Multiple Outliers in Multivariate Data", *Journal of Royal Statistical Society*, B, 54, pp. 761-771

Preparata, F. P. and Hong, S. J. (1977), "Convex Hulls of Finite Sets of Points in Two and Three Dimensions", *Communications of ACM*, 20, pp. 87-93

Rousseeuw, P.J. and Leroy, A.M. (1987), *Robust Regression and Outlier Detection*, John Wiley, New York

Rousseeuw, P.J. and van Zomeren, B.C. (1990), "Unmasking Multivariate Outliers and Leverage Points", *Journal of the American Statistical Association,* 85, pp.633-651.

Reducing Dimensionality Effects on Kernel Density Estimation: The Bivariate Gaussian Case*,**

Marco Di Marzio, Giovanni Lafratta
Department of Quantitative Methods and Economic Theory G. d'Annunzio University, Viale Pindaro 65127 Pescara, ITALY e-mail: [dimarzio | lafratta]@dmqte.unich.it

Abstract: It is well known that the kernel estimation of multidimensional densities is a difficult task due to the so-called "curse of dimensionality". The greater the data dimension, the greater is the sample size required to obtain efficient estimates. To reduce such dimensionality effects, we introduce further smoothing sources in addition to the usual bandwidth parametrization. In particular, preliminary kernel estimates are interpreted as smoothed samples and form the basis for successive density estimates, whose average (weights are given by empirical likelihoods of the observed sample) define the proposed sequential density estimator.

Keywords: Curse of dimensionality; Likelihood; Smoothed sample.

1. Introduction

In nonparametric density estimation, efficient estimates of multidimensional functions require the observation of larger and larger samples, rapidly increasing as the dimensions increase (Epanechnikov 1969, Scott-Wand 1991). This situation configures the so-called "curse of dimensionality" (Huber 1985; Härdle 1990).

To face this difficulty and to balance the number of dimensions with the available sample size, two approaches are possible. A first way is given by the statistical reduction of the number of dimensions, which can be obtained by eliminating redundant information (via principal component analysis, projection pursuit techniques, etc.; see Scott, 1992). From a theoretical point of view, a second way is obtained by increasing the sample in order to determine a sufficiently sized dataset. Incidentally, we observe that the two approaches are

---

* This work, though is the result of a close collaboration of the authors, has been specifically elaborated as follows: sections 1, 2, 4 by M. Di Marzio, sections 3,5,6 by G. Lafratta.

** The paper has been supported by a grant MURST 40% titled "Analisi dei dati spaziali", national coordinator Prof. Mauro Coli.

not symmetrical, since the application of the first one does not guarantee the effective balancing discussed above. In fact, the number of maintained dimensions can remain too high for successive efficient density estimations. As a consequence, in this paper we investigate the second approach and we develop a method to reduce dimensionality effects on kernel density estimation.

The paper is organized as follows. Section 2 introduces the concept of smoothed sample as a tool for increasing the size of available data. In Section 3 we discuss how to draw samples from estimated multidimensional densities. Section 4 contains a full description of the estimation algorithm we propose. Section 5 gives some evidence about the efficiency of the method via Monte Carlo simulations when sampling from standard bivariate Gaussian distributions. Finally, in Section 6 we report some concluding remarks.

2. Preliminary Sample Smoothing

Assume a sample $\mathbf{S} = \{\mathbf{X}_1,\ldots,\mathbf{X}_n\}$ of size n, with $\mathbf{X}_v = (\mathbf{X}_{v1},\ldots, \mathbf{X}_{vp})$, $v = 1,\ldots,n$, is drawn from an unknown density f of a p-dimensional random vector $\mathbf{X}= (\mathrm{X}_1,\ldots, \mathrm{X}_p)$. The conventional kernel estimator of $f(\mathbf{x})$ (Rosemblatt 1956, Cacoullos 1966) is

$$\hat{f}(\mathbf{x}; K, \mathbf{H}, \mathbf{S}) = n^{-1} \sum_{v=1}^{n} K_{\mathbf{H}}(\mathbf{x} - \mathbf{X}_v), \tag{1}$$

where

$$K_{\mathbf{H}}(\mathbf{u}) = |\mathbf{H}|^{-1/2} K(\mathbf{H}^{-1/2}\mathbf{u})$$

is a kernel function, with K a p-dimensional density and $\mathbf{H}$ a $p \times p$ symmetric positive definite bandwidth matrix. When $p > 1$, we use diagonal bandwidth matrices, so assuming an isotropic hypothesis. As a consequence, $\mathbf{H}$ must be seen as equal to $\mathbf{I}_p$ for some scalar h, with $\mathbf{I}_p$ the $p \times p$ identity matrix.

Kernel smoothing procedures determine an estimate $\hat{f}$ of f on the basis of the sample $\mathbf{S}$. In this way, the function $\hat{f}$ is treated as the final object of the analysis and no attempt is made to use $\hat{f}$ as a density from which *whatever sized* samples can be potentially drawn. Clearly, care has to be taken when choosing $\hat{f}$. Intuition, for example, suggests the use of small bandwidths in order to remain as close as possible to the observed sample. Furthermore, one should be aware that drawing a sample from $\hat{f}$ (no matter its size) does not guarantee effective results. The use of a sequence of preliminary kernel estimates from which samples are drawn in turn seems a better strategy. In order to apply such considerations, let us start with the following:

Definition 1 *Assume that* $K_{\mathbf{H}}$ *is a kernel function and g a p-dimensional density. Let y be a point in the support of g and* q: $\Re^p \rightarrow [0,+\infty[$ *the following function:*

$$\forall \mathbf{u} \in \Re^p : \mathrm{q}(\mathbf{u};\mathbf{y}) = \mathrm{K}_{\mathbf{H}}(\mathbf{y} - \mathbf{u}).$$

Then we define the Expected Proportion of Points for the pair $(K_{\mathbf{H}}, g)$ at **y** *as the integral:*

$$EPP(\mathbf{y};\mathbf{H}) = \int_{E(\mathbf{H},y)} g(\mathbf{z})\,d\mathbf{z} \tag{2}$$

where $E(\mathbf{H},\mathbf{y}) = supp(q)$ *if the support of q is bounded, while* $E(\mathbf{H},\mathbf{y}) = B(\mathbf{y},h)$*, i.e. the ball in* $\Re^p$ *centered at* **y** *having radius h, in the other case.*

Assume that, for a fixed value of **y**, we have to compute $EPP(\mathbf{y};\mathbf{H})$. Given a sample drawn from *g*, we are able to execute a kernel estimation of *g*, obtaining say $\hat{g}$. Thus, we estimate the required integral, say $E\tilde{P}P(\mathbf{y};\mathbf{H})$, by substituting $\hat{g}$ to *g* in (2).

The use we make of this concept is the following. We evaluate the *EPP* function at the sampled points $\mathbf{X}_1,\ldots,\mathbf{X}_n$ in **S**. The $E\tilde{P}P(\mathbf{X}_i;\mathbf{H})$ value can be interpreted as an estimate of the probability that a generic sample point belongs to the set $E(\mathbf{H},\mathbf{y})$, which represents approximately the region whose points, if sampled, will receive a non-zero weight through which they will contribute to the estimate of $f(\mathbf{X}_i)$. We interpret the mean $n^{-1}\sum_{i=1}^{n} E\tilde{P}P(\mathbf{X}_i;\mathbf{H})$ as the expected frequency of points which will belong to a non-zero area when estimating one of the values $f(\mathbf{X}_1),\ldots,f(\mathbf{X}_n)$. As a consequence, given a sample **S'** of size *n'*, we can estimate the expected proportion of points of **S'** which will belong to a non-zero area as the value $(n'/n)\sum_{i=1}^{n} E\tilde{P}P(\mathbf{X}_i;\mathbf{H})$. If, in particular, we refer to sample **S**, such expected proportion will be equal to the following index

$$SEPP(\mathbf{H},\mathbf{S}) = \sum_{i\in\{1,\ldots,n\}} E\tilde{P}P(\mathbf{X}_i;\mathbf{H}).$$

Finally, we determine the bandwidth h_{S} such that, for a real constant $0 < c \leq 1$,

$$h_{\mathrm{S}} = \sup\{h : SEPP(\mathbf{H},\mathbf{S}) \leq c\}. \tag{3}$$

As a consequence, if we use a bandwidth $h \in]0, h_S]$, our expectation is that, when estimating $f(\mathbf{X}_i)$ by means of $\hat{f}(\mathbf{X}_i; K, \mathbf{H}, \mathbf{S})$, no more than c points in **S** will have a non-zero contribute in determining the value of the estimate.
This enables us to state the following

Definition 2 ***(Smoothed samples)*** *Let* ***S*** *be a sample drawn from f:*

1. *every kernel smoothing* $\hat{f}(\cdot; K, \mathbf{H}, \mathbf{S})$ *such that* $h \le h_S$ *is a smoothed sample;*
2. *for every smoothed sample* $\hat{f}$ if **S'** *is a sample drawn from* $\hat{f}$, *and* $h \le h_S$, *then* $\hat{f}(\cdot; K, \mathbf{H}, \mathbf{S}')$ *is a smoothed sample;*
3. *the only smoothed samples are those given by* 1 *and* 2.

Smoothed samples are useful tools when increasing the size of available data as we will describe in Section 4, when introducing sequences of smoothed samples obtained with increasing bandwidths. In Section 3, instead, we discuss how to obtain effectively a smoothed sample.

3. Sampling from Multidimensional Densities

As the number p of dimensions increases, the frequency of points, sampled from f, which belong to the distribution tails, i.e. those areas whose probability is relatively small, will also increase. As a consequence, when estimating f on the basis of small samples it will be probable to overestimate the tails and to underestimate the relatively more probable areas. So, when we have to decide from what region W to execute the (re)sampling procedures from given estimates $\hat{f}$ i.e. that region which the sampled points will belong to, the above discussion must be applied in some way. In particular, we state for W the following choice:

$$W = \prod_{i}^{p} \left[\delta_{i,\alpha}, \delta_{i,100-\alpha}\right] \tag{4}$$

in this case $\prod$ is intended as the Cartesian product operator, $0 < \alpha \le 10$, and $\delta_{i,\beta}$ stands for the β-th percentile of the distribution $X_{1i}, \ldots X_{ni}$.
To obtain samples from $\hat{f}$, we apply the rejection method (see, for example, Ross 1996) as follows. We define the density

$$g(\mathbf{y}) = \frac{1}{\prod_{i=1}^{p}(\delta_{i,100-\alpha} - \delta_{i,\alpha})} w(\mathbf{y})$$

and the constant

$$\max_{\mathbf{y}\in D} \hat{f}(\mathbf{y})\prod_{i=1}^{p}(\delta_{i,100-\alpha}-\delta_{i,\alpha}),$$

where D is a p-dimensional grid on W. Thus it can be expected that, for all $\mathbf{y}\in W$, $\frac{f(\mathbf{y})}{g(\mathbf{y})}\leq a$, so that the method applies by generating $\mathbf{Y}$ from g, U from the uniform density $\frac{1}{a}1_{[0,1]}(u)$ and hence accepting $\mathbf{y}$ as generated from $\hat{f}$ if $U\leq\frac{\hat{f}(\mathbf{Y})}{ag(\mathbf{Y})}$ and rejecting it otherwise. This enables us to consider $\mathbf{Y}$ as distributed following $\hat{f}$ as required.

4. Sequential Kernel Density Estimation

We operate a set of density estimates ${}_j\hat{f}$, $j=1,\ldots,n$, each of which is obtained applying an iterated kernel smoothing procedure as follows. The algorithm consists of two steps.

The first step, referred to as *the smoothing step*, is intended to increase the size of available data. A sequence of say k smoothed samples ${}_j\hat{f}_i(\cdot\,;K,\mathbf{H}_i,\mathbf{S}_i)$ $i=1,\ldots,k$ with $h_i\leq h_S$, is generated given a vector of increasing bandwidths $h_1,\ldots,h_k$ and a vector of integers $n_1,\ldots,n_k$ representing increasing sample sizes. The number of smoothed samples k can be identified, for example, as the number of dimensions p, or as the number of modes in a pilot estimate of f. Our aim is to increase the sample size in some a way which minimizes the lack of information given when passing from a smoothed sample to the other. As a consequence, the h_i values need to be very small and increasing very slowly at first, while they need to increase more rapidly only in the last steps. So, let us define vector $\mathbf{z}$ whose i-th element, $i=1,\ldots,k$, is given as $z_i=h_S(i/k)$, and observe that the points z_i are equally spaced in the interval $I_S=\left[\frac{h_S}{k},h_S\right]$. In order to apply the discussion reported above, we choose to define h_i as the value assumed at z_i by a given convex function $v:I_S\rightarrow[0,h_S]$. When $i=1$, let us state $\mathbf{S}_1=\mathbf{S}$ and $n_1=n$. As a consequence, the first smoothed sample will be the same for all $j=1,\ldots,m$. For $1\leq i<k$, we ${}_j\hat{f}_i(\cdot\,;K,\mathbf{H}_i,\mathbf{S}_i)$, the i-th smoothed sample in the sequence, and, successively, a sample $\mathbf{S}_{i+1}$ of size n_{i+1} is drawn from ${}_j\hat{f}_i(\cdot\,;K,\mathbf{H}_i,\mathbf{S}_i)$, we obtain the sample $\mathbf{S}_k$, so we are able to compute the last smoothed sample ${}_j\hat{f}_i(\cdot\,;K,\mathbf{H}_k,\mathbf{S}_k)$.

In the second step, referred to *the estimation step*, we draw a sample ${}_j\mathbf{S}$ of size n_j^* from $\hat{f}_i(\cdot;K,\mathbf{H}_k,\mathbf{S}_k)$, with $n_k \le n^*$ compute ${}_j\hat{f}$ as the kernel estimate $\hat{f}(\cdot;K,\mathbf{H}^*,{}_j\mathbf{S})$, obtained on the basis of ${}_j\mathbf{S}$ given a bandwidth h^*.

Now, let ${}_jL(\mathbf{S}) = \prod_{i=1}^{n} {}_j\hat{f}(\mathbf{X}_i)$ be the likelihood that ${}_j\hat{f}$ assigns to sample $\mathbf{S}$. Then we suggest estimating the analyzed density through the weighted average of densities ${}_j\hat{f}$ as follows:

$$\tilde{f}(\cdot;\mathbf{S}) = \left(\sum_{j=1}^{m} {}_jL(\mathbf{S})\right)^{-1} \sum_{j=1}^{m} {}_j\hat{f}\,{}_jL(\mathbf{S}). \tag{5}$$

5. A Monte Carlo Study on Bivariate Gaussian Density Estimation

In this section we test the effectiveness of the proposed methodology through simulation experiments. We execute a Monte Carlo study in order to compare $\hat{f}(\mathbf{x};K,\mathbf{H}^*,{}_n\mathbf{S}_l)$, the conventional kernel estimator (1), with $\tilde{f}(\cdot;{}_n\mathbf{S}_l)$, the sequential kernel estimator (5). For n = 50,60,80,120, we draw 200 samples of size n, say ${}_n\mathbf{S}_1,\ldots,{}_n\mathbf{S}_{200}$, from a Gaussian standard bivariate distribution f. We compute, for all l = 1,…,200 and for $e = \hat{f}, \tilde{f}$, the integrated standard error when estimating f by means of e given ${}_n\mathbf{S}_l$:

$${}_nISE_l(e) = \int \left(e(x;{}_n\mathbf{S}_l) - f(\mathbf{x})\right)^2 d\mathbf{x}$$

where K is the Standard bivariate Gaussian kernel function and $\mathbf{H}^*$ is the bandwidth matrix which minimizes the MISE when estimating a Gaussian Standard distribution. We also employ $\mathbf{H}^*$ in the estimation step of the sequential algorithm. Since we introduce further sources of smoothing other than the bandwidth matrix, we decide to hold constant the bandwidth, so avoiding the problem of assessing the contribution on estimation performances of bandwidth selection procedures.

We set c = 1 in formula (3) and we select the convex function

$$v(z) = \exp\left(15(z - h_s)^2\right).$$

In addition, we set $m = 2$, $k = 2$, and we define $n_i = ni$, $i = 1,\ldots,k$, $n^* = n(k+1)$ and $\alpha = 5$ in formula (4).

In order to obtain a comparison which takes into account the sampling distribution as a whole, we determine, for both $e=\hat{f}$ and $e=\tilde{f}$, the following estimates of the Mean Integrated Standard Error for e:

$$ {}_nMISE_l(e)=l^{-1}\sum_{j=1}^{l}{}_nISE_l(e), $$

in which, for $l=1,\ldots,200$, the l-th estimate is performed using the first l samples in the considered sequence. Hence we compute, for $l=1,\ldots,200$, the following index:

$$ {}_nM_l=\frac{{}_nMISE_l(\tilde{f})}{{}_nMISE_l(\hat{f})}100. \tag{6} $$

Finally, for the purpose of comparing the relative performance of the two estimators sample by sample, we consider, for $l=1,\ldots,200$, the following ratio:

$$ {}_nD_l=\frac{{}_nISE_l(\tilde{f})}{{}_nISE_l(\hat{f})}100. \tag{7} $$

6. Discussion

Table 1 reports some of the simulation results. In particular, we record a value for ${}_{50}M_{200}$ equal to 54.43 which, shows that, with a sample size equal to 50, our method reduces the MISE of about 46% if compared with the conventional kernel estimator. Increasing the sample size, the ratio ${}_nM_{200}$ between the MISE's of the two methods decreases. For n = 60,80,120, we have, respectively, ${}_nM_{200}$ = 51.27, 46.32, 37.99, hence the relative overperformance of our method increases.

As reported in Table 2, MISE's decrease differently for the two methods. For example, observe that increasing the sample size of about 60% determine a MISE reduction of -30.96% for our method, while for the conventional kernel estimator it reduces of -18.87% only. This means that our method employ the increased sampled information on f in a more efficient way. Further sample by sample comparisons are described by means of the D_l index. In particular, we find that, for n=50,60,80,120, the condition D_l >100 holds true in correspondence, respectively, of 19,18,19 and 3 of the 200 considered samples.

As a result, the conventional kernel smoothing seems to suffer of relatively poor performance if compared with the sequential kernel smoothing when estimating bidimensional densities. In this paper we report some evidence about the fact that additional sources of smoothing can play an important role in facing ``the curse of dimensionality". In particular, the use of smoothed samples as a tool for increasing the sample size and the use of weighted averages of different sequential kernel estimates reduce consistently the Mean Integrated Standard Error. Further studies are obviously required in order to generalize thesc results to estimation problems involving dimensions higher than two.

Table 1: *Statistics of simulation results. The conventional kernel estimator is indicated by* CE, *while the sequential kernel estimator by* SE.

n	$ISE \times 10^3$	Mean	Median	Min	Max
	SE	4.2411	3.6577	0.6813	21.0798
50	CE	7.7920	7.2820	0.8698	20.6258
	Ratio	0.5443	0.5023	0.7832	1.0220
	SE	3.7148	3.1254	0.5013	16.6870
60	CE	7.2453	6.8353	1.1513	17.4455
	Ratio	0.5127	0.4572	0.4354	0.9565
	SE	2.9280	2.4583	0.4878	12.3037
80	CE	6.3215	6.0230	1.0313	15.0868
	Ratio	0.4632	0.4082	0.4730	0.8155
	SE	2.1147	1.7720	0.4501	7.0298
120	CE	5.5666	5.3470	1.6379	10.7521
	Ratio	0.3799	0.3314	0.2748	0.6538

Table 2: *Sample size increases and* MISE'*s decreases for the conventional* (CE) *and the sequential* (SE) *estimators. Variations are referred to the case n = 50.*

Sample size increases	CE MISE's decreases	SE MISE's decreases
20%	-7.02%	.41%
60%	-18.87%	.96%
40%	-28.56%	.14%

References

Cacoullos, T. (1966). Estimation of multivariate density, *Annals of the Institute of Statistical Mathematics* **18**, 179-189.

Epanechnikov, T. (1969). Nonparametric estimation of a multivariate probability density, *Theory of Probability and Applications* **14**, 153-158.

Härdle, W. (1990). *Applied Nonparametric Regression*, Cambridge University Press, Cambridge.

Huber, P. (1985). Projection pursuit (with discussion), *The Annals of Statistics* **13**, 435-475.

Rosenblatt, M. (1956). Remarks on some nonparametric estimates of a density function, *Annals of Mathematical Statistics* **27**, 832-837.

Ross, S. (1996). *Simulation*, Academic Press, London.

Scott, D. (1992) *Multivariate Density Estimation: Theory, Practice, and Visualization.* Wiley, New York.

Scott, D. & Wand, M. (1991).Feasibility of multivariate density estimates, *Biometrika* **78**, 197-205.

Shewhart's Control Chart: Some Observations

Massimiliano Giacalone
Department of Mathematics and Statistics, University of Naples "Federico II".
Complesso Monte S. Angelo, Via Cinthia, 80126 Naples.
e-mail: max@matstat.dms.unina.it

Abstract: Data Analysis in Shewhart's Control Chart, to use the original m samples n sized intensities, is the main subject of this paper. Given $m \times n$ intensities we examine three alternatives to sintetize the variability: a) arithmetic mean of m standard deviations $\left({}^{'}S\right)$; b) root mean square of m variances $\left({}^{''}S\right)$; c) global dispersion $\left({}^{'''}S\right)$. We prefer the global dispersion to estimate parent population σ^2.

As an alternative we suggest to analyze all the items of an unique random sample dimensioned in such a manner to have an efficient σ^2 estimate. A second introducted proposal is to use the Factory's needs: $\left(P_0, P_1, \alpha, \beta, L \text{ and } U\right)$. Some examples are given in the last session of the paper.

Keywords: Shewart's Control Chart, Sigma's Estimate, Data Analysis.

1. Introduction

Using S.C.C. (Shewhart's Control Chart) it is customary to operate 2 stages; a first stage devoted to data collection and limits $LCL_{\bar{x}}$, $UCL_{\bar{x}}$, LCL_s, UCL_s (Lower Control Limit and Upper Control Limit for mean and dispersion) computation.The second stage is devoted to chart's use.

In the first stage it is customary to produce $K = m \cdot N$ items, (in other words we have m lots N sized), to draw m single random samples n size from each lot N sized. The population is given by all items *produced and to be produced,* its mean is μ and its variance is σ^2; μ and σ^2 supposed stable in the first stage (*items produced*).

Let us call x_{ij} the ith intensity of the jth sample, so the jth sample mean is:

$$\overline{x_j} = \sum\nolimits_i x_{ij}/n \qquad (i = 1,2,\ldots n)$$

$$s_j^2 = \sum\nolimits_i \left(x_{ij} - \bar{x}_j\right)^2 / (n-1) \tag{1}$$

is the jth sample variance estimate;

$$\overline{\overline{X}} = \sum_j \bar{x}_j/m,$$

is the mean of the sample means.
Sample mean synthesis create no problem, not the same happens for s^2 or s.
Indeed some authors (W.A. Shewhart, 1931); (A. J. Duncan, 1965); (P.L. Piccari, 1974); (D.C. Montgomery, 1991) propose to compute:

$$'S = \sum s_j/m \tag{2}$$

Some other authors (Mittag-Rinne, 1993) propose to compute:

$$''S = \left\{\sum s_j^2/m\right\}^{1/2} \tag{3}$$

finally one may also compute:

$$'''S = \left\{\sum_i \sum_j \left(X_{ij} - \overline{\overline{X}}\right)^2 \Big/ (m \cdot n - 1)\right\}^{1/2}; \tag{4}$$

In this paper we study the rationale of each solution and we suggest an alternative proposal.

2. Synthesis analysis

Since root mean square is greater than or equal to arithmetic mean, we may write:

$$'S \leq {''S},$$

and declare that one of the introduced formulae can't be correct. Relation (2) is the main suspect because since:

$$E(s) \neq \sigma$$

The same may be said for (3), and this means $'S$ to be a biased σ estimate. Someone notes that, if the underlying population is normal, $'S$ actually estimates $\sigma \cdot c_2$; this is statistically correct but a little cumbersome. We remember that c_2 is a constant depending on the sample size n:

$$c_2 = \{2/(n-1)\}^{1/2} \cdot \Gamma\{(n-1)/2\};$$

tabulated values are presented in Duncan (1965).
Let us now consider the synthesis of sample variances (3). Kenney and Keeping

(1956), showed that:

$$E\left(s^2\right)=\sigma^2,$$

not only for simple samples, but also in presence of *m* simple samples. In case h independent samples are available from the universe, they suggest to use:

$$\hat{\sigma}^2 = Q/(U-h);$$

where

$$Q=n_1\, s_1^2+n_2\, s_2^2+\ldots\ldots+n_h\, s_h^2;$$

$$U=n_1+n_2+\ldots\ldots\ldots+n_h;$$

and s_1^2 is the variance in the ith sample consisting of n_1 variates.
If $n_i=n$ is the same for every sample, we have:

$$\hat{\sigma}^2=n\left(s_1^2+s_2^2+\ldots\ldots+s_h^2\right)/(U-h);$$

where $U=n\cdot h$. Clearly the last relation may be written in the form:

$$(n-1)/n\cdot\hat{\sigma}^2=\left(s_1^2+s_2^2+\ldots\ldots+s_h^2\right)/h$$

The constant $(n-1)/n$ is present because the authors started with $s_j^2=\Sigma_i\left(X_{ij}-\overline{X}\right)^2/n$ instead of s_j^2, but if the degrees of freedom are used, the result is correct and consistent with: $E\left(s^2\right)=\sigma^2$. This solution records time variations. In other words we have a trace of variability changes during data collection period.

Finally relation (4) is based on the whole group. It may be seen as the *total variance*, while $''S^2$ may be seen as *within variance.* Deviances are the same if *between variance* is equal to zero.

There is someone discouraging its use. For instance D.C. Montgomery (1991), affirms that the estimate of the process standard deviation σ used in constructing the control limits is calculated from the variability within each sample. Consequently, the estimate of σ reflects within-sample variability only. It is not correct the estimate of σ based on the usual quadratic estimator, say $'''S$, because if the sample means differ, then this will cause $'''S$ to be too large. Consequently, in this way ,σ could be overestimated.

A. J. Duncan (1965) shares the same opinion, and retaines that is not correct to estimate the process standard deviation from all the data (e. g. $'''S$) and use this in setting up limits for the $\overline{X}$-chart. The estimate of the process standard deviation to be used in setting up limits for the $\overline{X}$-chart must be computed from the within-sample variation to the exclusion of the between-sample variation.

Let us remember that if a production process presents stable between-sample variation it could be a good rule to look for the trouble and to remove it if possible.If the problem persist we do not see why to ignore it, computing the so called within variation. Another important remark is the difference between "first stage" and "second stage". In the second stage production must be monitored so that it is very useful to divide output into lots, let us say N sized, and investigate every single lot produced. If no trouble appears production can continue; on the contrary, if a trouble comes out it is much better to stop production and to look for happenings. In the second stage, points are regarded as independent events and O.C.C. (Operating Characteristic Curve) is computed under this assumption (G. Rouzet, 1957). In short, the division of production into lots N sized is a suitable procedure for the second stage as we said before.
The first stage problem is a different one. to estimate μ and σ^2 related to the character of interest. The subject involved is the *parent population* and its parameters. The division of items into lots N sized is not an essential operation. Perhaps the sample repetition is a mechanical consequence of the second stage technique, to some extent necessary if n=5, because μ and σ^2 estimates based on so a little sample should be extremely poor ones, so to have both ways saved some authors suggested to repeat the sample (and the lot) *m* times (Mittag-Rinne,1993). It seemed therefore a natural consequence to compute $\overline{X}_j$, s_j and ${}^{'}S$, ${}^{''}S$ and ${}^{'''}S$.

3. Simulation

In order to emphasize our opinion we consider a simulation. We shall use Wold's Random Normal Deviates divided into lots N=50 sized, one numbers column for lot. From each column we draw one sample n sized and this operation will be repeated m (=20) times as in the first stage practice. We compute m $\bar{x}$ and m σ^2,and the synthesis ${}''S^2$ is compared with ${}'''S^2$. We define: DifTot = $\sigma^2 - {}''S^2$ and DifUni = $\sigma^2 - {}'''S^2$. We also noted that here σ^2 is the population variance computed on $N \cdot m$ data = 1000 considering series of 100 samples. If DifTot<DifUni one point is given to ${}''S^2$, but if DifUni<DifTot then one point is given to ${}'''S^2$.
For series of samples $n = 5$ sized we found more than 75% points for DifUni, then for ${}'''S^2$.

4. Alternative proposals

The first stage procedure is a very expensive one. Infact after m samples we must revise the production process, therefore to save time and money we suggest to analyze all the items produced within the first stage and dimension this sample according to wanted protection.

Our suggestion seems particularly useful for destructive control analysis because with customary procedure if not analyzed items are out of tolerance, production-control costs increases.

Calling N the first stage lot size, we shall have: $\overline{\overline{X}}=\sum_i X_i/N$, and $\overline{S}^2=\sum_i (X_i-X)^2/(N-1)$, as an unbiased σ^2 estimate.

A different suggestion is based on the introduction of Factory's needs $(P_0, P_1, \alpha, \beta, L \text{ and } U)$. Many authors, use symbol L for *Lower specification limit* and symbol U for *Upper specification limit.*

Now let us call P_0 the well known *Acceptable Quality Level* and we underline that it seems suitable subdivide P_0 into to parts, the one on the left ${}_LP_0$ (fraction of too small items) and the other on the right ${}_UP_0$ (fraction of too large items), of course ${}_LP_0+{}_UP_0=P_0$. This is enought for the computation of:

$$\overline{X}_0=(Lz_U-Uz_L)/(Z_U-Z_L); \qquad (Z_L<0)$$

$$\sigma_0=(U-L)/(Z_U-Z_L);$$

where Z_L is the normal standardized fractile given ${}_LP_0$, and Z_U the one given ${}_UP_0$. $\overline{X}_0$ and σ_0 are the parameters to be used for Shewhart's variables Control Chart computation.

The SCC so obtained is a very different tool because it privileges Factory's needs, whereas customary procedure privileges process capability. Therefore, once obtained the new SCC $(UCL_{\overline{x}}, LCL_{\overline{x}}, UCL_s)$ we must look if production process is able to output material just as designer wants (L and U).

For this test we must collect N data related to the character of interest and compute $S^2=\sum(X_1-\overline{X})^2/(N-1)$, the variance of the last N items produced and compare S^2 with σ_0^2. If $S^2<\sigma_0^2$ the process is capable.

According to capability studies experience it is better to accept the process if $S^2/\sigma_0^2<1.23$. We did not use here the so called natural tolerance concept because it is enought to compare directly variances in order to have the test accomplished.

If production process is not able we suggest an innovative maintenance keeping into account cost embroiled with this operation.

To complete Factory needs list we remember P_1 known as *Lot Tolerance Percent Defective;* α, the producer's risk or first type error probability; β, the consumer's risk or second type error probability. All these values must be contractually chosen.

5. An example

Let us take some data based on the: "Inside diameter for automobile engine piston rings" (Montgomery, 1991, pag. 234).
We have: $\overline{\overline{X}} = 74.001$; $'S = 0.0090$; $''S = 0.0099$; $'''S = 0.0101$. The limits for the $\overline{x}$ chart are:

$$\mathrm{UCL}_{\overline{x}} = \overline{X} + A_3\,'S = 74.001 + (1.427)\,(0.009) = 74.014;$$

$$\mathrm{LCL}_{\overline{x}} = \overline{X} - A_3\,'S = 74.001 - (1.427)\,(0.009) = 73.988;$$

and for the S chart:

$$\mathrm{UCL}_S = B_4\,'S = (2.089)(0.009) = 0.019.$$

If we introduce : $L = 73.981$; $U = 74.021$; ${}_L P_o = 1\%$ ${}_U P_o = 1\%$, $(P = 2\%)$ we can compute the new parameters according to our proposal. We consider:

$$z_L = -2.32635;\quad z_L = 2.32635;\quad \chi^2_{0.002} = 16.92386;$$

$$\overline{X}_0 = (L \cdot z_U - U \cdot z_L)/(z_U - z_L) = 74.001;$$

$$\sigma_o = (U - L)/(z_U - z_L) = 0.0086.$$

Therefore:

$$\mathrm{UCL}_{\overline{x}} = 74.001 + 3\,(0.0086)/\sqrt{5} = 74.01254;$$

$$\mathrm{LCL}_{\overline{x}} = 74.001 - 3\,(0.0086)/\sqrt{5} = 73.98946;$$

$$\mathrm{UCL}_s = \sigma_0\,\sqrt{\chi^2/(n-1)} = 0.0086\sqrt{16.92386} = 0.0177.$$

We note that our UCL_s is based on the χ^2-distribution as suggested by Duncan (1965). Our limits are slightly narrower than Montgomery's ones, but if factory needs are the declared ones (L and U) we have a production process not capable. Here is very important the designer responsibility because a little larger tolerances would change the situation.
We repeat the observation outlined above. Customary control charts privileges process capability. In presence of a chart (UCL and LCL for $\overline{x}$ and s) we

must verify if designer's needs (L and U) are satisfied. On the contrary with our control chart designer needs are privileged but we do not know if production process is capable. In order to get this last peace of information we must compare $''S$ with σ_0. Of course if $''S < \sigma_0$ the process can satisfy designer's needs; on the contrary we must solve the trouble. For the process capability analysis many references are given by Montgomery (1991).

6. Use of the chart

In order to use the chart we draw a sample with $n = 5$ and compute $\bar{x}$ and s. With the following data ; 74.002; 73.990; 73.997; 74.003, 74.001, we obtain: $\bar{x} = 74.002$;s = 0.002588. The points are within limits so that production is good. Let us now try to use the s^2 chart as proposed by Duncan. We have: $s^2 = 0.0000067$ and $(\mathrm{UCL}_s)^2 = 0.000079 \cdot 4.230965 = 0.000313$; the point is within limits as before. If we suppose to have a point very near the limit e.g. s =0.0176, squaring it we get: $\sigma^2 = 0.00030976$ and it is also within limits.
If the point is just out of control e.g. s =0.0178, we get $s^2 = 0.00031684$ and the point is just out of control also in the new chart.

7. Conclusions

SCC (Shewart's Control Chart) is based on process ability to produce wanted items. Indeed, if control limits $(\mathrm{UCL}_{\overline{\mathrm{X}}}\ \mathrm{LCL}_{\overline{\mathrm{X}}}\ \mathrm{UCL}_{\mathrm{S}})$ are computed on either $''S^2$ or $'''S^2$, it is not worthy to insist on process ability. Clearly there is also the designer and their needs (L,U) to be considered so we must consider a capability study to test if they are in accordance.
We have seen how it is possible to get new control limits based on $\overline{\overline{X}}_0$ and σ_0 keeping into account designer's needs (L and U). Items must be output by production proccss so that now must look if it is able to do its work.
A different subject is the dispersion estimate related to SCC taking into account the presence of m lots. We have seen that 's is a biased statistic very cumbersome to be adjusted. A simulation leads to prefer $'''S^2$ but this means to use a σ^2chart instead of a σ chart. Nevertheless in our example we used $''S$ and $''S^2$ in order to simplify the discussion.
It seems worthy to remember that customary control chart construction privileges production process capability whereas our suggestion privileges designer needs. In every case we must verify the second coin's face to compare $''S$ with σ_0 or better $'''S$ with σ_0. Better else to compare $'''S^2$ with σ_0^2 because $'''S^2$ is a σ^2 unbiased estimate.

References

Duncan, A.J. (1965). *Quality Control and Industrial Statistics.* Richard D. Irwing, Inc. Homewood, Ill.

Kenney, A.F.& Keeping, E.S. (1956). *Mathematics of Statistics.* D. Van Nostrand Co. Inc., Princeton.

Mittag, H.G. & Rinne, H. (1993) *Statistical Methods of Quality Assurance.* Chapman & Hall, London.

Montgomery, D.C. (1991). *Introduction to Statistical Quality Control.* John Wiley & Sons. New York.

Piccari, P.L. (1974). *Manuale di Controllo di Qualità e Affidabilità.* ISEDI, Milano.

Rouzet, G. (1957). *Courbes d'Efficacité de la Carte de Contrôle pae Mesures en Fonction du Pourcentage de Pièces Défectueuses.* Revue de Statistique Appliquée, vol. V, 2, 19-32.

Shewart, W.A. (1931). *Economic Control of Quality of Manufactured Product.* D. Van Nostrand Co., Inc. New York.

Wold H., (1955) *Random Normal Deviates.* Department of Statistics University of London. Tracts for Computers Edited by E. S. Pearson D. Sc. n.XXV.

Projection Pursuit Regression with Mixed Variables

Annalisa Laghi
Laura Lizzani
Dipartimento di Scienze Statistiche, Università di Bologna
Via Belle Arti 41, 40126 Bologna
laghi@stat.unibo.it, lizzani@stat.unibo.it

Abstract: The aim of this paper is to extend projection pursuit regression to the case of mixed predictors, according to two different approaches. The former consists in converting each categorical regressor into dummy variables. The latter consists in preliminarily transforming the predictors by means of principal coordinate analysis. In presence of strongly non-linear regression functions and interactions between predictors, both procedures improve the results obtained by multiple linear regression, distance-based regression, MORALS and ACE. In particular, projection pursuit regression in conjunction with principal coordinate analysis shows very satisfactory performances.

Keywords: ACE, Distance-Based Regression Model, Mixed Predictors, MORALS, Principal Coordinate Analysis, Projection Pursuit Regression.

1. Introduction

Projection pursuit regression (PPR) is a non-parametric regression method (Friedman & Stuetzle, 1981; Friedman, 1984a) developed for continuous explanatory variables. We propose to extend it to the case of mixed predictors following two different approaches, exploited so far in the context of linear regression analysis. The former consists in converting each categorical regressor into dummy variables. The latter consists in preliminarily transforming the predictors by means of principal coordinate analysis (PCA) (Gower, 1966). The two approaches are tested on simulated and real data sets and compared with classical linear regression (CR) and three procedures purposely developed to handle the case of mixed data: the distance-based (DB) regression (Cuadras and Arenas, 1990; Cuadras et al., 1996), the multiple optimal regression by alternating least squares (MORALS) (Young et al. 1976; Young, 1981) and the alternating conditional expectation (ACE) method (Breiman and Friedman, 1985). The presented simulation studies stress the distinctive feature of PPR methods to model strongly non-linear regression surfaces and interactions between predictors.

The paper is structured as follows: in section 2 we briefly describe PPR, as developed by Friedman (1984a), in section 3 we present the two approaches proposed for the treatment of mixed predictors in PPR models. Examples and conclusions are discussed in the last section.

2. Projection pursuit regression

PPR, introduced by Friedman and Stuetzle (1981) and refined by Friedman (1984a), is a non-parametric regression method for modelling a q-dimensional random vector Y, of response variables, as a function of a p-dimensional random vector X, of predictors, on the basis of a sample of n matched observations of (Y, X). Each response variable is modelled as a different linear combination of smooth functions of different linear combinations of the predictor variables. The model takes the form:

$$E[Y_l|x] = \mu_{Y_l} + \sum_{m=1}^{M_0} \beta_{lm}\phi_m(\alpha_m' x); \quad l = 1,2,..,q \tag{1}$$

where $\mu_{Y_l} = E(Y_l)$, α_m are p-dimensional unit projection directions, and ϕ_m are univariate smooth functions, with zero mean and unit variance, of the projections $\alpha_m' x$, $m=1,2,.., M_0$.
Friedman's PPR algorithm (1984a) estimates the coefficients β_{lm} and α_m by least squares, and the smooth functions ϕ_m, for each selected projection direction, using a variable span smoother, called the supersmoother (Friedman, 1984b). The algorithm proceeds by finding a model with $M \geq M_0$ terms and then pruning the model back to a total of M_0 terms, where M_0 and M are user-specified parameters.
In the examples presented in section 4 we consider the case of a single response variable (q=1) and adopt the fraction of unexplained variance (FUV) as a measure of goodness-of-fit for PPR models (Friedman, 1984a):

$$FUV = \sum_{i=1}^{n}\left[y_i - \bar{y} - \sum_{m=1}^{M_0} \hat{\beta}_m \hat{\phi}_m(\hat{\alpha}_m' x_i)\right]^2 \Bigg/ \sum_{i=1}^{n}(y_i - \bar{y})^2 \tag{2}$$

where $\bar{y}$ is the sample mean of Y, $\hat{\beta}_m$, $\hat{\phi}_m$ and $\hat{\alpha}_m$ are the estimates of β_m, ϕ_m and α_m in model (1), respectively.

The introduction of linear combinations of smooth functions and of predictors allows PPR to model non-linear regression surfaces and interactions between explanatory variables, respectively.

3. Two approaches to the treatment of mixed regressors in PPR

The PPR model has been developed for continuous predictors (X). If X is composed of continuous, binary and categorical variables, we propose to transform the data in two different ways. The former consists in replacing, in model (1), each categorical predictor by dummy variables. The latter consists in preliminarily transforming the explanatory variables by means of PCA (Gower, 1966). This method starts by constructing an $(n \times n)$ Euclidean matrix of dissimilarities $D = \{d_{ij}\}$ $i, j = 1, \ldots, n$ between every pair of individuals in the sample. We adopt the similarity coefficient proposed by Gower (1971) to deal with mixed data, so defined:

$$s_{ij} = \left[\sum_{h=1}^{p_1} \left(1 - \left|x_{ih} - x_{jh}\right| / G_h\right) + a + c \right] \Big/ \left[p_1 + (p_2 - b) + p_3\right] \quad (3)$$

where p_1 is the number of continuous regressors, a and b are the number of positive and negative matches, respectively, for the p_2 dichotomous regressors, and c is the number of matches for the p_3 categorical regressors. G_h is the range of the h-th continuous regressor. In absence of missing values, $S = \{s_{ij}\}$ $i, j = 1, \ldots, n$ is positive semi-definite and this implies that the dissimilarity matrix $D = \{d_{ij}\}$, with $d_{ij} = \sqrt{1 - s_{ij}}$, is an Euclidean matrix in the sense that there is an exact configuration of points in $n\text{-}1$ dimensions or fewer, with exactly this matrix of Euclidean distances (Gower, 1971). The transformed predictors are obtained by the spectral decomposition of the symmetric matrix $B = -0.5 HD^2 H$, where $D^2 = \{d_{ij}^2\}$ and H is the $(n \times n)$ centring matrix. The solution is given by $X^* = V\Lambda^{1/2}$, where Λ is the diagonal matrix containing the non zero eigenvalues of B arranged in descending order and V is the matrix containing the corresponding eigenvectors. A small number k of columns of X^* are selected and inserted in model (1). In our examples we retain as many columns as the number of the original predictors, presenting the highest absolute correlation coefficient with the dependent variable, as suggested by Cuadras and Arenas (1990) and Cuadras et al. (1996).

In the following section we compare these two solutions with the CR model, the DB model, which assumes a linear relationship between the dependent variable and the k selected columns of X^*, the MORALS method, which consists in maximising the multiple correlation coefficient by using an algorithm based on the alternating least squares and optimal scaling principles, and, finally, with the ACE method, which finds smooth non-linear transformations, both of the response and independent variables, that produce the best fitting additive model.

4. Examples and conclusions

We test the performances of the two proposed procedures in handling the case of mixed predictors (PPR with the method based on converting categorical predictors into dummy variables, PPR-1; PPR in conjunction with principal coordinate analysis, PPR-2) both on simulated and real data sets, and compare them with the results obtained with classical linear regression (CR), distance-based regression (DB), MORALS and ACE. The S-Plus functions *ppreg*( ) and *ace*( ) are used for PPR (PPR-1, PPR-2) and ACE methods, respectively, while the SAS *proc transreg* for MORALS.
The performances of the different methods are evaluated on the basis of the fraction of unexplained variance (FUV).

Simulated data
Two samples, each of 100 observations, are generated according to the following models:

C1:
$$Y = 0.17X_1 \cdot 0.52X_3 + 0.26X_2 \cdot (0.2X_4 + 0.43X_5 + 0.64X_6) + \varepsilon$$
C2:
$$Y = sin(0.17X_1 \cdot 0.26X_2)^2 + sin(0.52X_3 \cdot (0.2X_4 + 0.43X_5 + 0.64X_6))^2 + \varepsilon$$

where X_1 and X_2 are normally distributed with zero mean and unit variance, X_3 is binary, X_4, X_5, X_6 are dummy variables representing a three-state categorical predictor and $\varepsilon \sim N(0,0.04)$.

Table 1: *Fraction of unexplained variance for cases C1 and C2.*

	C1	FUV	*C2*	FUV
CR		0.7083		0.9357
DB (*k=4*)		0.6106		0.7847
MORALS		0.4581		0.4411
ACE		0.5106		0.7160
PPR-1	$M_0=3$	0.2769	$M_0=3$	0.3720
PPR-2 (*k=4*)	$M_0=3$	0.1701	$M_0=4$	0.1890

PPR-2 shows the best results in both situations (see Table 1), but PPR-1 can be deemed very satisfactory, too. PPR-1 allows the interpretation of the solution in terms of the original regressors but it could become unreliable when the number of categorical predictors or categories increases. In such a situation the use of PPR-2 is advisable.

The poor performances of the DB model are due to its inability to model strongly non-linear regression surfaces. MORALS and ACE fail, as they are not able to capture interactions between predictors. CR performs the worst because it cannot deal with either non-linear relationships or interactions.

Real data

This example is taken from SAS/IML User's guide (1985, p. 67). The data come from an experiment in which nitrogen oxide emissions from a single cylinder engine were measured for various combinations of fuel, compression ratio and equivalence ratio. Only two kinds of fuel, ethanol and indolene, are considered, as in Cuadras and Arenas (1990) and Cuadras et al. (1996) where the same data set is used to test the performances of the DB model. The data set consists of 110 observations. Two predictors, compression ratio and equivalence ratio, are continuous, while the remaining one, fuel, is a two-state categorical variable.

All methods, but classical linear regression, show good performances (see Table 2). In particular our procedures yield very close results to MORALS and ACE.

Table 2: *Fraction of unexplained variance for fuel data.*

		FUV		FUV
CR		0.7709		
DB (*k=3*)		0.1126		
MORALS		0.0297		
ACE		0.0359		
PPR-1	$M_0=1$	0.0407	$M_0=2$	0.0222
PPR-2 (*k=3*)	$M_0=1$	0.0643	$M_0=2$	0.0300

Figure 1 shows that the residuals for the CR model are strongly structured, revealing that linear regression doesn't succeed in catching the bimodal shape of the regression function. This shape is clearly detected by the one-term PPR-1 model (see Figure 2). In Figure 3 the corresponding residuals, with their smooth representation, are plotted against the fitted values.

The smooth function determined by the one-term PPR-2 model shows a non-linear shape (see Figure 4). This is the reason why the DB model performs worse, resulting in structured residuals (see Figure 6). On the contrary no structure emerges from the plot of PPR-2 residuals (see Figure 5).

The results for both PPR-1 and PPR-2 are only slightly improved by adding one more term.

Figure 1: *Residuals Vs fitted values (CR)*

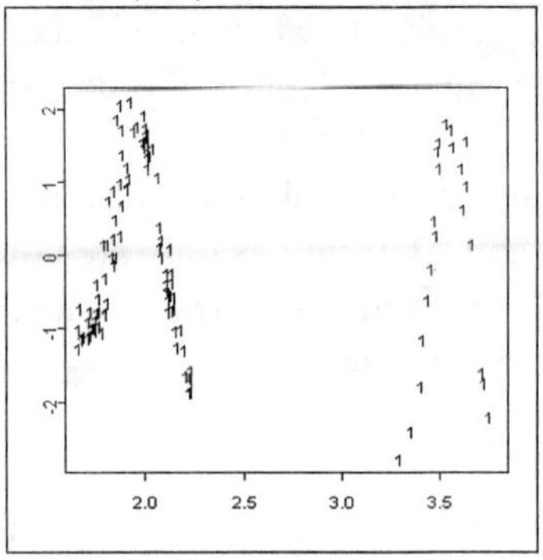

Figure 2: *Smooth function Vs linear projection (one-term PPR-1)*

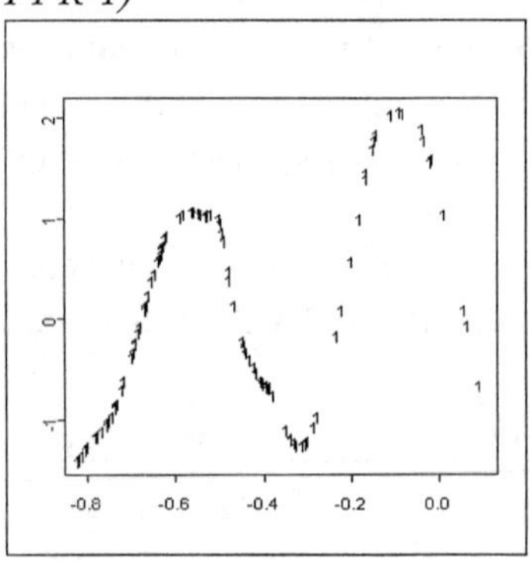

Figure 3: *Residuals* (1) *and smooth residuals* (2) *Vs fitted values (one-term PPR-1)*

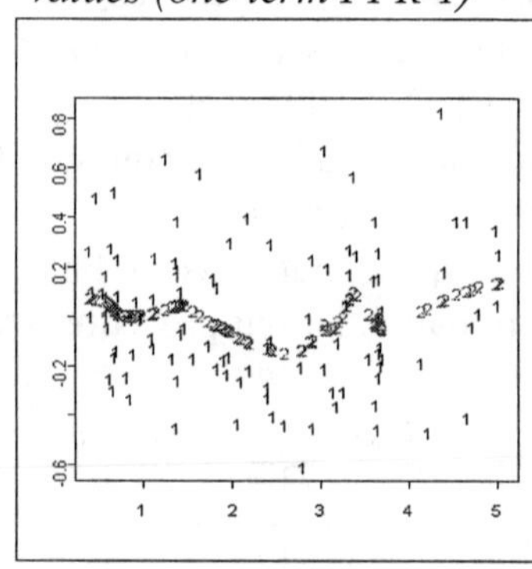

Figure 4: *Smooth function Vs linear projection (one-term PPR-2)*

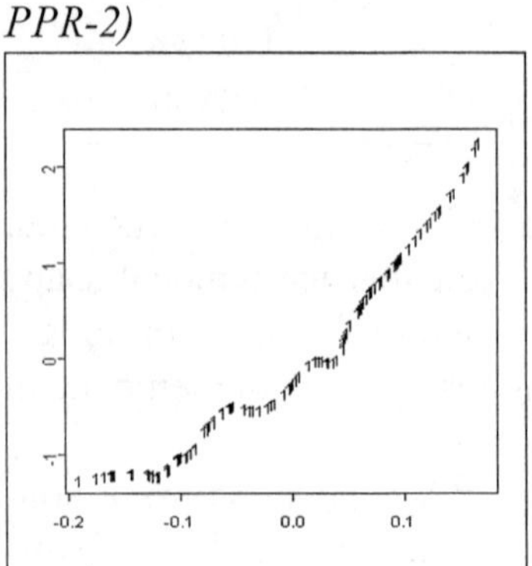

Figure 5: *Residuals* (1) *and smooth residuals* (2) *Vs fitted values (one-term PPR-2)*

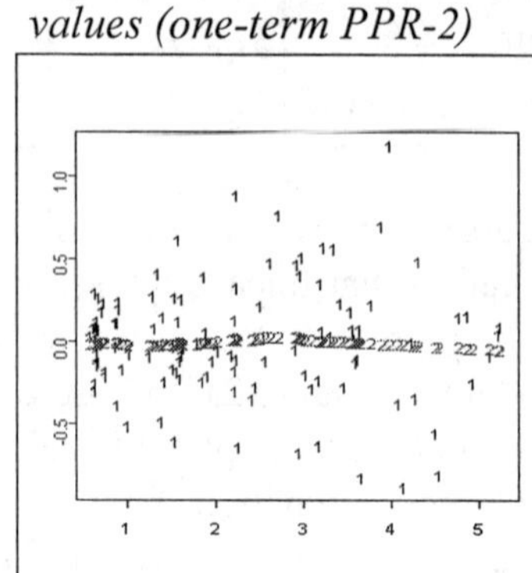

Figure 6: *Residuals* (1) *and smooth residuals* (2) *Vs fitted values (DB)*

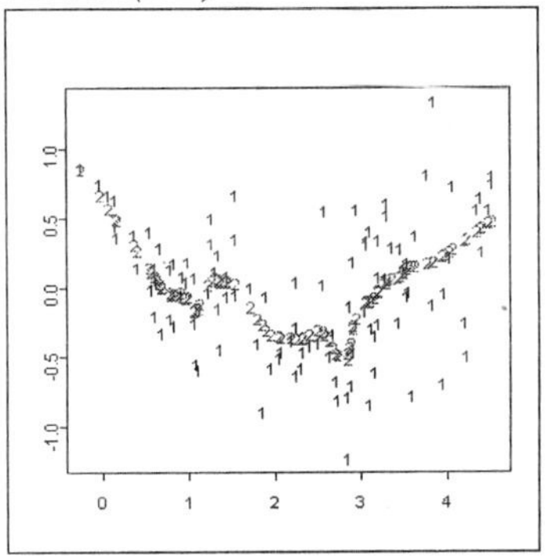

On the basis of these analyses and of further real and simulated examples, we can conclude that our proposed procedures can be deemed valid competitors to the parametric and non-parametric well-established regression methodologies developed to deal with mixed explanatory variables.
Projection pursuit regression based methods for the treatment of mixed regressors could be improved by inserting the predictor transformations in the model building stage rather than in a preliminary one. We are now working on this possibility following the same line of reasoning underlying MORALS and ACE procedures.

References

Breiman L. & Friedman J. H. (1985). Estimating optimal transformations for multiple regression and correlation, (with discussion), *Journal of the American Statistical Association*, 77, 580-619.

Cuadras, C. M. & Arenas, C. (1990). A distance-based regression model for prediction with mixed data, *Communications in Statistics - Theory and Methods*, 19, 2261-2279.

Cuadras, C. M., Arenas, C. & Fortiana, J. (1996). Some computational aspects of a distance-based model for prediction, *Communications in Statistics - Simulations*, 25, 593-609.

Friedman, J. H. (1984a). SMART: User's Guide, Technical Report no. 1, Department of Statistics, Stanford University, Stanford, CA.

Friedman, J. H. (1984b). A variable span smoother, Technical Report no. 5, Department of Statistics, Stanford University, Stanford, CA.

Friedman, J. H. & Stuetzle, W. (1981). Projection Pursuit Regression, *Journal of the American Statistical Association*, 76, 817-823.

Gower, J. C. (1966). Some distance properties of latent root and vector methods used in multivariate analysis, *Biometrika*, 53, 325-338.

Gower, J. C. (1971). A general coefficient of similarity and some of its properties, *Biometrics*, 27, 857-872.

SAS/IML (1985). User's guide, SAS Institute Inc., Cary, North Carolina.

Young, F. W. (1981). Quantitative analysis of qualitative data, *Psychometrika*, 46, 357-388.

Young, F. W., de Leeuw, J. & Takane, Y. (1976). Regression with qualitative and quantitative variables: an alternating least squares method with optimal scaling features, *Psychometrika*, 41, 505-529.

Recursive Estimation of System Parameter in Environmental Time Series Models

P. Mantovan, A. Pastore, S. Tonellato
Dipartimento di Statistica, Università Ca' Foscari di Venezia

Abstract: Dealing with high-frequency time series, such as environmental ones, raises important inferential and computational problems. Environmental monitoring and forecasting, for instance, require statistical procedures giving reliable estimates of unknown parameters and forecasts in real time. In this paper we consider dynamic linear models as a basic tool for the analysis of such kind of data and propose a recursive estimator for system parameter. A comparison of this estimator with some other estimation methods is provided via Monte Carlo simulations. The estimator we propose is computationally efficient and very easy to implement. Moreover, in our simulation study, it exhibits good asymptotic properties.

Keywords: Dynamic linear models, system parameter estimation, Kalman filter, method of moments estimators.

1. Introduction

Usually, time series of environmental data are collected with high frequency. Air pollution data, for instance, are usually collected in real time. In this context, a statistical model is required to give forecasts in a very short time. To this purpose, a useful tool is provided by dynamic models for which recursive estimation and forecasting procedures are available. For the concentration of each pollutant, Italian law fixes two thresholds: a warning threshold and a high risk threshold. Whenever the concentration of a pollutant becomes higher than the corresponding warning threshold, public authorities must implement some environmental policy interventions (such as traffic and house heating restrictions), in order to reduce pollution to an acceptable level. If the concentration of a pollutant becomes higher than the high risk threshold, citizens' health might be seriously damaged, so public interventions are even more urgently needed.

In the light of the considerations above, it is clear that environmental monitoring represents a great challenge for the statistician. In fact, we are required to provide forecasts about any crossing over of warning and high risk thresholds, and we should do it as early as possible, in order to allow prompt interventions. It is also obvious that reliable statistical models are required for this purpose. Since we usually have to deal with many response and explanatory variables, it follows that such models might be quite complex. Moreover, observations collected with a relatively high frequency are usually numerous, and this implies that data storing and handling might be quite problematic and time demanding. It is worth noting, finally, that environmental systems evolve over both space and time, so we have to deal with temporal and spatial dependences among observations and with many kinds of temporal and/or spatial changes in the relationships among the variables we are interested in. For all these reasons, highly flexible models and efficient estimation procedures are needed. In such a context recursive estimators are mostly useful, in that they permit the achievement of forecasts as soon as new observations are available.

In Section 2, we give a short description of dynamic linear models and propose a recursive method for the estimation of the system parameter. Section 3 recalls some other methodologies well known in the literature which will be compared with our estimator in a simulation study, in order to assess their performance in the operational context we have described above. The results we obtained will be shown in Section 4.

2. A recursive estimator for the system parameter of dynamic linear models

Dynamic linear models give a relationship between a sequence of c-variate, $c \geq 1$ observable random vectors $\mathbf{y}_t$ and a sequence of k-variate $k \geq 1$ unobservable random vectors $\mathbf{x}_t$ ($\mathbf{x}$ is usually referred as *state vector*), $t = 1, 2, \dots$. This relationship is determined by the following stochastic system:

$$\mathbf{x}_t = \mathbf{M}_t \mathbf{x}_{t-1} + \mathbf{u}_t \tag{1}$$

$$\mathbf{y}_t = \mathbf{H}_t \mathbf{x}_t + \mathbf{v}_t \tag{2}$$

where $\mathbf{H}_t$ is a known matrix for all t, $\mathbf{v}_t \sim WN(\mathbf{0}, \mathbf{R}_t)$ is the vector of measurement errors and $\mathbf{u}_t \sim WN(\mathbf{0}, \mathbf{Q}_t)$, indicates system noise. It is usually assumed that random errors $\mathbf{v}_t$ and $\mathbf{u}_t$ are mutually uncorrelated, that $\mathbf{u}_t$ is uncorrelated with $\mathbf{x}_j$ for all $j < t$, $t = 1, 2, \dots$, and $\mathbf{v}_t$ is uncorrelated with $\mathbf{x}_j$ for all j and t. Let $\mathbf{Y}_t = (\mathbf{Y}_{t-1}, \mathbf{y}_t)$ be the information set available at time t, with $\mathbf{Y}_0$ representing initial, information, and $(\mathbf{y}_1, \mathbf{y}_2, \dots, \mathbf{y}_t)$ be the series of the observed values. If $\mathbf{R}_t$, $\mathbf{Q}_t$ and $\mathbf{M}_t$ are known, the Kalman filter (Kalman and Bucy 1960), provides updating equations for both the estimate of the state vector, $\mathbf{x}_{t|t}$, and the associated error variance matrix $\mathbf{P}_{t|t}$. Moreover, if $\mathbf{u}_t$ and $\mathbf{v}_t$ are normally distributed, it is easy to define the predictive distribution of $\mathbf{y}_{t+s}$ and $\mathbf{x}_{t+s}$, $s > 0$ (a statistical development of this issue has been given recently by West and Harrison (1989)).

In most statistical applications, the system parameter $\boldsymbol{\theta}_t = \text{vec}(\mathbf{M}_t')$ is only partially known. To simplify the presentation, in this section we will assume that $\boldsymbol{\theta}_t$ is completely unknown and propose a recursive estimator. Assume that:

$$\boldsymbol{\theta}_t = \mathbf{A}\boldsymbol{\theta}_{t-1} + \boldsymbol{\eta}_t, \tag{3}$$

where $\mathbf{A}$ is a known matrix, and $\boldsymbol{\eta}_t \sim WN(\mathbf{0}, \mathbf{W})$ is a disturbance term, uncorrelated with measurement errors and system noise (when $\mathbf{A}$ is an identity matrix and $\boldsymbol{\eta}_t$ is null, the system parameter is called time invariant). In this case, the updating equations provided by the Kalman filter can be used only conditionally on $\Theta_t = (\boldsymbol{\theta}_1, \boldsymbol{\theta}_2, \ldots, \boldsymbol{\theta}_t)$, which is a reference trajectory of the values of the system parameter. The recursive algorithm we propose can be obtained as follows. Substituting $\mathbf{x}_t$ as defined by equation (1) into equation (2), we obtain:

$$\mathbf{y}_t = \mathbf{H}_t(\mathbf{I}_k \otimes \mathbf{x}'_{t-1})\boldsymbol{\theta}_t + \boldsymbol{\varepsilon}_t \tag{4}$$

with $\boldsymbol{\varepsilon}_t = \mathbf{H}_t\mathbf{u}_t + \mathbf{v}_t$. Equations (3) and (4) define a new dynamic linear model where $\boldsymbol{\theta}_t$ can be treated as the state vector and, conditionally on a trajectory $\mathbf{X}_{t-1} = (\mathbf{x}_0, \mathbf{x}_1, \ldots, \mathbf{x}_{t-1})$, it can be estimated using the Kalman filter. In order to obtain updating equations for both $\mathbf{x}_{t|t}$ and $\boldsymbol{\theta}_{t|t}$ we have to choose two reference trajectories: Θ_t and $\mathbf{X}_{t-1}$. Such trajectories are built step by step: given some fixed initial values, $\boldsymbol{\theta}_1$ and $\mathbf{x}_0$, at time t, $t \geq 1$, the estimate $\boldsymbol{\theta}_{t|t}$ can be obtained by applying the Kalman filter to equations (3) and (4), conditionally on $\mathbf{X}_{t-1}$; then, updating Θ_{t-1} by putting $\boldsymbol{\theta}_t = \boldsymbol{\theta}_{t|t}$, it is possible to get $\mathbf{x}_{t|t}$ applying the Kalman filter to equations (1) and (2) conditionally on Θ_t. So, the t-th value of each trajectory is given by: $\boldsymbol{\theta}_t = \boldsymbol{\theta}_{t|t}(\mathbf{X}_{t-1})$ and $\mathbf{x}_t = \mathbf{x}_{t|t}(\Theta_t)$. Assume you fixed the initial values $\theta_{1|1}$ e $x_{0|0}$, and the corresponding "prior" dispersion matrices $\boldsymbol{\Omega}_{1|1}$ and $\mathbf{P}_{0|0}$. At time $t = 1$ the estimate $\mathbf{x}_{1|1}$ can be obtained by applying the Kalman filter to eqs. (1) and (2). Hence, at time $t = 2$, assuming $\mathbf{x}_1 = \mathbf{x}_{1|1}$, the estimate $\boldsymbol{\theta}_{2|2}$ can be obtained by applying the Kalman filter to eqs. (3) and (4), conditionally on $\mathbf{x}_{1|1}$. This procedure is repeated at each time instant. The resulting algorithm is given by two mutually interacting Kalman filters and it is similar to the one proposed by Todini (1978). The updating equations for $\boldsymbol{\theta}_{t|t}$ and $\boldsymbol{\Omega}_{t|t}$ are given by:

$$\boldsymbol{\theta}_{t|t} = \mathbf{A}\boldsymbol{\theta}_{t-1|t-1} + \mathbf{K}_t\left[\mathbf{y}_t - \mathbf{H}_t(\mathbf{I}_k \otimes \mathbf{x}'_{t-1})\mathbf{A}\boldsymbol{\theta}_{t-1|t-1}\right] \tag{5}$$

$$\boldsymbol{\Omega}_{t|t} = \left[\mathbf{I}_{k^2} - \mathbf{K}_t\mathbf{H}_t(\mathbf{I}_k \otimes \mathbf{x}'_{t-1})\right]\boldsymbol{\Omega}_{t|t-1} \tag{6}$$

where $\mathbf{K}_t = \boldsymbol{\Omega}_{t|t-1}\left[\mathbf{H}_t(\mathbf{I}_k \otimes \mathbf{x}'_{t-1})\boldsymbol{\Omega}_{t-1|t-1}(\mathbf{I}_k \otimes \mathbf{x}'_{t-1})'\mathbf{H}'_t + \mathbf{H}_t\mathbf{Q}_t\mathbf{H}'_t + \mathbf{R}_t\right]^{-1}$, with $\boldsymbol{\Omega}_{t|t-1} = \mathbf{A}\boldsymbol{\Omega}_{t-1|t-1}\mathbf{A}' + \mathbf{W}$ (clearly, updating equations for a model with time invariant parameter can be easily obtained by putting $\mathbf{A} = \mathbf{I_k}$ and $\mathbf{W} = \mathbf{0}$).

The estimator we propose has the following properties: a) it is recursive; b) it is semiparametric (we did not make any assumption about data distribution); c) it does not require the solution of complex optimisation problems; d) it allows a quick intervention whenever structural changes happen (we will provide an example later).

3. Some other estimation methods

We compared our recursive estimator with three widely used methodologies: maximum likelihood estimation (ML), Bayesian posterior analysis and an estimation method proposed by Wojcik (1993), based on the method of moments. In the following, it will be assumed that $\boldsymbol{\theta}_t = \boldsymbol{\theta}$ for all t.

Maximum likelihood estimation. ML estimation is a very well known and popular methodology because of its asymptotic optimal properties (Gupta and Mehra 1974).

Bayesian method and Gibbs sampling. A Bayesian model might be more flexible than ML, but it generally requires heavy numerical computations. Monte Carlo Markov Chains are probably the most powerful tool a statistician can use to cope with highly complex models, but they are usually time demanding. We will give a brief description of the Gibbs sampler (Gelfand and Smith 1990). In our model the parameter of interest is $\boldsymbol{\psi} = [\mathbf{x}', \boldsymbol{\theta}']'$, with $\mathbf{x} = [\mathbf{x}_0', \ldots, \mathbf{x}_t']'$. The Gibbs sampler generates samples from the joint posterior distribution of $\boldsymbol{\psi}$ as follows. Given arbitrary starting values $\mathbf{x}_{(0)}$ and $\boldsymbol{\theta}_{(0)}$, we draw $\mathbf{x}_{(1)}$ from $f\left(\mathbf{x}|\boldsymbol{\theta}_{(0)}, \mathbf{y}\right)$, then $\boldsymbol{\theta}_{(1)}$ from $f\left(\boldsymbol{\theta}|\mathbf{x}_{(1)}, \mathbf{y}\right)$ to complete the first iteration. After s such iterations we obtain $\left(\mathbf{x}_{(s)}, \boldsymbol{\theta}_{(s)}\right)$. Under mild conditions, this random vector converges in distribution to the joint posterior as $s \to \infty$. Replicating the entire process in parallel G times provides a sample of random vectors $(\mathbf{x}^{(j)}, \boldsymbol{\theta}^{(j)})$, $j = 1, \ldots, G$, from the joint posterior distribution. These observations can be used to estimate any posterior marginal moments (whenever they exist) or quantiles, or posterior marginal densities. Now suppose that $\mathbf{R}_t$ and $\mathbf{Q}_t = \mathbf{Q}$ in the model defined by eqs. (1) and (2) are known for all t and assume we observed a sample $\mathbf{y} = (\mathbf{y}_1 \ldots \mathbf{y}_t)$. Generation of $\mathbf{x}$ from $f\left(\mathbf{x}|\boldsymbol{\theta}, \mathbf{y}\right)$ can be obtained using the method proposed by Carter and Khon (1994). As far as the generation of $\boldsymbol{\theta}$ is concerned, it can be easily shown that, if a Gaussian prior has been elicited on $\boldsymbol{\theta}$, i.e. $\boldsymbol{\theta} \sim N(\mathbf{m}_{\boldsymbol{\theta}}, \mathbf{V}_{\boldsymbol{\theta}})$ a priori, and defining $\mathbf{x}_i^* = \mathbf{I}_k \otimes \mathbf{x}_{i-1}'$, $\mathbf{x}^* = [\mathbf{x}_1^{*\prime}, \ldots, \mathbf{x}_t^{*\prime}]'$ and $\tilde{\mathbf{x}} = [\mathbf{x}_1', \ldots, \mathbf{x}_t']'$, $f\left(\boldsymbol{\theta}|\mathbf{x}, \mathbf{y}\right)$ will be the density of a normal random vector with variance $\mathbf{V}_{\boldsymbol{\theta}}^* = \mathbf{V}_{\boldsymbol{\theta}}^{-1} + \mathbf{x}^*(\mathbf{I}_t \otimes \mathbf{Q})^{-1}\mathbf{x}^{*\prime}$ and mean $\mathbf{m}_{\boldsymbol{\theta}}^* = \mathbf{V}^*{}_{\boldsymbol{\theta}}{}^{-1}\left[\mathbf{V}_{\boldsymbol{\theta}}^{-1}\mathbf{m}_{\boldsymbol{\theta}} + \mathbf{x}^{*\prime}(\mathbf{I}_t \otimes \mathbf{Q})^{-1}\tilde{\mathbf{x}}\right]$.

An estimator based on the method of moments The estimator we are introducing now has been defined by Wojcik (1993). It is considered here for ease of implementation and also because it is a consistent estimator. However, it requires some severe restrictions. Assuming that $\mathbf{y}_t$ (which, in this approach, is assumed to be a scalar quantity) is second order stationary, i. e. $\mathbf{R}_t = \mathbf{R}$, $\mathbf{Q}_t = \mathbf{Q}$ and $\mathbf{M}_t = \mathbf{M}$ for all t, with all eigenvalues of $\mathbf{M}$ strictly less than one in modulus, and exploiting the structure of the observability matrix and the Cayley-Hamilton theorem, Wojcik considers a relationship between the second order moments of $\mathbf{y}_t$ and the coefficients of the characteristic equation of $\mathbf{M}$. This relationship is:

$$\begin{bmatrix} \alpha_1 \\ \vdots \\ \alpha_k \end{bmatrix} = \begin{bmatrix} \boldsymbol{\Gamma}(1) & \cdots & \boldsymbol{\Gamma}(k) \\ \vdots & & \vdots \\ \boldsymbol{\Gamma}(k) & \cdots & \boldsymbol{\Gamma}(2k-1) \end{bmatrix}^{-1} \begin{bmatrix} \boldsymbol{\Gamma}(k+1) \\ \vdots \\ \boldsymbol{\Gamma}(2k) \end{bmatrix} \tag{7}$$

where $\boldsymbol{\Gamma}(i) = Cov(\mathbf{y}_t, \mathbf{y}_{t-i})$ and $\alpha_1, \ldots, \alpha_k$ are the coefficients of the characteristic equation of M. Substituting second order moments with their empirical estimates, we obtain consistent estimates of $\alpha_1, \ldots, \alpha_k$. The model can be suitably reparametrized via a non singular transformation of the state vector, leading to the so–called canonical representation, so that the transition matrix of the new model is the companion form of M, say $\mathbf{M}^*$:

$$\mathbf{M}^* = \begin{bmatrix} 0 & 1 & \cdots & 0 \\ \vdots & \ddots & \ddots & \vdots \\ 0 & \cdots & 0 & 1 \\ -\alpha_1 & -\alpha_2 & \cdots & -\alpha_k \end{bmatrix}^{-1}$$

which can be estimated consistently.

4. Results of a simulation study

We applied the three methods to simulated data generated from models for scalar observations with the following structure for all t:

$$\mathbf{x}_t = \mathbf{M}\mathbf{x}_{t-1} + \mathbf{B}u_t, \quad u_t \sim N(0, Q) \tag{8}$$

$$y_t = \mathbf{H}\mathbf{x}_t + v_t, \quad v_t \sim N(0, R), \tag{9}$$

$$\mathbf{M} = \begin{bmatrix} \phi_1 & \phi_2 & \phi_3 & \phi_4 \\ 1 & 0 & 0 & 0 \\ 0 & 1 & 0 & 0 \\ 0 & 0 & 1 & 0 \end{bmatrix}, \mathbf{H} = \begin{bmatrix} 1 & 0 & 0 & 0 \end{bmatrix}, \mathbf{B} = \begin{bmatrix} 1 & 0 & 0 & 0 \end{bmatrix}'.$$

We considered four models, denoted by **M1**, **M2**, **M3** and **M4**. In Table 1 the values of the parameters for each of them are reported together with the signal–to–noise ratio $Var(\mathbf{H}\mathbf{x}_t)/Var(y_t)$. For **M1**, **M2** and **M3** we generated 1000 independent time series of length 300. The ratio underlying the choice of these models is the following. It is well known that when the level of the signal–to–noise ratio is low we cannot achieve good estimates of the unobservable state, even when all parameters are known. We wanted to test how signal–to–noise ratio affects the performance of system parameter estimators. Moreover, we wanted to assess whether a high variability of the state (i.e a high value of Q in our simulation) implies or not a poor efficiency of the recursive estimator.

We wanted to test also the performance of the recursive estimator when the system parameter changes over time. Therefore we generated 100 independent time series of length 2000 from **M4**. The initial conditions in all the applications of the recursive estimator were fixed at: $x_{0|0} = 0$, $P_{0|0} = I_k, \theta_{0|0} = 0$, $\Omega_{0|0} = \begin{bmatrix} I_4 & \mathbf{0} \\ \mathbf{0} & \mathbf{0} \end{bmatrix}$.

Fig. 1 shows the comparison of the behaviour of the recursive estimator with that of ML estimator, which has been computed at

Table 1: *Parameter values and signal–to–noise ratios for models* **M1**, **M2**, **M3** *and* **M4**.

	ϕ_1	ϕ_2	ϕ_3	ϕ_4	Q	R	$\frac{Var(\mathbf{H}\mathbf{x}_t)}{Var(y_t)}$
M1	0.50	-0.10	0.07	0.01	0.50	0.03	0.955
M2	0.50	-0.10	0.07	0.01	0.50	0.15	0.811
M3	0.50	-0.10	0.07	0.01	2.50	0.03	0.991
M4 $t \leq 1000$	0.50	-0.10	0.07	0.01	0.50	0.03	0.955
M4 $t > 1000$	-0.50	-0.10	0.07	0.01	0.50	0.03	0.955

t = 20, 40, 60, 80, 100, 150, 200, 250, 300 (the curves relative to the quantiles of the latter have been obtained by linear interpolation). The 5-th, 50-th and 90-th percentiles of the Euclidean normi of $\boldsymbol{\theta} - \hat{\boldsymbol{\theta}}_{ML}$ and $\boldsymbol{\theta} - \boldsymbol{\theta}_{t|t}$ get quite close to each other at $t = 50$ and they tend to overlap for $t > 300$, indicating a good performance of the recursive estimator. Similar considerations can be made for the results about models **M2** and **M3**, which are reported in Table 2 for t = 20, 50, 300. Bayesian analysis via Gibbs sampling gave satisfactory results as well, but it does not seem adequate to the application field we are interested in, because of the computational complexity of the Gibbs sampler. In fact, the analysis of a single time series of length 300 requires about 20 minutes on a PC with a 75 Mhz Pentium coprocessor. This is a very long time if compared with the speed of the recursive estimator reported below. As can be seen in Table 2, moderate changes in the value of signal–to–noise ratio do not seriously affect the behaviour of the recursive and ML estimators. The variability of one step ahead forecast errors is instead sensitive to the values of the variances of the disturbances appearing in eqs. (8) and (9). The behaviour of the estimator based on the method of moments was very unsatisfactory. This result is not surprising since, whenever the matrix appearing in eq. (7) is close to singularity, numerical instability problems may arise. For this reason the estimates assume values quite distant from the true ones even for time series of length 2000. As far as computing time is concerned, ML estimation requires about 2 minutes for a time series of length 300 using scoring algorithm, whereas the recursive estimator produces its output in 0.005 minutes.

We find very interesting the performance of the recursive estimator when a structural change happens. In the estimation of the system parameter for model **M4**, we followed two alternative strategies: in one case we used the recursive estimator as defined in Section 2 (no correction), in the other one we increased the dispersion matrix of $\boldsymbol{\theta}_t$ at $t = 1100$ by putting $\boldsymbol{\Omega}_{1100|1100} = 0.5\,\boldsymbol{\Omega}_{0|0}$, in order to accelerate the learning process of the recursive estimator. The result is shown in Fig. 2: the good performance of the estimator in adapting to the new situation is apparent. The behaviour of innovations is shown in Fig. 3, and it suggests that a careful investigation of innovations behaviour might give useful insights as far as the detection of model changes is concerned.

Table 2: *Comparison of 5-th, 50-th and 95-th quantiles of* $||\hat{\boldsymbol{\theta}} - \boldsymbol{\theta}||$ *obtained through recursive* (**RE**) *and maximum likelihood* (**ML**) *estimators.*

		M1		**M2**		**M3**	
		RE	ML	RE	ML	RE	ML
$t = 20$	$q_{0.50}$	0.435	0.351	0.463	0.365	0.432	0.325
	$q_{0.95}$	0.777	0.542	0.822	0.580	0.822	0.580
$t = 50$	$q_{0.50}$	0.287	0.219	0.326	0.281	0.270	0.216
	$q_{0.95}$	0.526	0.426	0.586	0.461	0.489	0.383
$t = 300$	$q_{0.50}$	0.114	0.095	0.147	0.123	0.113	0.100
	$q_{0.95}$	0.212	0.186	0.282	0.266	0.205	0.197

Figure 1: *Model M1: (a) 5-th, 50-th and 95-th percentiles of* $||\boldsymbol{\theta} - \hat{\boldsymbol{\theta}}_{t|t}||$ *(bold lines)* $||\boldsymbol{\theta} - \hat{\boldsymbol{\theta}}_{ML}||$ *(dotted lines); (b) 5-th, 50-th and 95-th percentiles of innovations when RE is used.*

(a) (b)

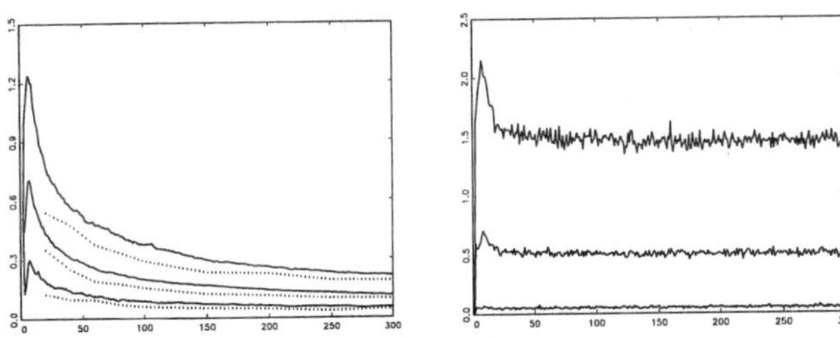

Figure 2: *Parameter change: plot of 5-th, 50-th and 95-th percentiles of* $||\boldsymbol{\theta}-\hat{\boldsymbol{\theta}}_{t|t}||$ *with and without correction.*

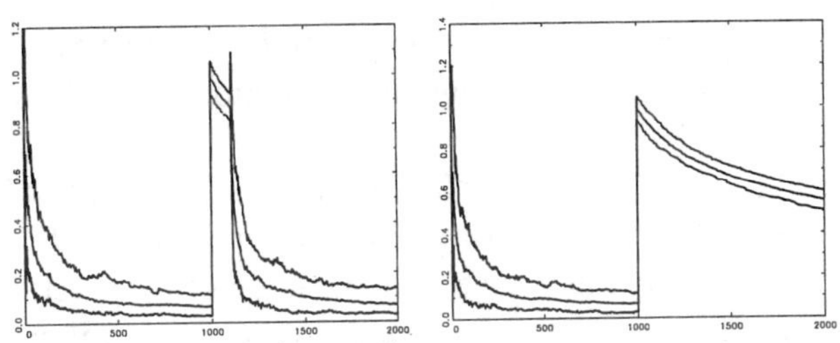

5. Conclusions

The results of our simulation study shed some light on the applicability of the recursive algorithm we propose for the estimation of the system parameter. In the examples we examined signal–to–noise ratio did not seem to affect strongly the performance of the estimator. Moreover, we can hypothesise that the asymptotic

Figure 3: *Parameter change: plot of 5 th, 50-th and 95-th percentiles of innovations with and without correction.*

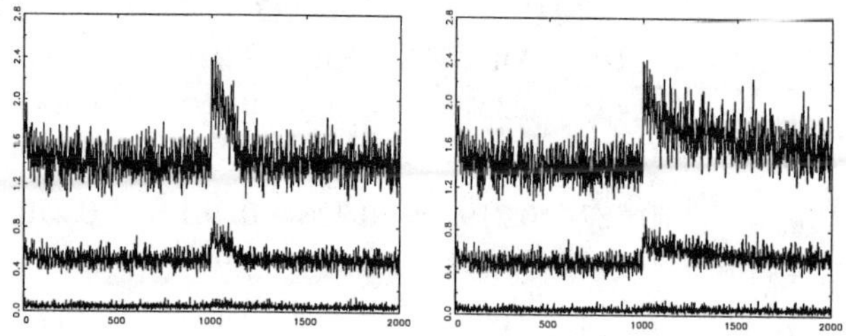

behaviour of our recursive procedure might be very close to ML estimation. In the context of environmental monitoring our method has some advantages over ML. In fact, ML requires some complex optimisation procedures which need to be carefully checked in order to avoid convergence to local maxima. Moreover, it is difficult to define the likelihood of the model when some parameter changes happen if change points are unknown. Our method, instead, being recursive, is very quick to implement and may adapt rapidly to parameter changes throug a very simple intervention. Moreover, it does not require any assumption about data distribution.

References

Carter, C. K. and R. Khon (1994). On Gibbs sampling for state–space models. *Biometrika 81*, 641–553.

Gelfand, A. E. and A. F. M. Smith (1990). Sampling based approaches to calculating marginal densities. *Journal of the American Statistical Association 85*, 398–409.

Gupta, N. K. and R. K. Mehra (1974). Computational aspects of maximum likelihood estimation and reduction of sensitivity function calculations. *IEEE Trans. Aut. Contr. 19*, 774–783.

Kalman, R. E. and R. S. Bucy (1960). New results in linear filtering and prediction theory. *Journal of Basic Engineering Transaction of the ASME*, 95–108.

Todini, E. (1978). Mutually interactive state parameter (misp) estimation. In *Proceedings of the Conference on Application of Kalman Filter to Hydrology, Hydraulics and Water Resources*, Pittsburg, pp. 135–151.

West, M. and J. Harrison (1989). *Bayesian Forecasting and Dynamic Linear Models*. Springer & Verlag.

Wojcik, P. J. (1993). On-line estimation of signal and noise parameters and the adaptive kalman filtering. In G. Chen (Ed.), *Approximate Kalman Filtering*. In Approximate Kalman Filtering, Chen, G. (ed.). Singapore, World Scientific Publishing Co., Inc.

Kernel Methods For Estimating Covariance Functions From Curves

Andrea Pallini
Dipartimento di Scienze Statistiche, Università di Bologna,
Via delle Belle Arti 41, I-40126 Bologna (Italy).

Abstract. We propose kernel methods for estimating covariance functions, when the data consists of a collection of curves. Every curve is modelled as an independent realization of a stochastic process with unknown mean and covariance structure. We consider a kernel density estimator, which has the positive semi-definiteness property on the "time" points and also in the continuum. We describe a cross-validation procedure, which leaves out an entire curve at a time, to choose the bandwidth (smoothing parameter) automatically from the observed collection of curves.

Key words: Covariance, Cross-validation, Functional Data Analysis, Kernel Density Estimation, Smoothing.

1. Introduction

In many research areas, experiments typically produce response which are curves, rather single data points. For instance, curves arise naturally as observations in the investigation of growth, in survival analysis, in signal processing, and more generally in the interpretation of automated on-line data, where a separate curve is observed for each individual/unit in a sample. A more detailed account of the statistical analysis of curves (named Functional Data Analysis) may be deduced from Ramsay (1982), Ramsay & Dalzell (1991) and Rice & Silverman (1991).

Suppose that a collection of n curves is available, where each curve is observed at "time " points $t_1, \ldots, t_p$. Assume that the sample curves are independent realizations of a stochastic process $X(t)$, with unknown mean $\mu(t) = E\{X(t)\}$ and unknown covariance function

$$\rho(s,t) = COV\{X(s), X(t)\} ,$$

for general (s,t) and without structural assumptions about ρ. The t_i's are not necessarily evenly spaced. The jth point on the ith curve will be indicated by $X_i(t_j) = X_{ij}$. We wish to estimate the covariance function ρ, nonparametrically, from the n observed curves.

Related work may be considered, for instance, Azzalini (1984), Hart & Wehrly (1986) and Diggle & Hutchinson (1989), when the curves arise from time series or repeated measurement data, and Diggle *et al.* (1987), Berman & Diggle (1989) and Sampson & Guttorp (1992), when the curves exhibit some spatial interdependence.

In particular, in section 2 we describe our basic kernel density estimator of the covariance function ρ. In section 3, we describe a cross-validation procedure for an automatic choice of the corresponding bandwidth. In section 4, we obtain a version of this kernel density estimator with the positive semidefiniteness property in the continuum. Finally, in section 5 we show the effectiveness of the entire procedure on a real example.

2. Kernel estimation of covariance functions

Hall *et al.* (1994) have proposed a kernel method for estimating the covariance function ρ of a stochastic process. We try to extend their basic estimator to the case of n curves.

For every $i = 1, \ldots, n$, we denote by

$$\overline{X}_j = n^{-1} \sum_{i=1}^{n} X_{ij}, \qquad \widehat{X}_{jk} = n^{-1} \sum_{i=1}^{n} \left(X_{ij} - \overline{X}_j\right) \left(X_{ik} - \overline{X}_k\right),$$

the sample mean of $X_1(t_j)$ and the standard covariance function between $X_1(t_j)$ and $X_1(t_k)$, for every $j, k = 1, \ldots, p$.

Let K denote a bivariate kernel function, which is taken to spherically symmetric (cf. Wand & Jones (1995), section 4.2). Let $H \equiv (h_1, h_2)^T$ be the bandwidth or smoothing parameter. A kernel estimator of the covariance function ρ may be defined as

$$\widehat{\rho}(s,t) = \left\{ \sum_{j \neq k} \widehat{X}_{jk} \, K\left(\frac{s - t_j}{h_1}, \frac{t - t_k}{h_2} \right) \right\} \left\{ \sum_{j \neq k} K\left(\frac{s - t_j}{h_1}, \frac{t - t_k}{h_2} \right) \right\}^{-1} . \tag{1}$$

Note that $\widehat{\rho}(s,t)$ is defined, without assuming that $\rho(s,t)$ admits an orthogonal expansion in terms of a specific set of eigenfunctions (cf. Rice & Silverman (1991) and Ramsay & Dalzell (1991)).

Assume that $t_i = \lambda u_i$, where λ is a positive number, such that $\lambda = \lambda(p) \to \infty$ as $p \to \infty$, and $u_1, \ldots, u_p$ are observed values of independent random variables $U_1, \ldots, U_p$, all having distributions not depending on p and the process X. Alternatively, we may take $t_1, \ldots, t_p$ to be evenly spaced on an interval of width λ. For simplicity, we take a bandwidth H with equal components, i.e.

$$h_1 = h_2 \equiv h .$$

Under suitable regularity conditions on the distributions of $U_1, \dots, U_p$ and the kernel K, it follows that

$$E\{\widehat{\rho}(s,t) - \rho(s,t)\}^2 = q(s,t)\,(h^4 + \lambda^{-1}) + o(h^4 + \lambda^{-1}), \tag{2}$$

where

$$q(s,t) = \frac{1}{4}\left\{\frac{\partial q(s,t)}{\partial s\,\partial t}\right\}^2 .$$

Estimator $\widehat{\rho}(s,t)$ converges to the true covariance function $\rho(s,t)$ in L^2, for every (s,t), as $ph\lambda^{-1} \to 0$. Under the condition that the process is observed over an increasingly wide range of time points, the kernel density estimator $\widehat{\rho}(s,t)$ consistently estimates the covariance function $\rho(s,t)$, for every (s,t), provided the bandwidth is properly selected.

A more complicated asymptotic setting is needed, if we include the $j = k$ terms in sums defining the kernel density estimator (1). However, the resulting alternative kernel density estimator would have a similar asymptotic behaviour, without remarkable advantages in terms of mean square error.

3. Bandwidth selection by cross-validation

Bandwidth h may be chosen by a cross-validation procedure, which works by leaving out an entire curve at a time, instead of leaving out a single "time" point. A general introduction to cross-validation methods may be found in Wand & Jones (1995), chapter 3.

For every $j, k = 1, \dots, p$, with a slight abuse of notation, we denote by

$$\widehat{\rho}_h^{(-i)}(t_j, t_k)$$

the estimate (1) of the covariance function $\rho(t_j, t_k)$, obtained by applying (1) to all the curves, but the ith curve. The cross-validation score may be then defined as

$$S(h) = \sum_{i=1}^{n} \sum_{j \neq k} \left\{\widehat{X}_{jk} - \widehat{\rho}_h^{(-i)}(t_j, t_k)\right\}^2 . \tag{3}$$

A bandwidth selector $\widehat{h}$ for h in (1) is the minimizer of (3).

Note that $\widehat{X}_{jk}$ is a standard estimator of the covariance between time points t_j and t_k, where $j, k = 1, \dots, p$, borrowed from classical multivariate analysis. Two different time points t_j and t_k are regarded as labels of two dependent n-variate random variables. Choice of h determines the smoothed surface $\widehat{\rho}$, which interpolates points $\widehat{X}_{jk}$, where $j, k = 1, \dots, p$. The minimization of the cross-validation score (3) produces the surface $\widehat{\rho}$, which turns out to be, in a certain sense, optimal.

Typically, this cross-validation procedure demands an efficient numerical minimization algorithm (cf. Press *et al. (1992)*, chapter 10) as a necessary ingredient. If a unique minimizer does not exist, $\widehat{h}$ may subjectively be chosen by plotting S given by (3), as a function of h, for a convenient set of candidates. A subjective choice for h could be, for instance, the value $\widehat{h}$ producing a desired level of smoothing in surface $\widehat{\rho}(s,t)$. A subjective choice for h coul also be the value $\widehat{h}$ able to produce a surface $\widehat{\rho}(s,t)$, which well approximates the behaviour of the standard estimator $\widehat{X}_{jk}$, from a specified region of points (s,t) of interest.

We do not discuss here the question of selecting an optimal bandwidth, able to minimize the mean square error in (2) and maintain the consistency of $\widehat{\rho}(s,t)$ for every (s,t) as $ph\lambda^{-1} \to 0$.

4. Ensuring the positive semi-definiteness property

Estimator $\widehat{\rho}(s,t)$ given by (1), is not necessarily itself a covariance function, since it does not typically satisfies the positive semi-definiteness property

$$\int\int \widehat{\rho}(s,t)\, w(s) w(t)\, ds\, dt \geq 0\,, \tag{4}$$

for all integrable functions w. Hall *et al.* (1994) suggest a procedure for making a kernel estimator of the covariance function ρ itself a covariance function. The rationale behind that procedure can be applied to estimator (1) as well.

By Bochner's theorem (cf. Bhattacharya & Waymire (1990), pages 663-664), property (4) is equivalent to nonnegativity of the Fourier Transform $\widehat{\rho}^{\dagger}$ of $\widehat{\rho}$, that is

$$\widehat{\rho}^{\dagger}(\theta_1, \theta_2) \geq 0\,,$$

for all θ_1 and θ_2, where

$$\widehat{\rho}^{\dagger}(\theta_1, \theta_2) = \int_{-\infty}^{\infty}\int_{-\infty}^{\infty} \widehat{\rho}(s,t)\, e^{i\,(\theta_1 s + \theta_2 t)}\, ds\, dt\,. \tag{5}$$

To obtain (5) from (1), we first define the function

$$\widehat{\rho}_1(s,t) = \begin{cases} \widehat{\rho}(s,t) & 0 \leq s \leq S_1,\, 0 \leq t \leq T_1 \\ \widehat{\rho}(s,T_1)\, \frac{t-T_1}{T_2-T_1} & 0 \leq s \leq S_1,\, T_1 < t \leq T_2 \\ \widehat{\rho}(S_1,t)\, \frac{s-S_1}{S_2-S_1} & S_1 < s \leq S_2,\, 0 \leq t \leq T_1 \\ \widehat{\rho}_1(S_1,T_1)\, \frac{s(T_1-T_2)+t(S_1-S_2)+S_2T_2-S_1T_1}{S_1T_1-S_1T_2-S_2T_1-S_2T_2} & S_1 < s \leq S_2,\, T_1 < t \leq T_2 \\ 0 & \textit{elsewhere}\,, \end{cases} \tag{6}$$

where S_1, S_2, T_1 and T_2 work as truncation points. Surface $\widehat{\rho}(s,t)$, outside of the region $\{0 \le s \le S_1,\, 0 \le t \le T_1\}$, is substituted with planes (whose equations are given in (6)) or with the value 0.

We assume the conditions for points $t_1, \dots, t_p$, stated above in section 2. Under further regularity conditions on the distributions $U_1, \dots, U_p$ and the kernel K, it can be shown that $\widehat{\rho}_1(s,t)$ consistently estimates the true covariance function $\rho(s,t)$, for every (s,t).

We put

$$\widehat{\rho}^{\dagger}(\theta_1,\theta_2) = 2\int_0^{\infty}\int_0^{\infty} \widehat{\rho}_1(s,t)\,\cos(\theta_1 s + \theta_2 t)\,ds\,dt\,,$$

and finally define

$$\widetilde{\rho}_1(s,t) \;=\; (2\pi)^{-1}\int_{-\theta_1^*}^{\theta_1^*}\int_{-\theta_2^*}^{\theta_2^*} \widehat{\rho}_1^{\dagger}(\theta_1,\theta_2)\,d\theta_1 d\theta_2\,, \tag{7}$$

where, for all θ,

$$\theta_i^* = \inf\left\{\theta_i > 0 : \widehat{\rho}_1^{\dagger}(\theta_i,\theta) \ge 0\right\}\,,$$

$i = 1, 2$.

Estimator (7) is a version of (1), which always satisfies the positive semi-definiteness property (4).

5. An example

We consider a data set on the weights (kg) of 48 pigs in 9 successive weeks presented in Diggle *et al.* (1994), pages 34-46. In figure 1, we plot the corresponding $n = 48$ growth curves, where the lines connect the repeated observations for each pig.

Note that the "time" points, $t_1, \dots, t_9$ say, are evenly spaced, satisfying conditions for consistency of estimators (1) and (7). The underlying stochastic process is apparently nonstationary, with increasing mean function $\mu(s)$ and variance function $\rho(s,s)$. Again see figure 1.

In order to determine the basic kernel estimator (1) of the covariance function ρ, we have used the product Gaussian kernel function given by

$$K(s,t) = (2\pi)^{-1} e^{-s^2/2} e^{-t^2/2}\,.$$

Numerical minimization of the cross-validation score (3) has yielded the bandwidth selector $\widehat{h} = 0.75$. The truncation points chosen to define estimator $\widehat{\rho}_1$ given by (6) are $S_1 = T_1 = 2.5$ and $S_2 = T_2 = 3.1$. Estimator (7) has been obtained by standard FFT and numerical integration algorithms.

Figure 1: *Weights (kg) of 48 pigs in 9 successive weeks.*

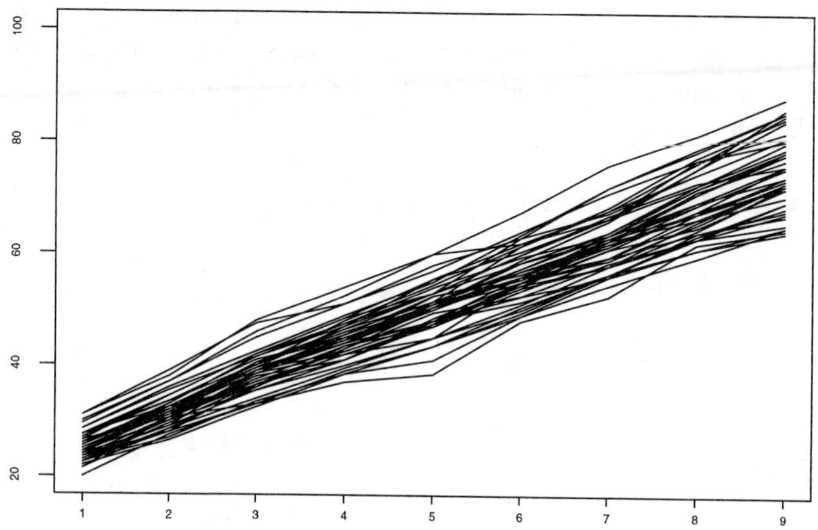

The behaviour of estimator $\widetilde{\rho}_1(s,t)$ given by (7) is finally displayed in figure 2, where points (s,t) are from the region $[1,9] \times [1,9]$, the bottom vertex being the point $(1,1)$. Note that the covariance estimate $\widetilde{\rho}_1(s,t)$ is given as a continuous function of the two variables s and t, where $(s,t) \in [1,9] \times [1,9]$.

The covariance function estimate exhibits an increasing variance function $\widetilde{\rho}_1(s,s)$ (cf. figure 1), as s increases. Set

$$I \equiv \{1,2,3,4,5,6,7,8,9\} .$$

We have estimates of the covariance function $\rho(s,t)$ also for points $(s,t) \in [1,9] \times [1,9]$, which have not been observed, namely, which are different from $(t_j,t_k) \in I \times I$. For small values of s and t in (s,t), estimator $\widetilde{\rho}_1(s,t)$ produces values, which are relatively far (with the present example) from values $\widehat{X}_{jk}$, for every $j,k = 1,\ldots,q$. For successive values of s and t in (s,t), estimator $\widetilde{\rho}_1(s,t)$ has a better performance, in this sense.

A value for $\widetilde{\rho}_1(s,t)$ means estimate for the covariance between $X(s)$ and $X(t)$. As shown in figure 2, the covariance point estimates are nearly symmetrical, that is, it turns out that

$$\widetilde{\rho}_1(s,t) \approx \widetilde{\rho}_1(t,s) ,$$

Figure 2: *Covariance function estimator* $\widetilde{\rho}(s,t)$, $(s,t) \in [1,9] \times [1,9]$.

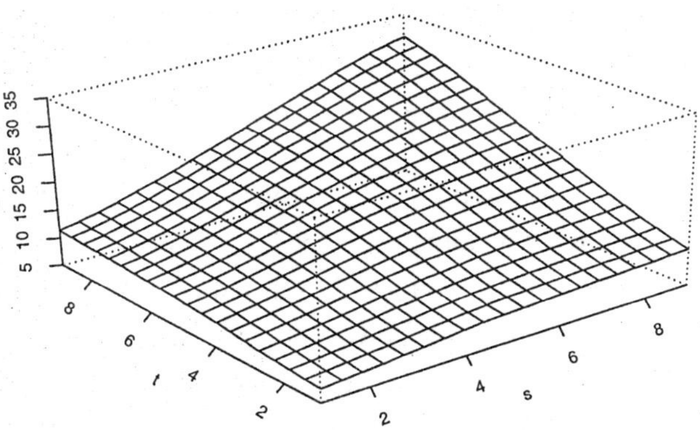

for every $(s,t),(t,s) \in [1,9] \times [1,9]$.

6. Conclusions

We have proposed a kernel density estimator for the unknown covariance function of a stochastic process. The kernel density estimator does not need settings, where stationarity of the process is crucial. This point partly explains our choice of a surface, instead of a curve, to represent the estimated covariance function of a stochastic process of interest.

It is well known that the so-called positive semidefiniteness property characterizes a covariance function. An estimator for a covariance function is typically preferable, when it satisfies this property, simply because it is itself a covariance function. We have applied the procedure by Hall *et al.* (1994), in order to make our basic kernel density estimator fulfil the positive definiteness property.

The proposed estimator is very computationally intensive and may not be competitive with others. We are thinking about examples, whereas a more standard estimator could be effective as well. A faster estimation procedure may probably be achieved by applying a different and more efficient cross validation technique.

References

Azzalini, A. (1984). Estimation and hypothesis testing for collections of autoregressive time series. *Biometrika*, 71, 85-90.

Bhattacharya, R.N. and Waymire, E.C. (1990). *Stochastic Processes With Applications.* Wiley and Sons, New York.

Berman, M. and Diggle, P. (1989). Estimating weighted integrals of the second-order intensity of a spatial point process. *J. Roy. Statist. Soc. B*, 51, 81-92.

Diggle, P.J., Gates, D.J. and Stibbard, A. (1987). A nonparametric estimator for pairwise-interaction point processes. *Biometrika*, 74, 763-770.

Diggle, P.J. and Hutchinson, M.F. (1989). On spline smoothing with autocorrelated errors. *Aust. J. Statist.*, 31, 166-182.

Diggle, P.J., Liang, K.-Y. and Zeger, S.L. (1994). *Analysis of Longitudinal Data.* Oxford University Press, Oxford.

Hall, P., Fisher, N.I. and Hoffmann, B. (1994). On the nonparametric estimator of covariance functions. *Ann. Statist.*, 22, 2115-2134.

Härdle, W. (1991). *Smoothing Techniques. With Implementation in S.* Springer-Verlag, New York.

Hart, J.D. and Wehrly, T.E. (1986). Kernel regression estimation using repeated measurement data. *J. Amer. Statist. Assoc.*, 81, 1080-1088.

Press, W.H., Teukolski, S.A., Vetterling, W.T. and Flannery, B.P. (1992). *Numerical Recipes.* Cambridge University Press, Cambridge.

Ramsay, J.O. and Dalzell, C.J. (1991). Some tools for Functional Data Analysis (with discussion). *J. R. Statist. Soc. B*, 53, 539-572.

Rice, J.A. (1984). Bandwidth choice for nonparametric regression. *Ann. Statist.*, 12, 1215-1230.

Rice, J.A. and Silverman, B.W. (1991). Estimating the mean and covariance structure nonparametrically when the data are curves. *J. R. Statist. Soc. B.*, 53, 233-243.

Sampson, P.D. and Guttorp, P. (1992). Nonparametrci representation of nonstationary spatial covariance structure. *J. Amer. Statist. Assoc.*, 87, 108-119.

Silverman, B.W. (1985). Some aspects of the spline smoothing approach to non-parametric regression curve fitting (with discussion). *J. R. Statist. Soc. B*, 47, 1-52.

Wahba, G. (1990). *Spline Models For Observational Data.* CBMS 59, SIAM, Philadelphia, Pennsylvania.

Wand, M.P. and Jones, M.C. (1995). *Kernel Smoothing.* Chapman and Hall, London.

Detection of Subsamples in Link-Free Regression Analysis

Giovanni C. Porzio*
Dipartimento di Matematica e Statistica, Università di Napoli
Via Cintia, 80126 Napoli, Italy
porzio@dmsna.dms.unina.it

Abstract: A regression analysis could fail if the sample is actually composed of more subsamples. We show that the regression function plot is a powerful tool to detect such a feature in the data. Its behaviour when more subpopulations are present is investigated in the framework of link-free regression analysis. A dynamic graphics procedure to detect the coexistence of more subsamples in the data is proposed.

Key words: Dynamic graphics, Fisher consistency, Regression graphics.

1. Introduction

In regression analysis, it is usually assumed that an unique model describes the whole data set. However, sometimes regression data allow several fits due to the presence of more subpopulations in the sample. Each of them could require different models with different parameter values, and they will cause the failure of an overall analysis. Hence, given a sample for a regression analysis, the point is to assess if it is composed of more than one subsample. It is aim of this paper to give tools to explore if our data cloud is actually composed of more data clouds in some way mixed.

Several numerical approaches have been considered to address this problem. Morgenthaler (1990) proposed a method in the framework of robust analysis, while some kind of regression tree techniques could be used as well (e.g. RECPAM, Ciampi (1994)). Both, however, use regression models with a given link function.

In this paper, we propose a dynamic graphics procedure that performs an exploratory analysis within the broader environment of link-free regression model. As graphical tools, we suggest to use the regression function plot along with the scatterplot matrix. We call regression function plot the plot of the response against a linear combination of the predictors. In linear regression, it becomes the response against fitted values plot, while in simple regression it corresponds to the scatterplot of the response variable against the predictor. Cook and Weisberg (1994) call it a Summary Plot in the regression problem, and stress its use as an important tool in regression graphics. This plot could be

* Work supported by ex-40% MURST Research Project "Nuovi Metodi di Classificazione e Analisi dei Dati".

used to assess the correct functional form required to describe the data, and it can highlight the presence of outliers. In this paper, we propose to use such a plot to detect subsamples in link-free regression analysis.

In Section 2 we present the regression function plot and show how subpopulations could appear in its population version. In Section 3 the sample version of the plot is investigated, and in Section 4 a dynamic graphics procedure to detect subsamples in regression data is proposed. In Section 5 an example is presented, while in Section 6 some concluding remarks follow.

2. Subpopulations in the regression function plot

Given a response random variable $\mathcal{Y} \in \Re^1$ and a set of p random predictors $\mathcal{X} \in \Re^p$, the goal of a regression analysis is to study the conditional distribution $F(y \mid \mathcal{X} = x)$, and in particular the regression mean function $E(\mathcal{Y} \mid \mathcal{X} = x)$. A general model for this distribution, that includes many regression models, could be:

$$F(y \mid \mathcal{X} = x) \equiv F(y \mid \mathcal{X} = x'\beta), \tag{1}$$

with β a $p \times 1$ vector of parameters. A corresponding model for the regression mean function could be:

$$E(\mathcal{Y} \mid \mathcal{X} = x) \equiv E(\mathcal{Y} \mid \mathcal{X} = x'\beta) = g(x'\beta), \tag{2}$$

where $g(\bullet)$ is an unknown link function.

This kind of modelling has been called link-free regression analysis because, unlike the standard regression model, we want to make inference not only on the parameter vector β, but on the link function g as well. The standard linear regression analysis is included in such a framework with $g(x'\beta) = \alpha + x'\beta$.

Given model (1), the distribution of $\mathcal{Y} \mid \mathcal{X} = x$ is the same as the distribution of $\mathcal{Y} \mid \mathcal{X} = x'\beta$, for each x. Hence, only one linear combination of the predictors is needed to extract from $\mathcal{X}$ all of the information about the distribution of $\mathcal{Y} \mid \mathcal{X}$.

As a consequence, the plot $\{\mathcal{Y}, \mathcal{X}'\beta\}$ visualize the regression mean function (2), even in the case of more than one predictor. Geometrically, this plot represents a projection of the data in the plane from which the regression hyperplane appears as a straight line. Here, by regression hyperplane we mean the one that "best" fit the data, even if the regression function is not linear.

This feature enables us to visualize the unknown link function and gives tools to infer about it - see Cook and Weisberg (1994, Chap. 7, 10) for more details. This is the reason why we will call such a plot the "regression function plot". Conditions on the reliability of regression plots investigated by Cook (1994) still hold for the regression function plot.

First, we will describe such a plot for population data (we refer to it as population plot in the follow), and we will investigate its appearance when more populations are present.

For the sake of simplicity, let us consider the two populations case. The more than two populations case can be addressed similarly. Using the above notation, we will assume respectively the models and the mean functions:

$$F_i(y \mid \mathcal{X}=x) \equiv F_i(y \mid \mathcal{X}=x'\beta_i), \qquad i=1,2 \qquad (3)$$

$$E_i(\mathcal{Y} \mid \mathcal{X}=x) \equiv E_i(\mathcal{Y} \mid \mathcal{X}=x'\beta_i) = g_i(x'\beta_i), \qquad i=1,2 \qquad (4)$$

where subscript i=1, 2 denotes respective populations and E_i expected values with respect to the i-th distribution function.

Two possible cases can arise and the way they could appear on the population regression function plot $\{\mathcal{Y}, \mathcal{X}'\beta\}$ is different:

i) The two populations parameters are equal up to a proportionality constant c ($\beta_1 = c\beta_2$).

In this case, (unless $g_1(\bullet)=g_2(\bullet)$, and $\beta_1=\beta_2$) the two populations will just appear along two different regression functions in the plot $\{\mathcal{Y}, \mathcal{X}'\beta\}$.

It is worth noting that as a special case (c=1, $g_1(\bullet)=\alpha+g_2(\bullet)$) we have shifted regression function: two mean functions of the same shape but with different intercept will explain the data at hand. If both these regression functions are linear, the two data clouds will lie along two parallel hyperplanes. In this case, looking at the $\{\mathcal{Y}, \mathcal{X}'\beta\}$ plot, an overall hyperplane will be a straight line lying between the parallel ones. The specific position of the data, of course, will depend on the population sizes and variances.

ii) The two populations have $\beta_1 \neq c\beta_2$.

In such a case we do not have an unique projection direction to visualize the regression functions. However, if β_1 is close to β_2, for some directions β^* between β_1 and β_2, the plot $\{\mathcal{Y}, \mathcal{X}'\beta^*\}$ could give an interesting view of the data as the populations can still show a different shape in this projection direction.

3. Subsamples in the regression function plot

As a matter of fact, to visualize the regression function shape is enough to consider a value proportional to β: the regression function plot $\{\mathcal{Y}, \mathcal{X}'\beta\}$ shows the same regression function shape as the plot $\{\mathcal{Y}, \mathcal{X}'\gamma\beta\}$, where γ is any proportionality constant. Hence, for our purposes, an estimate for β is equivalent to an estimate for $\gamma\beta$. Now, even if the link function is unknown, under elliptically distributed predictors, the ordinary least squares estimator is Fisher- consistent for $\gamma\beta$ in model (1) (Li and Duan, 1989).

Let y and X be the $n \times 1$ vector and the $n \times p$ matrix of observed values, independent distributed observations on the multivariate random vector $(\mathcal{Y}, \mathcal{X})$. We will denote as $\hat{\beta}$ an estimate for $\gamma\beta$ in model (1), and the regression function plot in its sample version will be $\{y, X\hat{\beta}\}$.

When more than one population is present, the regression function sample plot $\{y, X\hat{\beta}\}$ shows different features. From now on, assume that data are drawn from two different populations and an unique overall estimate $\hat{\beta}$ from our sample is available.

Let us consider first the case when the populations parameters are equal up to a proportionality constant ($\beta_1 = c\beta_2$).

Using on the whole sample a Fisher-consistent estimator for $\gamma\beta$ in (1), will yields a suitable estimate $\hat{\beta}$ for the direction $\beta_1 = c\beta_2$: the two subsamples will appear along two different regression functions in the plot $\{y, X\hat{\beta}\}$.

If we have two shifted linear regression function the two data clouds will lie along two parallel hyperplanes. In this case, looking at the $\{y, X\hat{\beta}\}$ plot, the overall estimated hyperplane will be a straight line lying between the ones that could be fitted on the two subsamples. Hence, it is reasonable to expect that data from two subsamples will be split approximately, in the $\{y, X\hat{\beta}\}$ plot, above and under the single estimated line. Their specific position, of course, will depend on the unknown subsample sizes and variances.

When the two subsamples come from populations with $\beta_1 \neq c\beta_2$ (case *ii)* previously analysed), the β estimated from the whole sample will be neither a suitable estimate for the direction β_1 nor for $c\beta_2$.

Therefore, the plot $\{y, X\hat{\beta}\}$ does not give a 'correct' representation of both the regression functions as the direction $X\hat{\beta}$ is any of the possible directions in the predictors space. Notwithstanding, this plot could give an interesting view of the data as the subsamples can still show a different shape in this projection direction even if less clearly than in the previous case. Closer β_1 is to $c\beta_2$, better the overall estimate should in some way discriminate the two regression functions.

To summarize, given a sample derived from two populations, two regression functions appear more or less clearly in the regression function plot $\{y, X\hat{\beta}\}$. However, since which observations belong to the subsamples is unknown, it could be very difficult to identify even approximately the subsamples.

In order to give tools to detect and identify subsamples, in the next section a dynamic graphics procedure is proposed.

4. A dynamic graphics procedure

Goal of the proposed procedure is to find data subsets such that their regression functions are different from each other. Since few data points with such a feature can be detected by standard outlier analysis (Chatterjee and Hadi, 1988), the proposed procedure search for an appropriate broad partition in the predictors space or along variables not used in the analysis (both referred as splitting variables in the following).

From the practical point of view, since we are analyzing real data observed on the same variables, it seems reasonable that we will have both approximately the same shape in the regression functions (i.e. $g_1(\bullet)\cong g_2(\bullet)$, perhaps with different parameter values), and β_1 close to $c\beta_2$.

Hence, at first we suggest to linearize the whole sample regression function by a unique inverse transformation of the link function, as in the standard link-free regression analysis (Cook and Weisberg 1994, Chap. 10). Then, to check for subsamples using the regression function plot in concert with the splitting variables scatterplot matrix. Once the data have been divided into subsamples, two regression analyses on the two subsamples should be performed: their comparison will evaluate the effectiveness of such a broad partition.

The proposed procedure is then as follows:

Step 1. Get an estimate $\hat{\beta}$ from the whole data set.

Step 2. Linearize, as possible, the regression function by inverse link function transformation.

Step 3. Display jointly the plot $\{y, X\hat{\beta}\}$ and the scatterplot matrix of the splitting variables.

Step 4. Iteratively and interactively: *i*) Look for cluster and/or patterns in the plot $\{y, X\hat{\beta}\}$. *ii*) If any, use dynamic selection to check if they correspond to cluster in the splitting variables. *iii*) Select data subsets by slicing (for different slice window width) or brushing in the scatterplot matrix cells. *iv*) Look for a pattern of the selected points in the plot $\{y, X\hat{\beta}\}$.

Step 5. If in the plot $\{y, X\hat{\beta}\}$ such a pattern is found, try to model it by a dummy variables for the identified subsamples or try to fit different models.

5. An example

As an example for the proposed procedure, we will use a data set from the Minnesota Dep. of Children, Families and Learning referred to 46 schools in Minnesota for the 1994-1995. (Source: Minneapolis Star and Tribune, March 18, 1996). The response variable is the percent of graduating students who attend four year college (P4Y), while the predictors are characteristics of the school: score at the ACT test, percent of minority students, cost of the student, and percent of students with free lunch (Pfree-lunch). Ordinary least squares estimation and link function inverse transformation were performed. As a result, P4Y^2 is linearly and positively related with the first three predictors and negatively with the forth. Numerical results are given in Table 1.

Table 1: *Minnesota Schools data. Regression on the whole sample.*

Coefficient Estimates	Estimate	Std. Error	t-value
Constant	-4972.39	3830.84	-1.298
ACTscore	339.227	164.798	2.058
Cost	0.725155	0.259484	2.795
log[Pfree_lunch]	-1492.20	219.716	-6.791
log[PMinority]	364.570	179.583	2.030

Summary Analysis of Variance Table			
Source	df	SS	MS
Regression	4	73226848.	18306712.
Residual	41	25478270.	621421.
R^2: 0.741875			
F: 29.46			*p*-value: 0.0000

The *p*-value for the F test is pratically equal to 0. However, the regression function plot $\{y, X\hat{\beta}\}$ shows an S-shape pattern of the data (Fig.1(a), top): the model does not completely describe the data. A further power transformation does not work, while a more complex nonlinear function $g(\bullet)$ is not worth as it could require too many parameters in the model.

Fig.1: *Minnesota Schools data. Predictors scatterplot matrix and plot $\{y, X\hat{\beta}\}$.*

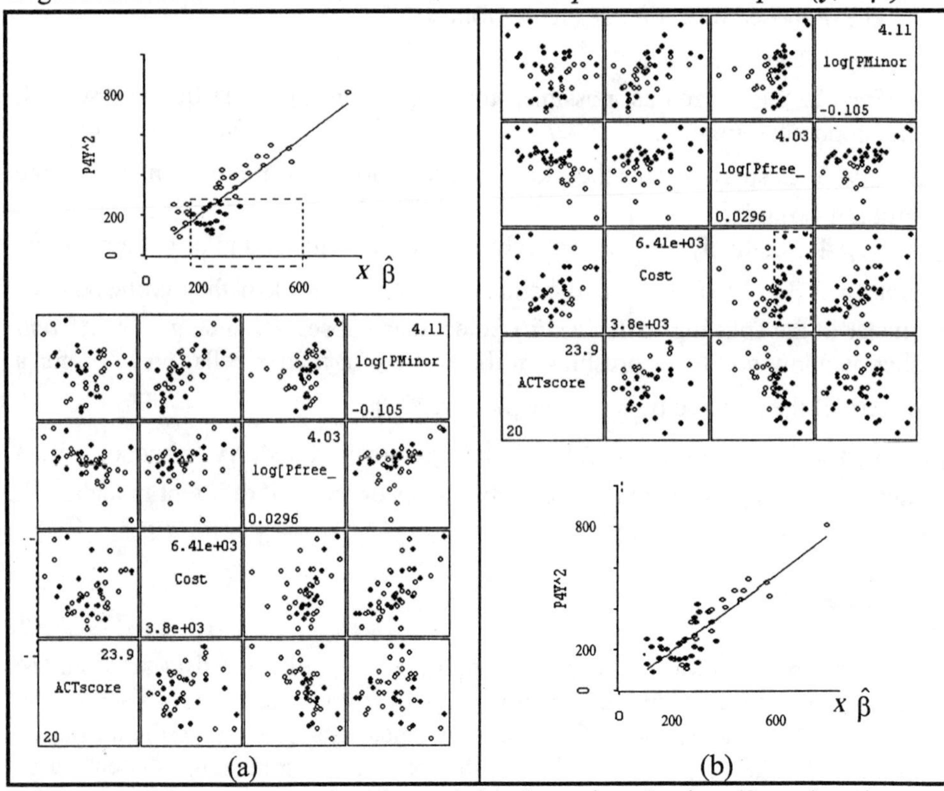

Instead, when our procedure is used, by dynamic selection (Step 4, *i-ii*), we note that the bottom of the S-shape in the plot $\{y, X\hat{\beta}\}$ corresponds to values in the upper half range of the variable Pfree-lunch (Fig.1(a)). Hence, according to our procedure (Step 4, *iii-iv*), we check this suspected structure and note that higher values of this variable correspond to a pattern in the plot $\{y, X\hat{\beta}\}$ (Fig.1(b)). This pattern could be then partially explained by two different models for lower and higher values of the variable Pfree-lunch, a variable revealing the economical and social level of the students in the school. Refitting the model separately for the subsets of data with lower and higher value of Pfree-lunch we found that the dependence of P4Y on all the predictors is stronger when the level of Pfree-lunch is low, whereas for higher values of Pfree-lunch the F-test for regression yields a p-value of 0.1082 (Table 2).

Table 2: *Minnesota Schools data. Two regressions for Pfree-lunch values.*

	Pfree-lunch lower value 23 cases.			*Pfree-lunch higher value 23 cases.*		
	Estimate	Std. Error	t-value	Estimate	Std. Error	t-value
Constant	-18079.4	5310.89	-3.404	3375.46	5270.19	0.640
ACTscore	898.663	246.780	3.642	26.6400	206.643	0.129
Cost	0.718061	0.332047	2.163	0.429684	0.413189	1.040
log[Pfree_lunch]	-1298.91	269.546	-4.819	-1482.39	646.305	-2.294
log[PMinority]	450.179	209.713	2.147	340.745	301.846	1.129
R^2:	0.88077			0.329808		
Anal. of Variance:	Source	df SS	MS	Source	df SS	MS
	Regression	4 59074818.	14768705.	Regression	4 5429068.	1357267.
	Residual	18 7997004.	444278.	Residual	18 11032255.	612903.
	F: 33.24		p-value: 0.0000	F: 2.21		p-value: 0.1082

At a first glance, it seems that the percent of graduating students who attend college does not depend globally on the recorded characteristics of the school, if the economical and social level of the students is low. As a conclusion, the coexistence of two subpopulations gives reason to the contradictory presence in the whole sample of a 0 p-value for the F test and a non linear pattern in the regression function plot.

6. Remarks

We have introduced the regression function plot as a tool to detect subsamples in link-free regression analysis. As a particular case, the proposed procedure -omitting Step 2- works in standard linear regression analysis as well.

For visual enhancement, at a first moment the plot $\{X\hat{\beta}, y\}$ could be used in the suggested procedure, instead of the regression function plot.

Only subsamples with different regression mean function $E(y|\mathcal{X} = x'\beta)$ have been considered. However, the ideas are straightforwardly extendable to the other regression moment functions such as the regression variance function $Var(y|\mathcal{X}= x'\beta)$.

It is worth noting the case when the subsamples regression functions are not dependent on the same predictor variables. Let $\mathcal{X}_i$ be a subset of $\mathcal{X}$: for the i-th subsample we can have $E_i(\mathcal{Y} \mid \mathcal{X}) = E_i(\mathcal{Y} \mid \mathcal{X}_i) = E_i(\mathcal{Y} \mid \mathcal{X}_i = \mathbf{x}'\beta_i)$ with some values in β_i equal to zero. In such a case $\beta_1 \neq c\beta_2$, unless c is close to zero and then $E_i(\mathcal{Y} \mid \mathcal{X}) \cong E_i(\mathcal{Y})$. For our procedure to work, we only need $(\bigcup_i \mathcal{X}_i) \subset \mathcal{X}$.

In the case we have more than two subsamples in the data, our procedure could work as well. On the other hand, to investigate if any subsample is in turn composed by more 'sub-subsamples', the iteration of the procedure is recommended.

Acknowledgments. Thanks are due to J. Antoch and G. Galmacci for many helpful comments on an earlier version of this paper.

References

Ciampi, A. (1994). Classification and Discrimination: the RECPAM Approach, in: *Compstat, Proceedings in Computational Statistics*, Dutter, R. & Grossmann, W. (Eds.), Physica-Verlag.

Chatterjee, S. & Hadi, A.S. (1988), *Sensitivity Analysis in Linear Regression*, Wiley, New York.

Cook, R. D. (1994). On the Interpretation of Regression Plots, *Journal of the American Statistical Association*, 89, 177-189.

Cook, R. D. & Weisberg, S. (1994). *An Introduction to Regression Graphics*, Wiley, New York.

Li, K. C. & Duan, N. (1989). Regression Analysis Under Link Violation, *The Annals of Statistics*, 17, 1009-1052.

Morgenthaler, S. (1990). Fitting Redescending M-Estimators in Regression, in: *Robust Regression*, Lawrence, K. D. & Arthur, J. L. (Eds.), Dekker, New York.

Asymptotic Prior to Posterior Analysis for Graphical Gaussian Models

Alberto Roverato
Dipartimento di Economia Politica, Università di Modena
e-mail: roverato@unimo.it

Abstract: In this paper we derive the asymptotic posterior distribution, in a conjugate analysis, for the marginal and partial correlation coefficients in a graphical Gaussian model. An example of prior to posterior analysis is given and the problem of the specification of the hyper parameters discussed.

Keywords: Conditional Independence, Conjugate Prior, Correlation Coefficient, Graphical Model, Isserlis Matrix, Matrix Completion.

1. Introduction

Recent work established a class of statistical models, known as graphical models, that exploit the close relationship between conditional independence and separation in undirected graphs (see Lauritzen, 1996).

The conditional independence structure of a multivariate Normal distribution is dictated by the zero pattern of its concentration matrix Σ^{-1}. The use of the concentration matrix in the parametrisation of the normal distribution has many advantages, however non-zero elements of Σ^{-1} are difficult to interpret since they are unnormalised quantities. Consequently, when the interest is on the strength of the association structure, this parameters have to be transformed to obtain partial correlation coefficients.

In this paper we apply the results of Roverato and Whittaker (1996, 1998) to derive the asymptotic posterior distributions, in a Bayesian conjugate analysis, of the marginal and of the partial correlation coefficients. We give an example of prior to posterior analysis based on real data where we discuss the difficulties concerned with the specification of the hyper parameters.

The notation is given in Section 2. In Section 3 we present some basic theory relating to graphical Gaussian models and in Section 4 we describe the HIV data used in the application. In Section 5 we derive the required asymptotic distributions. Finally in Section 6 we carry out the application and discuss the problem of the specification of the hyper parameters.

2. Notation

Let V be a finite set with $|V| = p$, and let Γ be a $p \times p$ symmetric invertible matrix. The rows and columns of Γ are indexed by the elements of V, so that Γ itself is indexed by $V \times V$. When $V = \{1, \ldots, p\}$, Γ is indexed by row and column numbers.

The Isserlis matrix of Γ, $\text{Iss}(\Gamma)$, (Isserlis, 1918; Roverato and Whittaker, 1998) is the symmetric matrix indexed by $\mathcal{W} \times \mathcal{W}$ where $\mathcal{W} = \{(i,j) : i, j \in V, i \leq j\}$, with elements $\{\text{Iss}(\Gamma)\}_{(i,j),(r,s)} = \gamma_{ir}\gamma_{js} + \gamma_{is}\gamma_{jr}$.

The graph theory requisite for graphical models may be found in Lauritzen (1996). We use the convention that, for an arbitrary undirected graph $G = (V, \mathcal{V})$ where V is the vertex set and $\mathcal{V}$ is the set of edges, for all $i \in V$ the pair (i,i) is included in $\mathcal{V}$ and that if $(i,j) \in \mathcal{V}$ then $i \leq j$. The set $\mathcal{W}$ is therefore the edge set of the complete graph and we denote by $\bar{\mathcal{V}} = \mathcal{W} \backslash \mathcal{V}$ the set of edges not in G.

EXAMPLE 1 With $V = \{1, 2, 3\}$ and graph G ①——②——③ the edge set is $\mathcal{V} = \{(1,1), (2,2), (3,3), (1,2), (2,3)\}$ while $\bar{\mathcal{V}} = \{(1,3)\}$. □

For any undirected graph $G = (V, \mathcal{V})$ the pair $(\mathcal{V}, \bar{\mathcal{V}})$ is a partition of $\mathcal{W}$. To this correspond the submatrices $\text{Iss}(\Gamma)_{\mathcal{V}\mathcal{V}}$, $\text{Iss}(\Gamma)_{\mathcal{V}\bar{\mathcal{V}}}$ and $\text{Iss}(\Gamma)_{\bar{\mathcal{V}}\bar{\mathcal{V}}}$ as well as the partial matrix $\text{Iss}(\Gamma)_{\mathcal{V}\mathcal{V}|\bar{\mathcal{V}}} = \text{Iss}(\Gamma)_{\mathcal{V}\mathcal{V}} - \text{Iss}(\Gamma)_{\mathcal{V}\bar{\mathcal{V}}}[\text{Iss}(\Gamma)_{\bar{\mathcal{V}}\bar{\mathcal{V}}}]^{-1}\text{Iss}(\Gamma)_{\bar{\mathcal{V}}\mathcal{V}}$.

For a set $\mathcal{C} \subseteq \mathcal{W}$ we define the $\mathcal{C}$-incomplete matrix $\Gamma^{\mathcal{C}}$ as the symmetrised matrix indexed by $V \times V$ with elements $\{\gamma_{ij}\}$ for all $(i,j) \in \mathcal{C}$, and with the remaining elements unspecified. In the Example 1 above the incomplete matrices corresponding to the sets $\mathcal{V}$ and $\bar{\mathcal{V}}$ are respectively

$$\Gamma^{\mathcal{V}} = \begin{pmatrix} \gamma_{11} & \gamma_{12} & * \\ \gamma_{21} & \gamma_{22} & \gamma_{23} \\ * & \gamma_{32} & \gamma_{33} \end{pmatrix} \quad \text{and} \quad \Gamma^{\bar{\mathcal{V}}} = \begin{pmatrix} * & * & \gamma_{13} \\ * & * & * \\ \gamma_{31} & * & * \end{pmatrix},$$

where asterisks denote unspecified elements. The matrix $\Gamma^{-\mathcal{V}}$ is a shorthand for $(\Gamma^{-1})^{\mathcal{V}}$. If it is possible to fill an incomplete matrix $\Gamma^{\mathcal{C}}$ to obtain a (full) positive definite matrix we say that $\Gamma^{\mathcal{C}}$ admits a positive completion.

Let $\Gamma^{\mathcal{V}}$ be a $\mathcal{V}$-incomplete matrix, with $G = (V, \mathcal{V})$, which admits a positive completion. We say that Γ_G is *the completion of* $\Gamma^{\mathcal{V}}$ if it is the unique positive definite matrix such that

$$(\Gamma_G)^{\mathcal{V}} = \Gamma^{\mathcal{V}} \qquad \text{and} \qquad \{\Gamma_G^{-1}\}_{ij} = 0 \ \text{ for all } \ (i,j) \in \bar{\mathcal{V}}. \tag{1}$$

See Grone *et al.* (1984) for a proof of the existence and uniqueness of such matrix.

We denote by $\text{diag}(\Gamma^{\mathcal{V}})_{\mathcal{V}\mathcal{V}}$ the matrix indexed by $\mathcal{V} \times \mathcal{V}$ with the distinct specified elements of $\Gamma^{\mathcal{V}}$ in the main diagonal and zero elsewhere.

For an undirected graph $G = (V, \mathcal{V})$, we denote by $\mathcal{M}_*(G)$ the set of all $\mathcal{V}$-incomplete matrices and by $\mathcal{M}_*^+(G)$ the set of all $\mathcal{V}$-incomplete matrices that admit positive completion. Furthermore we denote by $\mathcal{M}_0(G)$ the set of all symmetric matrices indexed by $V \times V$ with element (i,j) equal to zero whenever $(i,j) \in \bar{\mathcal{V}}$. The intersection of $\mathcal{M}_0(G)$ with the set of all positive definite matrices is denoted by $\mathcal{M}_0^+(G)$.

The trace of the product of two square matrices is $\mathrm{tr}(\Phi\Gamma) = \sum_{i=1}^{p}\sum_{j=1}^{p}\phi_{ij}\gamma_{ij}$. If $\Phi \in \mathcal{M}_0(G)$ only the specified elements of $\Gamma^{\mathcal{V}}$ enter into this sum, and so we can write $\mathrm{tr}(\Phi\Gamma) = \mathrm{tr}(\Phi\Gamma^{\mathcal{V}})$.

3. Graphical Gaussian models

We consider graphical models for a random vector X_V with distribution P_V on an undirected graph $G = (V, \mathcal{V})$. We say that the distribution P_V is Markov with respect to G if X_A is independent of X_B given X_S, $X_A \perp\!\!\!\perp X_B | X_S\,[P_V]$, whenever S separates A from B in G. (For the Gaussian case that interests us, the global, local and pairwise Markov properties are identical, see Lauritzen 1996.)

A graphical Gaussian model is defined as the intersection of the set of Markov distributions relative to G and the set of p-variate normal distributions with mean equal to zero and variance matrix Σ, which we assume positive definite. An off-diagonal element σ^{ij} of Σ^{-1} is zero if and only if $X_i \perp\!\!\!\perp X_j | X_{V\setminus\{i,j\}}\,[P_V]$. Thus in graphical Gaussian models the pairwise conditional independence structure of X_V is dictated by the zero structure of Σ^{-1} so that $\Sigma = \Sigma_G$ is a G-completed matrix.

A natural measure of the interaction represented by the edge (i,j) of the graph is given by the partial correlation coefficient

$$\rho_{ij.V\setminus\{i,j\}} = -\frac{\sigma^{ij}}{\sqrt{\sigma^{ii}\sigma^{jj}}} \tag{2}$$

which is zero if and only if $(i,j) \notin \mathcal{V}$.

The moment parameter of the distribution is $\Sigma^{\mathcal{V}} \in \mathcal{M}_*^+(G)$ while the canonical parameter is $\Sigma^{-1} \in \mathcal{M}_0^+(G)$, the inverse of the completion of $\Sigma^{\mathcal{V}}$. The likelihood function of Σ^{-1} based on a random sample $X^{(n)} = (X^1, \ldots, X^n)$ from P_V is

$$L(\Sigma^{-1}) \propto \exp\left\{-\frac{n}{2}\mathrm{tr}(\Sigma^{-1}S) + \frac{n}{2}\log|\Sigma^{-1}|\right\}, \tag{3}$$

where S is the sample variance matrix. In our notation, the maximum likelihood estimate $\widehat{\Sigma}^{-1}$ (Speed and Kiiveri, 1986) is the inverse of the completion of $S^{\mathcal{V}}$.

4. The HIV study

Table 1 presents some summary statistics for six variables measured in Genoa

Table 1: *Summary statistics for the HIV data: sample variances (main diagonal), correlations (lower triangle) and partial correlations (upper triangle).*

X_1	8.8374	0.479	-0.043	-0.033	0.356	-0.236
X_2	0.483	0.1919	0.068	-0.084	-0.224	-0.110
X_3	0.220	0.057	8924231.9	0.085	0.552	-0.330
X_4	-0.040	-0.133	0.149	20392.4	0.091	0.013
X_5	0.253	-0.124	0.523	0.179	1952795.2	0.384
X_6	-0.276	-0.314	-0.183	0.064	0.213	1.378
	X_1	X_2	X_3	X_4	X_5	X_6

and Padua paediatric hospitals on 107 three month old babies. These data come from a larger Italian study investigating early diagnosis of HIV infection in children from HIV positive mothers. The variables are related to various measures on blood and its components: X_1 and X_2 immunoglobin G and A, respectively; X_4 the platelet count; X_3, X_5 lymphocyte B and T4, respectively; and X_6 the T4/T8 lymphocyte ratio. (For a detailed description of these data see Boccuzzo, 1991.)

Discussion with the experts running the study suggests the presence of a strong association between variables X_1, X_2 and between variables X_3, X_5, X_6; together with an association structure of these variables compatible with the graph of Figure 1.

An essential quantity to be computed in a Bayesian analysis of the proposed graphical model is a posterior distribution for the relevant association coefficients which, in the Gaussian case, are the marginal and the partial correlations corresponding to the edges of the graph. In the next Section we consider an asymptotic approach to this problem.

5. Asymptotic distributions

Exact prior to posterior analysis for the correlation coefficients of an arbitrary graphical Gaussian model still poses difficulties concerned with the algebric derivation of a tractable posterior distribution. In this Section we derive the asymptotic distributions for the marginal and the partial correlations in a conjugate analysis.

The standard conjugate prior density for Σ^{-1}, with respect to the product of Lebesgue measures on the diagonal and non-zero super-diagonal elements of Σ^{-1}, can be deduced from (3) (see Bernardo and Smith, 1994, p.269) and has form

$$\pi(\Sigma^{-1}|A^{\mathcal{V}}, h) \propto \exp\left\{-\frac{h}{2}\mathrm{tr}(\Sigma^{-1}A^{\mathcal{V}}) + \frac{h}{2}\log|\Sigma^{-1}|\right\}, \tag{4}$$

for $\Sigma^{-1} \in \mathcal{M}_0^+(G)$. The hyper parameters are an incomplete matrix $A^{\mathcal{V}} \in$

$\mathcal{M}_*^+(G)$ and a positive constant h. The posterior distribution has density function of the same form with hyper parameters $T^{\mathcal{V}} = (hA^{\mathcal{V}} + nS^{\mathcal{V}})/(h+n)$ and $m = h + n$. By the similarity between (3) and (4) it can be shown that the posterior density has maximum at $\Sigma^{-1} = T_G^{-1}$, the inverse of the completion of $T^{\mathcal{V}}$

Let $R^{\mathcal{V}}$ be the incomplete matrix with main diagonal equal to that of $\Sigma^{\mathcal{V}}$ and with marginal correlations in the specified off-diagonal positions. Clearly $R^{\mathcal{V}}$ can be written as a (bijective) function of $\Sigma^{\mathcal{V}}$, $R^{\mathcal{V}} = g(\Sigma^{\mathcal{V}})$, so that, by (2), $P^{\mathcal{V}} = -g(\Sigma^{-\mathcal{V}})$ is the incomplete matrix with main diagonal equal to that of $-\Sigma^{-\mathcal{V}}$ and partial correlations in the specified off-diagonal positions. Roverato and Whittaker (1998) showed that the asymptotic posterior distributions for $\Sigma^{-\mathcal{V}}$ and $\Sigma^{\mathcal{V}}$ are

$$\Sigma^{-\mathcal{V}} \overset{a}{\sim} N\left(T_G^{-\mathcal{V}}, m^{-1}\mathrm{Iss}(T_G^{-1})_{\mathcal{V}\mathcal{V}|\bar{\mathcal{V}}}\right) \quad \text{and} \quad \Sigma^{\mathcal{V}} \overset{a}{\sim} N\left(T^{\mathcal{V}}, m^{-1}\mathrm{Iss}(T_G)_{\mathcal{V}\mathcal{V}}\right) \quad (5)$$

respectively. The asymptotic posterior distribution of the transformed parameters can be obtained by applying standard delta-type methods (see Bernardo and Smith, 1994, p.295). The transformation $g(\cdot)$ can be shown to have Jacobian

$$J(\Sigma^{\mathcal{V}})_{\mathcal{V}\mathcal{V}} = \frac{\partial g(\Sigma^{\mathcal{V}})}{\partial \Sigma^{\mathcal{V}}} = \mathrm{diag}(g(\Sigma^{\mathcal{V}}))_{\mathcal{V}\mathcal{V}}\, H_{\mathcal{V}\mathcal{V}}\, \mathrm{diag}(\Sigma^{\mathcal{V}})^{-1}_{\mathcal{V}\mathcal{V}} \quad (6)$$

where $H_{\mathcal{V}\mathcal{V}}$ is the matrix indexed by $\mathcal{V}\times\mathcal{V}$ with elements $\{H_{\mathcal{V}\mathcal{V}}\}_{(i,j),(r,s)}$ equal to 1 if $i = r$ and $j = s$, to $-1/2$ if either $i = r = s \neq j$ or $j = r = s \neq i$ and 0 elsewhere. As usual, in taking the derivative in (6) the denominator is assumed to be a row vector, the numerator a column vector and the off-diagonal elements are considered only once. Applying $J(\cdot)$, evaluated at the mode of the posterior density, to (5) we obtain

$$P^{\mathcal{V}} \overset{a}{\sim} N\left(-g(T_G^{-\mathcal{V}}),\ m^{-1}J(T_G^{-\mathcal{V}})_{\mathcal{V}\mathcal{V}}\mathrm{Iss}(T_G^{-1})_{\mathcal{V}\mathcal{V}|\bar{\mathcal{V}}}J(T_G^{-\mathcal{V}})'_{\mathcal{V}\mathcal{V}}\right) \quad (7)$$

and

$$R^{\mathcal{V}} \overset{a}{\sim} N\left(g(T^{\mathcal{V}}),\ m^{-1}J(T^{\mathcal{V}})_{\mathcal{V}\mathcal{V}}\mathrm{Iss}(T_G)_{\mathcal{V}\mathcal{V}}J(T^{\mathcal{V}})'_{\mathcal{V}\mathcal{V}}\right) \quad (8)$$

which are the required asymptotic distributions.

7. Application

In this Section we carry out a prior to posterior analysis of the HIV data based on the results of Section 6.

A critical point in the conjugate Bayesian analysis of the graphical Gaussian model is the specification of the prior hyper parameters h and $A^{\mathcal{V}}$. The posterior distribution is very sensitive to different prior specifications. Furthermore note that the number of hyper parameters exceeds the number of parameters.

The positive constant h can be thought of as the number of imaginary data points establishing the prior belief, consequently in absence of genuine prior expert views it should be small.

The incomplete matrix $A^{\mathcal{V}}$ reflects the prior belief on the real value of $\Sigma^{\mathcal{V}}$. Dealing with the saturated model, $A^{\mathcal{V}} = A$, some authors (Chen, 1979; Dickey *et al.*, 1985) proposed to reduce the number of hyper parameters by constraining A to have lower dimension structure such as $A = \Delta^{1/2}\Phi\Delta^{1/2}$, where $\Phi = (1-\phi)I + \phi 1_p 1_p'$ has intraclass correlation structure (with $-1/(p-1) < \phi < 1$ so as to assure $\Phi > 0$) and $\Delta = \mathrm{diag}(\delta_1, \ldots, \delta_p)$.

We remark that a similar choice of $\Phi^{\mathcal{V}}$ induces a systematic bias in the mean of the asymptotic posterior distributions here considered. To see this, consider two sample correlations r_{ij} and r_{ls} such that $r_{ij} > 0$ and $r_{ij} = -r_{ls}$. In this case the evidence provided by the data is of equal strength in the associations relative to ρ_{ij} and ρ_{ls}. In such a situation it is desirable, when no otherwise specified in the prior, that this characteristic is maintained in the posterior distribution. However, it can be easily checked in $g(T^{\mathcal{V}})$ that any choice of $\phi > 0$ leads to a asymptotic posterior mean of ρ_{ls} closer to zero than that of ρ_{ij}. Furthermore a similar behaviour can be empirically observed in $-g(T_G^{-\mathcal{V}})$. Therefore an intraclass correlation structure of $\Phi^{\mathcal{V}}$ implies an asymmetrical inference depending on the signs of the single correlation coefficients. Note that setting ϕ to 0 would overtake this difficulty but at the price of a strong prior information of independence between all variables, which is seldom justified.

Our proposed solution to this problem is to set the specified (i,j)-element $(i \neq j)$ of $\Phi^{\mathcal{V}}$ to $\phi \times \mathrm{sign}(r_{ij})$. Although this is not a pure Bayesian approach to the problem, it keeps the hyper parameter dimension low without introducing a systematic bias in the analysis. In order to make the prior distribution not too informative, for the HIV data we set $h = 1$ and $\phi = 0.3$ (this is an admissible value for ϕ since $\Phi^{\mathcal{V}}$ has positive completion). The hyper parameter Δ is set to $\mathrm{diag}(10, 10^0, 10^7, 10^5, 10^7, 10)$.

The resulting asymptotic marginal posterior densities are presented in Figure 1. The given marginal distributions are completely specified by their mean and variance values. Nevertheless the plot of the posterior densities in the $[-1, 1]$ interval is useful since it allows an immediate visualisation of the densities behaviour around zero. For instance, it can be noted that all the correlations involving variable X_4 (forth row and column in the picture) have high density values at zero. A more detailed analysis showed that all of the four corresponding 90% confidence intervals include zero, suggesting that the model may be overparametrised.

Figure 1: *Independence graph of the hypothesised model for HIV diagnostic data and marginal asymptotic posterior densities for variances (main diagonal), partial correlations (upper triangle) and marginal correlations (lower triangle).*

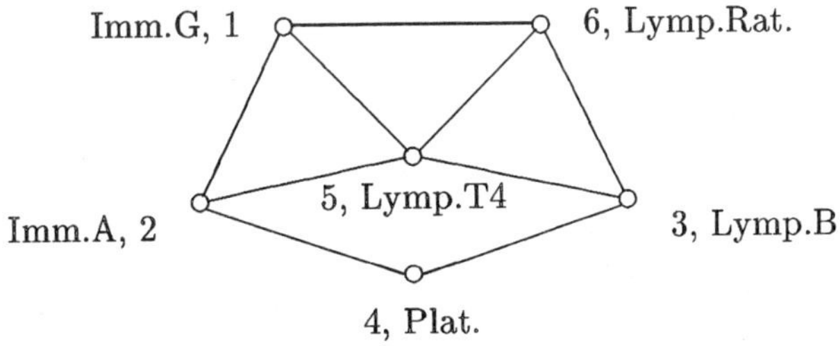

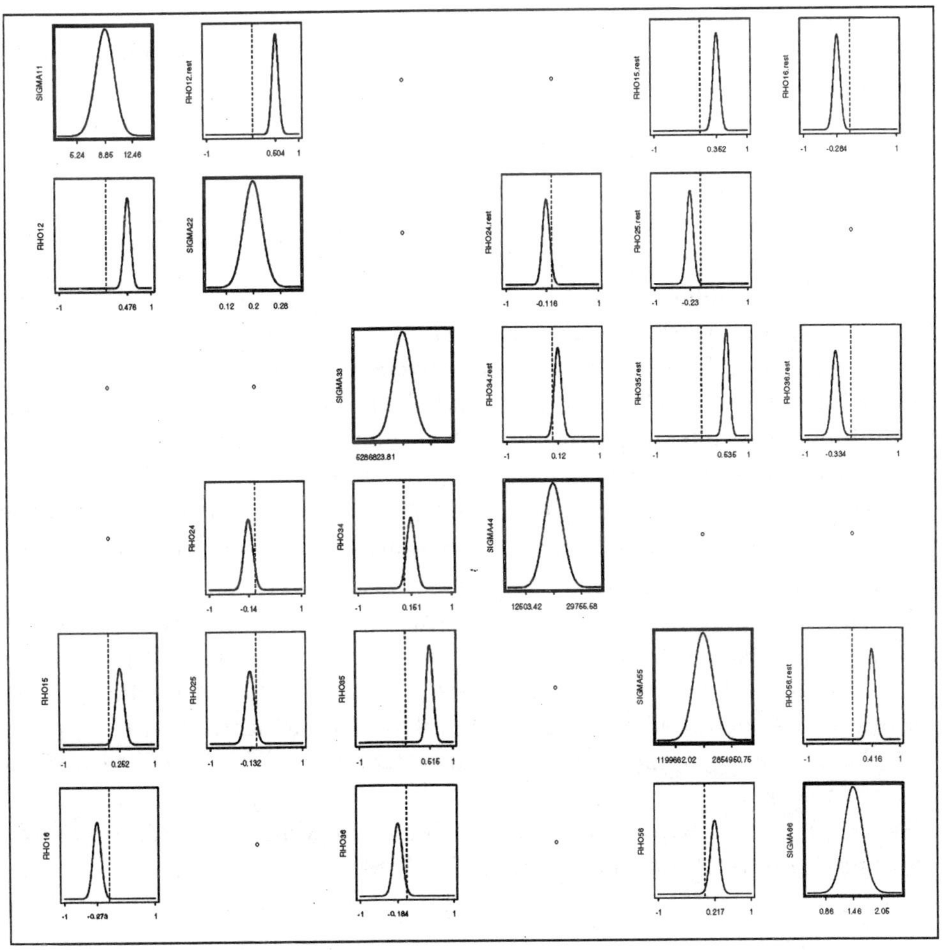

8. Conclusions

Graphical models are specified by set of pairwise conditional independencies and this leads to a parametrisation in terms of partial association coefficients. Nevertheless also the marginal independence pattern is of relevance in the comprehension of the problem under consideration. Figure 1, including both marginal and partial correlations, seems to be an effective device to summarise the data structure under a graphical Gaussian model.

The Splus functions relative to the work of this paper are available from the author.

Acknowledgements: We are grateful to G. Boccuzzo for making the HIV data available and helpful discussions concerning its analysis.

References

Bernardo, J.M. and Smith, A.F.M. (1994). *Bayesian Theory*, Wiley, Chichester.

Boccuzzo, G. (1991). *Funzioni Diagnostiche dell'Infezione Connatale da HIV*, Thesis, Department of Statistics, University of Padua.

Chen, C. (1979). Bayesian inference for a Normal dispersion matrix and its application to stochastic multiple regression analysis, *Journal of the Royal Statistical Society*, ser. B, 41, 235-248.

Dickey, J.M., Lindley, D. V. & Press S. J. (1985). Bayesian estimation of the dispersion matrix of a multivariate normal distribution, *Communications in Statistics, Part A - Theory and Methods*, 14, 1019-1034.

Grone, R., Johnson, C.R., Sà, E.M. and Wolkowice, H. (1984). Positive definite completions of partial Hermitian matrices, *Linear Algebra and its Applications*, 58, 109-124.

Isserlis, L. (1918). On a formula for the product-moment correlation of any order of a normal frequency distribution in any number of variables, *Biometrika*, 12, 134-139.

Lauritzen, S. L. (1996). *Graphical Models*, Oxford University Press, Oxford.

Roverato, A. and Whittaker, J. (1996). Standard errors for the parameters of graphical Gaussian models. *Statistics and Computing*, 6, 297-302.

Roverato, A. and Whittaker, J. (1998), The Isserlis matrix and its application to non-decomposable graphical Gaussian models, *Biometrika*, 85, 3, to appear.

Speed, T.P. and Kiiveri, H. (1986). Gaussian Markov distributions over finite graphs, *Annals of Statistics*, 14, 138-150.

PART IV

Case Studies

- **Applied Classification and Data Analysis**

Using Qualitative Information and Neural Networks for Forecasting Purposes in Financial Time Series

Simone Borra, Agostino Di Ciaccio
Faculty of Economics, University of Urbino
P.za della Repubblica 13, Urbino, Italy
e-mail: borra@econ.uniurb.it , diciaccio@econ.uniurb.it

Abstract: In the Italian financial market, stock fluctuations are highly dependent on political and economic events. For this reason, any realistic forecast should consider this kind of information. In this paper we show a way to include economic and political events in order to forecast a financial time series. Then we applied neural networks, econometric analysis and some recent non-parametric regression models to empirical data observed over a period of 61 weeks. The respective performances of the different approaches were then compared.

Keywords: Forecast, Neural Networks, Non-parametric Models, Financial Time series.

1. Introduction

In the Italian financial stock market, fluctuations are highly dependent on political and economic events. This feature has precluded the construction of reliable forecasting models.
In this paper we analyze the prices of listed shares traded on the Italian Stock Exchange (Borsa Valori di Milano), with the aim of forecasting the performance of the Italian Stock Index-MIB (basis 3/1/94=1000).
The presence of stock price dependence on political and economic events was verified at first glance by scanning the MIB series. Large changes in stock prices occur in correspondence to the most relevant events of Italian political life.
Such events could be of a political (i.e. the fall of a government) or a political-economic nature (i.e. a reduction in the discount rate). Furthermore, they can be classified as national or international events. Usually they have an immediate effect on stock prices and most traders use such news to speculate on the market.
Such considerations suggest that a valid prediction of the MIB should be obtained by adding current political and economic information to previously observed values in the series (see i.e. Tivegna, 1996).

The data we used refer to the period from April 5, 1994 to June 31, 1995 (61 weeks). To avoid anomalies associated with the weekly opening and closing of the market we considered the Tuesday MIB quotation for each week.
As predictive variables we took into account: the 10 years BTP Futures, the Lira/Dollar exchange rate, the Lira/German Mark exchange rate and the mean of the variation rate of the main foreign stock indices (London, N.Y., Frankfurt, Tokyo).
It is true that more detailed data, considering daily indices, over a longer period of time could have been analyzed. On the other hand, the aim of this paper is to evaluate the feasibility of inserting qualitative exogenous information jointly to the use of modern non-parametric methods. The results obtained can be a useful starting point towards the construction of a more complex analysis.

2. Indices of political and economic events

Qualitative information on national and international economic events was derived by skimming the front page of the most important Italian economic newspaper "Sole 24 ore". This newspaper provides a reliable review of the main economic and political events in Italy.
Qualitative indices were constructed by ranking each political or economic news reported on the front page according to previously defined criteria. The order in the ranking was based only on the journal's point of view, to obtain data independent from the observer's opinion.
Each news item was first classified according to 4 categories:
1) domestic politics, 2) international politics, 3) domestic economics, 4) international economics.
Then the items were scored according to:

i) position on the front page: 4=top of the page; 2=center; 1=bottom;
ii) space taken up on the page, measured as the number of columns occupied (score 0-2);
iii) emphasis given to its title (score 0-4).

The three different scores were added and the resulting score was multiplied by a coefficient (+1, 0,-1) according to the journalist's opinion on the reported event (positive, neutral or negative).
Weekly indices were obtained for each news category by summing up the scores reported for each day from Tuesday through to Monday of the previous week (omitting Saturdays and Sundays).
After analyzing the performance of the four political-economic indices, only three of them (dropping the "international economy" index) seemed to be relevant in estimating the MIB index. Thus, we considered the domestic politics (DP) and economic (DE) indices and the international politics (IP) index referred to the events which occurred between time t and time t-1, together with the following lagged (t-1) variables: BTP futures, ITL/US$ rate, MIB index and the average of the main stock markets (ASM).

The analysis was performed on a data set spanning 61 weeks, using the data of the first 55 weeks to define the model and the successive 6 weeks to test the forecasts.

3. Econometric analysis, parametric and non-parametric regression models

The univariate analysis of financial time series presents particular difficulties, due to the latter's specific features: very high frequency of observed times (weekly, daily, intraday); a wide observation period; high and heteroskedastic variability; non-significant autocorrelation coefficients. Some empirical analysis of financial variables identified a random walk process according to the efficient market hypothesis, so the past history of variables could not be used to predict future events. In this case the last observation available is the best unbiased estimator for future values.

The econometric analysis of the logarithm of MIB (LMIB) on our data confirms this results, highlighting a random walk process. In fact, the autocorrelation and partial autocorrelation functions of LMIB identify an AR(1) and the first differences LMIB leads to a stationary model (the diagnostics tests Weighted Symmetric = -2.1 and Dickey-Fuller = -3.36) with the residual's homoskedasticity (Arch test = 0.49; LR heteroskedasticity test = 0.18; WHITE test = 0.68). Furthermore all cointegration tests between LMIB and the other variables are not significant, although we have to warn that the sample size was insufficient to conduct a correct test.

In the multiple linear regression model and in the other regressive models we considered the MIB to be a dependent variable and the other 7 variables to be explicative variables. We applied the following semi-parametric or non-parametric models: Projection Pursuit Regression (Friedman and Stuetzle, 1981), Local Regression model (Cleveland & Devlin 1988), Generalized Additive model (Hastie & Tibshirani 1990), Regression Tree (Breiman et al. 1984).

Most of these methods can be considered to be a special case of the general expression:

$$Y = \alpha_0 + \sum_{m=1}^{M} \alpha_m \, \mathrm{B}_m(\mathbf{X}, \vartheta_m) + \varepsilon \tag{1}$$

where $\mathbf{X}$ is a vector of K explicative variables. From this point of view, the methods differ with respect to the definition of the set $\{B_m\}$ of basis functions, allowing a unified view of the different approaches (Borra & Di Ciaccio 1998).

To apply these models we had to fix the smoothing parameters, to avoid overfitting the data. To tune the parameters the 55 weeks used were repeatedly and randomly split into 50 weeks (for the estimation phase) and 5 weeks (for the tuning).

Surprisingly, we found that the multivariate linear regression model has a good fit ($R^2 = 0.94$, Durbin Watson = 1.90). This performance was also obtained by means of the contribution of the political-economic indices which were useful in reducing heteroskedasticity and to explain the sudden variations in the MIB. In the last paragraph we reported a table with the performance of the models with respect to the forecast.
Projection Pursuit Regression appears to be the most flexible model. This method uses the following approximation:

$$y = \sum_{m=1}^{M} f_m \left(\sum_{i=1}^{k} \alpha_{im} x_i \right) + \varepsilon \qquad (2)$$

that is, the dependent variable y is fitted by a sum of M additive functions of linear combinations of the explicative variables.
The functions f_m are required to be smooth but are otherwise arbitrary. In this way any smooth function of k variables $(x_1,x_2,\ldots,x_k)$ can be well represented by a large M. On the other hand, a bad forecasting performance is usually obtained for a large M due to the overfitting of training data. Other parameters affect the behavior of the model, for example the smoothness level of the functions f_m, and sometimes it is not simple to find the right model. In general, with non-parametric models, it is advisable to have a large data-set to allow a reliable tuning of the smoothing parameters.
In our application, using the validation set, we selected M=2 which gave a good forecast, indeed a decisively better one than the other regressive models.

4. Identification of the neural network model

The advantages and the limits of the Artificial Neural Networks (ANN) are well known in comparison with consolidated statistics methods (Cheng, Titterington, 1994; Ripley, 1994; Borra & Di Ciaccio 1998). ANN are suitable to the description of non linear problems and they have been applied with success in numerous forecasting applications.
ANN can be seen as a methodology to find a universal approximator with the following expression:

$$y = f_0 \left(\mu + \sum_{m=1}^{M} \gamma_m g_m \left(\alpha_m + \sum_{i=1}^{K} \alpha_{im} x_i \right) \right) + \varepsilon \qquad (3)$$

This expression represents a neural network with K input nodes, M hidden nodes and one output node. In formula (3), x_i is i-th input variable, y is the dependent variable (output node), α_{im} denotes a weight of the arc linking input variable i to hidden node m, and γ_m denotes a weight of the arc linking hidden node m to output node y. The g_m's are fixed functions (generally chosen to be monotone, in particular sigmoidal functions) and f_o is a linear or sigmoidal or threshold function.

Identifying the appropriate ANN is a critical problem, as it is necessary to consider a wide number of choices: the individuation of the input variables; the individuation of network architecture; the values to be assigned to the parameters of learning.

One of the problems that occurs with back-propagation and associated networks is the problem of over-fitting. In this case, the network performs well on the training data, but poorly on independent test data. To deal with this problem we reduced the network size and split the data as described in the previous paragraph, testing the network prediction during the learning phase on the 5 test weeks.

We used 2 types of neural networks: the multilayer feedforward neural network (FNN) and the recurrent neural network (RNN). In the former case, input data come "forward", from nodes of hidden layers to nodes of the output layer; in the latter case, the input layer's activity patterns pass through the network more than once before generating a new output pattern.

We used a multilayer feedforward neural network with 7 nodes in input, 3 nodes for the first hidden layer and 2 nodes for the second hidden layer. Several alternatives regarding the parameters have been tested. The best results were provided by standard backpropagation, by the sigmoid activation function with range [0, 1] (learning rate initially was set at 0.3 and momentum term fixed at 0.7).

Figure 1: *Recurrent Neural Network 7-4-2-1*

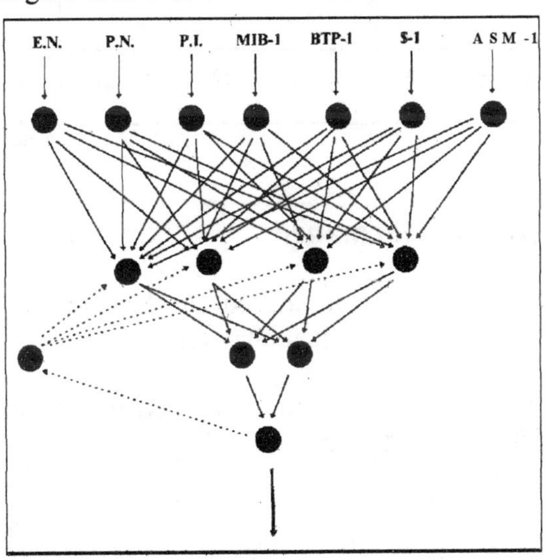

Many empirical works have demonstrated that when the time series is non linear, the prediction performance of RNN is better than FNN. Indeed, by

introducing time-lagged model components, the neural networks may respond to the same input in a different way at different times, depending on the sequence of inputs.
Our final neural architecture is composed by two hidden layers, the first layer with 4 nodes and the second layer with 2 nodes, and 1 node in a recurrent layer. The output layer is fed back into the first hidden layer by means of the node of the recurrent layer (in this manner the nodes of the first layer can see their own previous output, so that their subsequent behavior can be characterized by previous responses). The activation function, learning rate and momentum are the same as in the previous feedforward neural network.

Figure 2: *Comparison of fit of two models*

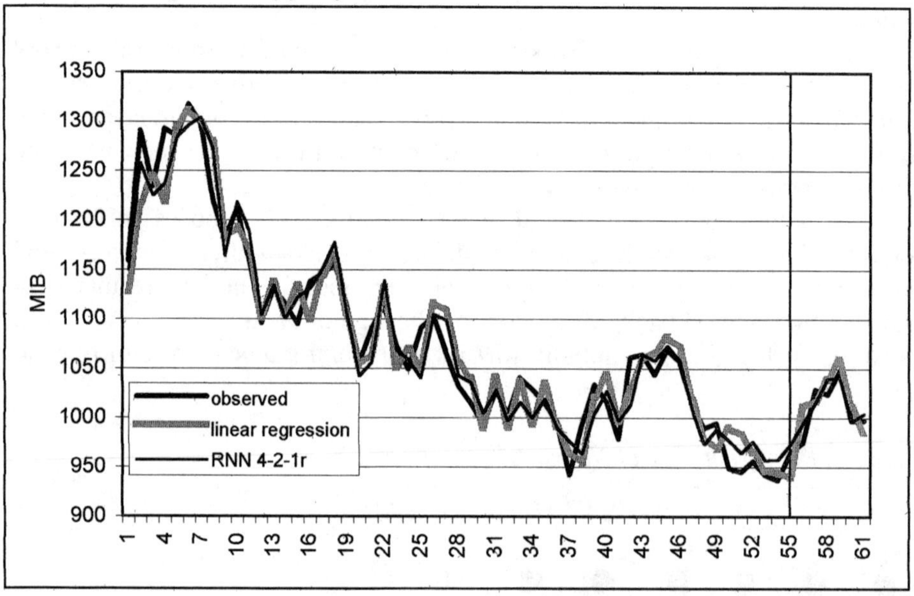

5. Comparison of the forecasting performance

To compare the performance of the models introduced in the previous paragraph, it is possible to adopt several forecast evaluation criteria.
We considered the following indices:

1) the linear correlation between the observed variation rate of MIB and the variation rate on the predicted MIB, denoted by rv;
2) the index defined as:

$$PR^{2^*} = 1 - \frac{\sum_{t \in V} (y_t - \hat{y}_t)^2}{\sum_{t \in V} (y_t - y_{t-1})^2} \tag{4}$$

where V is the validation set, y_t is the observed value at time t, $\hat{y}_t$ is the value predicted by the model at time t. Substantially, the PR^{2^*} index is the rate of the total forecasting error of the model, with the total error obtained assuming a random walk process;

3) the mean square error, MSE, which, being a quadratic loss function, is especially useful in the presence of large errors;
4) the adjusted mean absolute percentage error index, AMAPE, corrects the problem of asymmetry between the observed and forecast values:

$$\text{AMAPE} = \frac{1}{n(V)} \sum_{t \in V} \left| \frac{y_t - \hat{y}_t}{y_t + \hat{y}_t} \right| \tag{5}$$

where $n(V)$ is the total number of observations of V;

5) the standard index MAPE which is obtained from AMAPE by putting a denominator equal to y_t;
6) the percentage of correct sign predictions, CSP, which measures the potential profitability of forecasting model in a market trading strategy:

$$\text{CSP} = \frac{1}{n(V)} \sum_{t \in V} z_t \cdot 100 \tag{6}$$

where z_t is 1 when $(y_t - y_{t-1})(\hat{y}_t - \hat{y}_{t-1}) > 0$ and zero otherwise.

Table 1 shows the values of performance indices obtained by each method. We could note that the regression tree and local regression are quite incapable of forecasting the validation sample. Linear regression and GAM reach about the same results although the latter model is a generalization of a linear approach. This can be due to a bit of overfitting of GAM. Neural networks and Projection Pursuit give the best performances, with a slight preference for RNN.

Table 1: *Comparison with performance's indices*

Model	PR^{2^*}	rv	MSE	AMAPE%	MAPE%	CSP
Regression	0.675	0.635	329.0	0.769	1.552	66.67
RNN 7-4-2-1	0.875	0.871	126.9	0.475	0.956	83.33
FNN 7-3-2-1	0.839	0.749	163.2	0.538	1.080	83.33
Projection Pursuit	0.843	0.775	158.6	0.563	1.128	66.67
Local Regression	0.492	0.377	513.6	0.870	1.751	50.00
GAM	0.635	0.542	369.6	0.845	1.703	66.67
Regression Tree	0.147	0.361	863.2	1.086	2.203	16.67

The results obtained offer interesting suggestions for a comparison of the different methods and underline the importance of the use of exogenous qualitative information to improve the forecast of MIB. Of course, these results can not be easily generalized to other financial series or to other periods of time. In conclusion, we believe that this research can constitute a good starting point for the analysis of more extensive data-sets.

References

Borra, S., Di Ciaccio, A. (1998). Non-parametric regression models for the conjoint analysis of qualitative and quantitative data, *Proceedings of the 6-th Conference of IFCS*, Rome, to appear.

Breiman, L., Friedman, J.H., Olshen, R.A., Stone, J. (1984). *Classification and regression trees*. Monterey:Wadsworth and Brooks/Cole.

Brooks, C. (1997). Linear and non-linear (non-)forecastability of high-frequency exchange rates, *Journal of Forecasting*, 16, 125-145.

Cleveland, W.S., Devlin, S.J. (1988). Locally-weighted Regression: An Approach to Regression Analysis by Local Fitting, *J. Am. Statist. Assoc.*, 83, 596-610.

Cheng, B., Titterington, D.M. (1994). Neural Networks: a review from a statistical perspective, *Statistical Science*, n.1, 2-54.

Friedman, J.H., Stuetzle, W. (1981). Projection pursuit regression, *J.A.S.A.*, 76, 817-823.

Hastie, T., Tibshirani, R. (1990). *Generalized Additive Models*, Chapman & Hall, London.

Ripley, B.D. (1994). Neural Networks and related methods for classification, *J.R.S.S.*, B, 56.

Tivegna, M. (1996). News politiche ed economiche nelle fluttuazioni della lira. L'esperienza recente: 28 marzo 1994 - 28 dicembre 1995, Working Paper n.9, *Quaderni del CEIS*, Università Tor Vergata.

A New Approach to the Stock Location Assignment Problem by Multidimensional Scaling and Seriation

Angelo M. Mineo*, Antonella Plaia**•

*Dep. of Technology and Mechanical Production - University of Palermo
e-mail: amineo@unipa.it

** Institute of Statistics - Faculty of Economics - University of Palermo
e-mail: plaia@unipa.it

Abstract: The problem of the best stock location assignment in a warehouse has a fundamental role while optimising picking activities. In the present paper, this problem has been faced by considering seven variables to compute similarity between items. In this context, the problem of the choice of the most adequate similarity (or dissimilarity) measure between units while applying Multidimensional Scaling (MDS), has been examined. Besides the right metric, the possibility of applying a Seriation algorithm has been also considered. By using both MDS and seriation not just a single target can be considered, but we are able to manage with a plenty of variables; on the contrary with techniques used in literature, proper to Operational Research, just a single variable is under observation, and therefore just a single goal can be achieved. A wide discussion on the results is presented.

Key words: multidimensional scaling, seriation, similarity measures, stock location assignment.

1. Introduction

The problem of the best stock location assignment in a warehouse has a fundamental role while optimising picking activities. Among the different components of times constituting this activity (Mineo, Plaia, 1997 a), times to reach the first picking position from I/O (and return) and to reach all the picking positions in the picking list, represent the most relevant component: an improvement in stock location assignment can reduce significatively this component of time.

In the quoted paper, the use of Multidimensional Scaling (MDS) (Schiffman, Reynolds, Young, 1981) has been proposed in order to optimise stock location

• Authors' names are listed in alphabetical order.

assignment; by using MDS not just a single target can be considered, but we are able to manage with a plenty of variables; on the contrary with techniques used in literature, proper to Operational Research (Mineo, Plaia, 1997 b), just a single variable is under observation (i.e. the picking rate), and therefore just a single goal can be achieved.
The reported results show the usefulness of such a methodology, even if it has been applied to the data of a warehouse (of a Sicilian network of super- and hyper-markets) where a sub-optimal solution has been gained thanks to human experience and know-how. The aim of that paper was to show the adequacy of MDS to solve such a problem. From this point of view, the solution reached by Mineo & Plaia (1997 a) can be considered promising. Nevertheless we think that some of the simplistic solutions adopted while applying MDS, justified by the aim of that paper, require much deepness, and this will be the object of the present paper. Moreover we shall compare the MDS location assignment with the one gained by applying another multivariate technique, the Seriation.

2. The application

In order to better deepen the problem mentioned in the precedent section, data from a warehouse of a Sicilian chain of super- and hyper-markets have been considered. In the warehouse about 2000 different goods are stocked; these can be clustered in respect of goods affinity in about 25 classes, according to what is suggested by the management of the Private Label to whom the Sicilian chain belongs.
Actually, a lack in homogeneity has been perceived inside some classes. Therefore, two different solutions are considered in the present paper in order to obtain more homogeneous classes.
On one side, we have considered just the number of item[1]-per-good picked in the period of time 1.7.1996 to 31.12.1996, so obtaining 44 smaller classes by cutting off some of the goods whose behaviour, according to the considered variable, was quite different from the other goods in the class.
On the other side, a hierarchical agglomerative clustering algorithm (single linkage algorithm, Hartigan, 1975) has been applied to the 7 variables listed below (the same variables, recorded in the same period mentioned above, will be used while applying MDS and seriation) inside each class. The resulting dendrograms have been used in order to obtain a new set of 44 classes, generally different from the former.
Besides the number of item-per-good picked from the warehouse in the above mentioned period (variable 1) the following 6 variables have been considered:

2. average volume of the item in the class;
3. item fragility;
4.-6. average number of items per good per order, for each of the three

[1] Item: smallest pickable quantity of good.

different kind of supermarket in the chain;

7. number of goods in the class.

Figures 1 and 2 show two different situations obtained by applying the two approaches. As it is possible to notice, on the left side of figure 1 we find a good whose behaviour is really different from the other goods in the class (according to the number of picked items). On the contrary, on the right side of the figure, no isolable goods are present; therefore we get two classes from the histogram on the left, while we leave all the goods of the right histogram in a single class. Similarly, we maintain a single class for the goods on the left of figure 2, while we get 4 classes from the dendrogram on the right.

A 45th fictitious class (Mineo, Plaia, 1997 b) has been later added to the 44 in order to take into account the necessity to allocate the classes with the highest values of the seven variables next to I/O position in the warehouse, as these are composed by the most handled items: therefore 45th class represents the I/O position.

In the following sections a similarity matrix among classes is considered in order to compare the solutions gained by applying both MDS and seriation.

Figure 1: *Two different situations by applying the 1st approach*

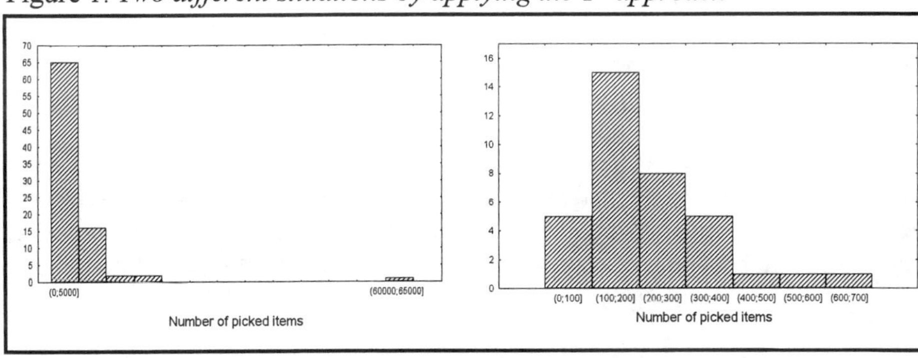

Figure 2: *Two different situations by applying the 2nd approach*

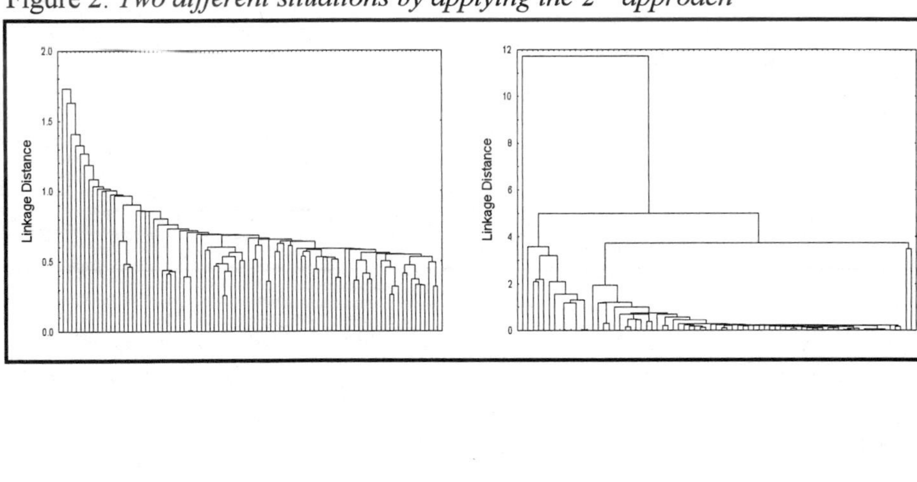

3. New choices in the application of the MDS

In order to select the proper measure of similarity the opportunity to use Gower's index of similarity (Gower, 1971) has been considered. The problem raises because one of our variable «fragility» is not quantitative; nevertheless we need to maintain the ranking order (according to the meaning of the variable). The index proposed by Gower is consequentially not suitable in our case, as it would reduce our variable to a categorical one. So we prefer to treat it as a quantitative one and non-metric MDS algorithm is applied to the two Euclidean dissimilarity matrices between the 45 classes gained by the two approaches described in section 2.

Criticisms could also be carried out to the use of the 45th fictitious class, used from the authors to represent the I/O place of the warehouse, because of the way this class has been constructed, i.e. by assigning to each variable the corresponding highest value observed on the 44 classes: indeed it could be thought that this class introduces a distortion in the final configuration, carrying therefore to a procedure of allocation of the goods strongly conditioned just from this class; this is due to the fact that MDS is a no-robust technique in presence of outliers (Spence, Lewandowsky, 1989). Actually, if we compare the final configurations with and without the fictitious class, just a rotation of the axes is observable with the first approach (based on variable 1, fig. 3), while a distortion in the final configuration is present by using the second approach (based on the clustering algorithm, fig. 4) to get the set of 44 classes. Therefore just the first approach has been considered to get the solution.

Moreover, referring to the diagram on the right in fig. 3, dimension 1 seems to be the most important to position classes along the aisles of the warehouse, considering the meaning of class 45 and the way it has been built (Mineo, Plaia, 1997 b).

Figure 3: *MDS solutions with 44 (left) and 45 (right) classes by the 1st approach*

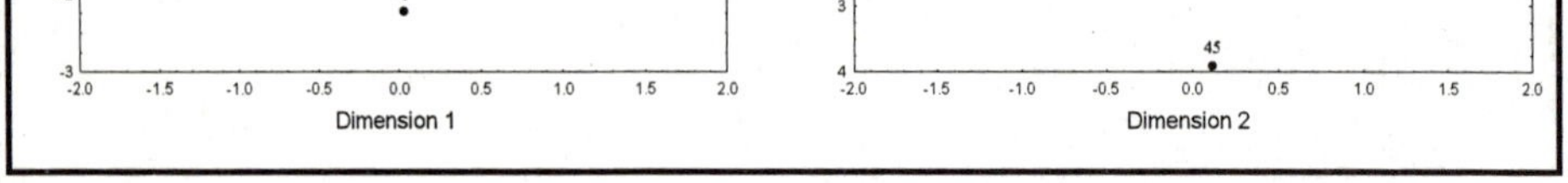

Figure 4: *MDS solutions with 44 (left) and 45 (right, the class 45 has co-ordinates D1=6.633, D2=-0.000019) classes by the 2nd approach*

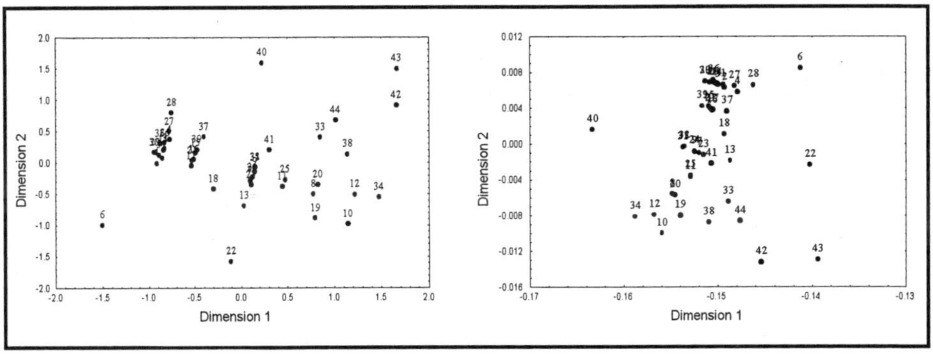

4. Seriation

In our opinion, further attempts have to be done to validate the choice of a non-metric MDS algorithm. We have used the one implemented in the STATISTICA® package (StatSoft, Inc. 1995) consisting of minimising, by means of the steepest descent method, the so-called raw-stress function, defined as

$$\text{raw-stress} = \sum_{i,j}^{n} \left[d_{ij} - f(\delta_{ij}) \right]^2 \qquad (1)$$

where d_{ij} are the reproduced distances, given the number of the dimensions, and $f(\delta_{ij})$ represent the monotonous transformation of the dissimilarities δ_{ij} computed on the input data.
Actually, if we think to the way aisles are travelled in the warehouse, i.e. to the 'traversal' travel policy (Caron, Marchet, 1994) which provides for the complete crossing of each aisle where at least one item has to be picked (fig. 5), a seriation algorithm (Wright, 1985) is applied: as a matter of fact, our problem could be presented as placing objects along a continuum.
A loss function as:

$$LF(\mathbf{x}) = \sum_{i<j} (\delta_{ij} - |x_i - x_j|)^2 \qquad (2)$$

has to be minimised over $\mathbf{x} = (x_1, x_2, \ldots, x_n)$, where x_i is the coordinate of object i on the axis (think to a single aisle made by the 20 real aisles of the warehouse) and δ_{ij} has the usual meaning.

Figure 5: *Traversal travel policy*

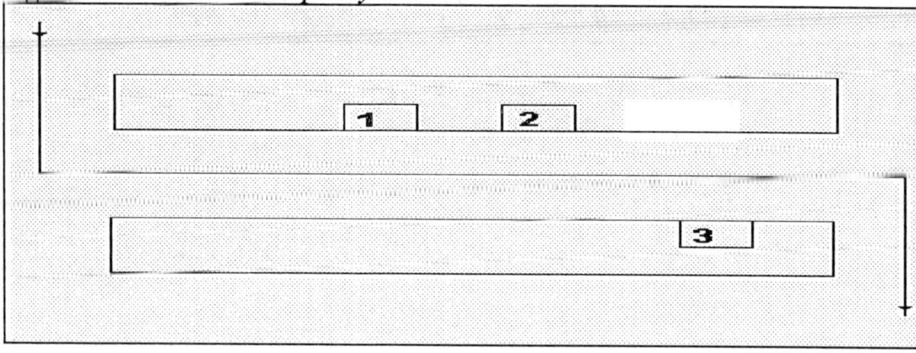

Both the approaches to get the two sets of 44 classes presented in section 2 have been considered in order to apply the algorithm.

5. Results

In order to compare all the solutions, a C language simulator has been developed, which computes the distances travelled in order to satisfy a sample of 10 orders drawn for each of the 39 stores of the supermarket chain.
In fig. 6 the following solutions are compared:

1. the current stock location assignment;
2. the non-metric MDS solution;
3. seriation solution applied to the 44 classes gained by the first approach;
4. seriation solution applied to the 44 classes gained by the second approach.

It is immediately evident that the 4th solution is absolutely not competitive, being worse than the other seriation solution: so we can conclude that the clustering algorithm is not suitable to get our final set of classes.
Both the 2nd and the 3rd configurations seem to be better than the current one and approximately equivalent one another: it is important to highlight that the current assignment is not a starting one for the warehouse, being actually the results of years of experience and improvement.
Concluding, both non-metric MDS and seriation seem to be suitable to improve the stock location assignment of a warehouse, being also better than the COI assignment (Mineo, Plaia, 1997 b) which results to be the best solution found in literature.
The comparison between non-metric MDS and seriation will be repeated as soon as the data of a whole year will be available, in order to better deepen possible differences in their solutions, by introducing other variables such as a seasonal component, which could influence stock location assignment.

Figure 6: *Distances travelled by the operator to satisfy a sample of orders for each of the 39 stores of the chain with the four techniques*

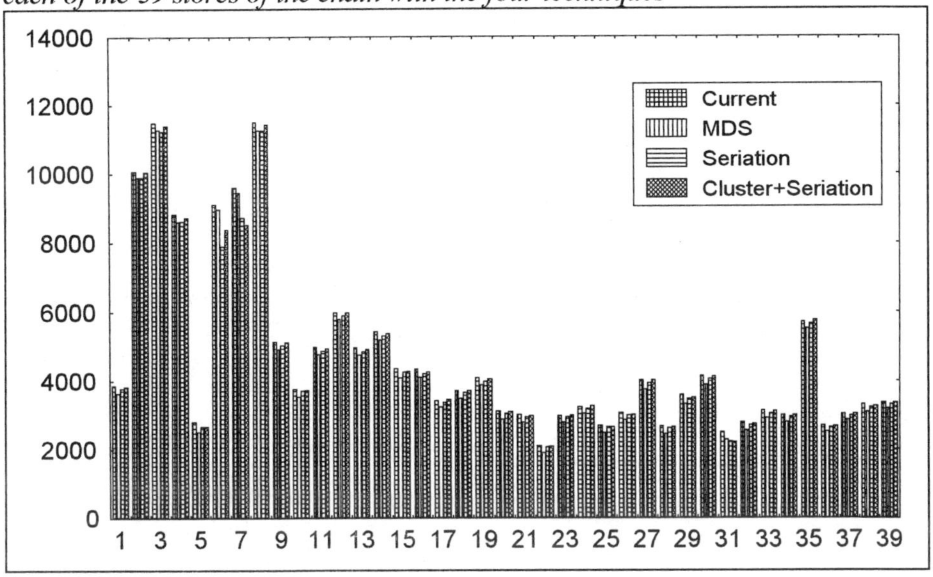

Acknowledgements

We would like to thank U. La Commare, Professor of Mechanical Technology at the Dep. of Technology & Mechanical Production of Palermo, and A. Solofra, Informative System Manager at MAR S.p.A. (Palermo), for their precious cooperation, especially for giving us the data for the application in the paper.

References

Caron, F., Marchet, G. (1994). Politiche di gestione dei sistemi picking, *Advanced topics on production system design and management*, PSD&M Varenna (Lecco) 1-4 giugno 1994, 237-265.

Gower, G. (1971). A general coefficient of similarity and some of its property, *Biometrics*, 27, 857-874.

Hartigan, J. A. (1975). *Clustering algorithms* John Wiley and Sons, New York.

Mineo, A. M., Plaia, A. (1997 a). L'allocazione fisica dei materiali in un magazzino: un approccio basato sul multidimensional scaling, Ati del Convegno SIS - *La Statistica per le imprese* - Torino , 2-4 April 1997, Vol II, 351-358.

Mineo A. M., Plaia A. (1997 b). Multidimensional Scaling and Stock location assignment in a warehouse: an application, *Proceedings of the VIII*

International Symposium on Applied Stochastic Models and Data Analysis (ASMDA), Anacapri (Napoli), 11-14 June 1997, 297-302.

Schiffman, S. S., Reynolds, M. L., Young, F. W. (1981). *Introduction to Multidimensional Scaling - Theory, Methods and Applications*, Academic Press Inc. Orlando.

Spence, I., Lewandowsky, S. (1989). Robust multidimensional scaling, *Psychometrika*, 54, 3, 501-513.

StatSoft, Inc. (1995). *STATISTICA for Windows [Computer program manual for rel. 5.0]*. Tulsa, OK,USA.

Wright, R. V. S. (1985). Detecting pattern in tabled archaeological data by principal components and correspondence analysis: programs in BASIC for portable microcomputers, *Science and Archaeology*, 27, 35-38.

Food Coding in Nutritional Surveys

Aida Turrini
Istituto Nazionale della Nutrizione, Via Ardeatina, 546 - 00178 Rome, Italy

Abstract: Nutritional studies are aimed at evaluating implications of food behaviour in order to detect possible health problems. Nevertheless, the results can be used to plan educational campaigns, regulatory interventions, and so on. In this context, food classification can vary according to different criteria. Therefore, food coding systems must be flexible enough in order to satisfy the various requirements. This approach has been utilised in the INN-CA study carried out by the Istituto Nazionale della Nutrizione (INN) in 1995, the characteristics and first results are discussed in the present paper.

Key words: Food coding, Nutritional studies.

1. Introduction

Food behaviour represents a complex topic for both the number of variables involved in its definition and the number of factors by which it is influenced by. Therefore, the study of this phenomenon can be performed according to different purposes: economic, socio-cultural, health and other aspects.

In the nutritional approach food behaviour patterns are analysed in order to estimate the "food components" intake, that means, nutrients and non-nutrients substances conveyed by foods, and the explanatory factors. This leads to a variety of methodologies for approaching the study of nutritional patterns. Particularly, nutritional surveys can be classified by "completeness" (number of explicative variables included) and "precision" (in measuring foods and food components intake) (Bingham, 1987,1991; Cialfa *et al.*, 1991; Fidanza, 1984; Marr, 1971; Pekkarinen, 1970; Saba *et al.*, 1990, 1992; Turrini, 1991 *et al.*, 1993, 1995; Willet, 1990). Generally, studies aimed at outlining food behaviour patterns for the whole population provide the basic information to plan further studies to examine in-depth specific aspects (population groups with specific problems, single nutrients or non-nutrients intake, etc.), to project educational campaigns, to define food policy interventions, and so on (Turrini, 1993).

In this context, the food data collection can be performed by utilising either "free-writing" forms, such as inventories and diaries (open-ended section) or fixed food items list (close-ended section). The first technique provides a detailed picture of consumed food products, but it poses some problems in data processing.

2. Food coding

Rationale

The first conceptual problem in food coding is the definition of the properties to be considered in data processing. In fact, food sets can be partitioned into many subsets according to different points of view (Figure 1), each of them is variously related to aspects of food behaviour and its consequences on nutritional and health status of individuals (Saba *et al*, 1990, 1992).

Figure 1: *Factors influencing food classification in nutritional surveys*

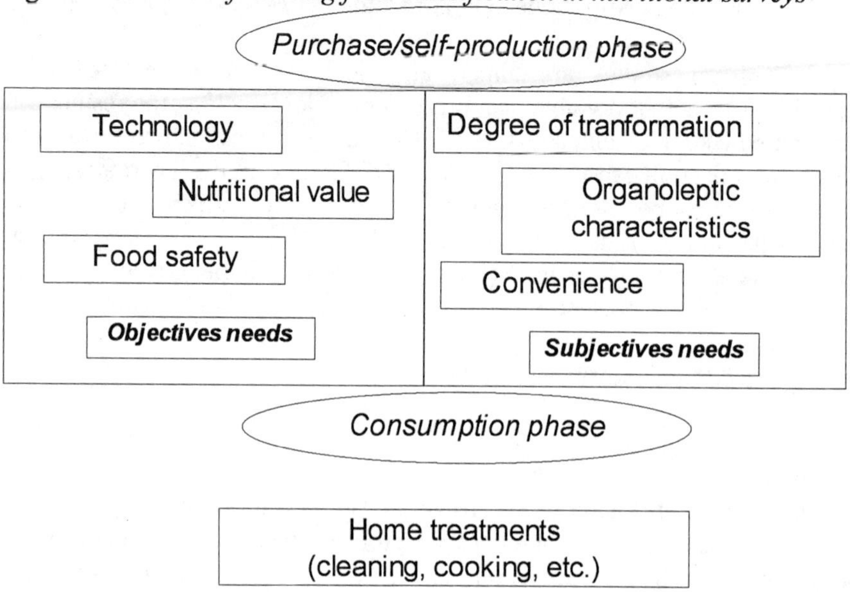

As consequences, two levels of requirements are detected: the identification for nutritional evaluation (attribution of composition data) and some level of description both for aggregating food items to match categories defined by different criteria and to pick out certain sources of undesirable food components (additives, residues, contaminants, etc.).

Furthermore, aggregation is indispensable for reaching statistical significativity in consumed quantities. In fact, it is practically impossible to reach the sample size sufficient for each single food product.

Therefore, the set of p food products $P = \{P_1, P_2, \ldots, P_p\}$ is aggregated in order to obtain the variables $\text{food}^{(s)}$ (s=1,...,k) that will be constructed as following:

$$\text{food}^{(s)} = P_1 + P_2 + \ldots + P_i + \ldots + P_{i_s} = \Sigma_{i \in I_s} P_i \qquad s=1,\ldots,k \qquad (1)$$

where the number of categories k depends on the purpose of the analysis (source of food components, total diet, comparison with other studies, etc.) and $I_p = I_1 \cup I_2 \cup \ldots \cup I_s \cup \ldots \cup I_p$.

According to characteristics such as origin, packaging, convenience, preservation method, etc., each food products P_i belongs to different subsets:

$$^{(c)}P_i^{(v)} = \{P_i \colon P_i \in F_v^{(c)},\ i=1,\ldots,p\} \qquad (2)$$

where $F_v^{(c)}$ is the subset of food products with the attribute v of the characteristic c and $F_1^{(c)} \cup F_2^{(c)} \cup \ldots \cup F_{vc}^{(c)} = P$.

Fixed the characteristic c different list of foods will be obtained for each attribute v:

$$^{(c)}\text{food}_v^{(s)} = \sum_{i \in Is} {}^{(c)}\text{P}_i^{(v)} \qquad s=1,\ldots,k;\ v=1,\ldots,v_c. \tag{3}$$

Structure

Food coding systems can be classified in two types: hierarchical codes, mixed codes (hierarchical and crosswise). In nutritional studies, we utilise a mixed code composed by two parts: the first is aimed at identifying the type of products (food group, subgroup, other levels of detail) and the second part is dedicated to the description of the characteristics (packaged or not, fresh or preserved, etc.). The first part is hierarchically organised while the second part is crosswise. This organisation is common to systems adopted for food items thesaurus (LanguaL) and systems studied for coding food surveys data (Eurocode 2). Besides, specific food coding systems are developed depending on objectives, research field, national requirements, and so on (Turrini *et al.*, 1992).

The systems differ from each other mainly for the extension of the second part that is strongly related to the study purpose.

3. A study case

Approach

In general food coding is applicable to three different types of food data (Figure 2):

1) food products (source: surveys);
2) food composition (source: chemistry laboratories);
3) regulatory statements (source: law).

Figure 2: *Typology of food data*

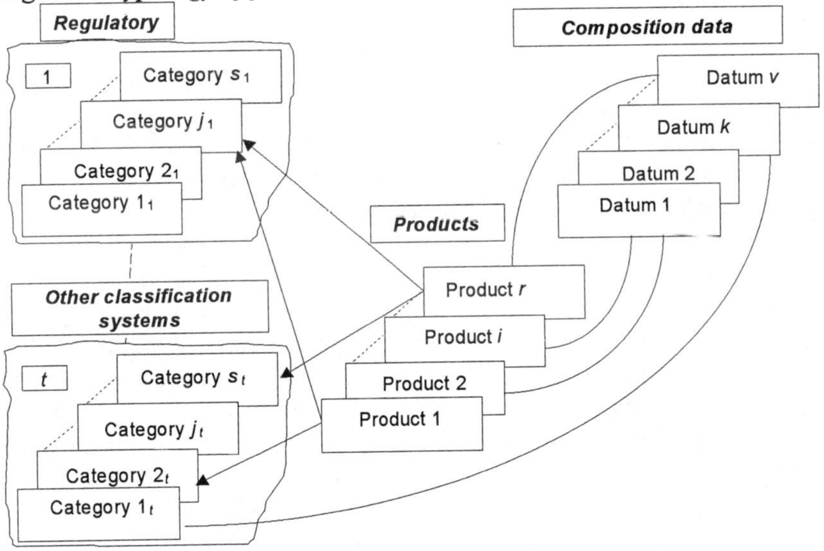

Major problems rise in the practical application of coding systems to surveyed food products. In fact, this operation requires a knowledge of the food products that

individuals do not always have. Furthermore, the food products world is continuously evolving in order to adapt the supply to the consumer's requests.
Consequently, it is very important to define a flexible coding procedure and to create a detailed documentation.
In 1995 a nation-wide food behaviour study, named INN-CA 1995, was carried out by the INN. The food section was open-ended and it included 1) household inventory, 2) purchase/wastage diary, 3) recipes form, 4) individual diaries.
The food code was framed in 9 fields as illustrated below:

Hierarchical part

1. Group
2. Subgroup
3. Detail 1 (variety)
4. Detail 2 (brand)

Crosswise part

5. Origin (vegetable, animal, mineral, composed)
6. State (raw, packaged, ready-to-eat)
7. Treatment (fresh, type of preservation)
8. Other information
9. Cooking method (prepared at home)

The objective was to obtain a food products database by recording single food names (including varieties and brand name). Because the study was conducted in 16 different centres, to avoid duplicated codes for diverse foods, a preliminary list of coded items was distributed. Therefore, when the food products were not present in the previous list all fields were assigned except detail 2 or detail 1, conventionally put to 99.

<u>*Results*</u>

Food items have been revised at two levels: *a*. alphabetical correction of writing errors, *b*. food coding standardisation.
In table 1 it is clearly shown that the application of the standardised food coding caused a strong reduction in the number of the items by itself. Therefore, the system is efficient with regards to the standardisation in food description.
It is also evident that the food groups basically originated according to a nutritional criterion (main nutrients provided) are not homogeneous according to the number of items. In fact, they contain different typologies of food products. For example, "Cereals and cereal products" comprehends bread, pasta, rusks, cornflakes, etc., "Meat" includes beef, pork, poultry, rabbit, lamb, offal, salami, etc. and the preparation could be industrial or not. For this reason food groups are divided into subgroups that are more homogeneous and details 1 and 2 indicate single variety or packaged food products.

Table 1: *Number of food items before (A) and after (B) alphabetical correction and codes standardisation (C) by food groups*

Group	*Name*	*A*	*B*	*C*
01	**Cereals and cereal products**	4938	3924	448
02	**Vegetables**	3371	2524	701
03	**Fruit**	1159	865	231
04	**Meat**	2356	1774	691
05	**Sea products**	1097	858	284
06	**Eggs**	87	60	18
07	**Milk and dairy products**	2410	1809	305
08	**Oil and fats**	853	649	56
09	**Sugar and sweets**	3456	2751	652
10	**Beverages**	2433	1878	229
11	**Miscellaneous**	1345	1102	283
12	**Dishes**	2102	1809	812
	Total	*25607*	*20003*	*4710*

Codes standardisation caused a stronger reduction than the alphabetical correction because it was based on objective characteristics of food products, while the latter was affected by the "writing style". It provided the minimum number of items necessary to defined the category surveyed.

4. Conclusion

The problem of defining aggregation criteria is central in the elaboration of nutritional survey. In surveys that utilise pre-coded food sections the problem must be solved *a priori*, in surveys based on open-ended food section this issue must be tackled at the preliminary phase of data processing.

The food coding system proposed seems to answer to the informatic requirements for the organisation of a food products database starting from survey data.

Further analyses will be performed to test its ability in outlining food sets according to different criteria (e.g. grouping by preservation methods, origin, etc.).

The expected results will be the identification of general food coding procedures in relation to different classification criteria.

References

Bingham SA (1987). The dietary assessment of individuals; methods, accuracy, new techniques and recommendations. *Nutr. Abs. Rev. (Series A)*, **57**, N. 10.

Bingham SA (1991). Limitations of the various methods for collecting dietary intake data. *Ann. Nutr. Metab.* **35**, 117-127.

Cialfa E, Turrini A & Lintas C (1991). A national food survey. Food Balance Sheets and other methodologies: a critical overview, in: *Monitoring dietary intakes*, ILSI Monographs, I Macdonald (ed.), chap. 4. Springer-Verlag, Berlin-Heidelberg-New York-London-Paris-Tokyo-Hong Kong-Barcelona.

Fidanza F (1984). Tecniche di rilevamento delle abitudini e dei consumi alimentari, in: *Nutrizione umana*, F Fidanza & G Liguori (a cura di), 346-378. Idelson, Napoli.

Marr JW (1971). Individual dietary surveys, purposes and methods. *World Rev. Nutr. Diet.* **13**, 110-139.

Pekkarinen M (1970). Methodology in the collection of food consumption data. *World Rev. Nutr. Diet.* **12**, 148-164.

Saba A, Turrini A, Mistura G, Cialfa E & Vichi M (1990). Indagine nazionale sui consumi alimentari delle famiglie 1980-84 alcuni principali risultati. *Riv. Soc. It. Sci. Alimen.* anno 19, **4**, 53-65.

Saba A, Turrini A & Cialfa E, (1992). Estimates of intakes: methodology and results of some studies carried out in Italy. *Food Add. Contam.* **5**, 527-534.

Turrini A, Saba A & Lintas C (1991). Study of the Italian reference diet for monitoring food constituents and contaminants. *Nutr. Res.* **11**, 861-874.

Turrini A, Carnovale E, Giangiacomo R, Pettinelli A (1992). Application of different coding systems to Italian dairy products, in: *Proceedings of the 2° meeting FLAIR-Eurofoods-Enfant*, Dublin (EIRE) 10-12/6/92.

Turrini A (1993). Indagini alimentari su scala nazionale: metodologia e possibilità di utilizzazione, in: *Atti della XXV Riunione Generale della Società Italiana di Nutrizione Umana*, Roma 23-25/9/1992, Giornale Europeo di Nutrizione Clinica, suppl. al n. 3, pp. 61-69

Turrini A, D'Amicis A (1995). Elementi di valutazione della qualità dei dati rilevati nelle indagini alimentari, in: *Atti del convegno "Metodi di misura nella ricerca per lo studio dell'Obesità"* , Roma 15-16/3/1995.

Willet W (1990). *Nutritional Epidemiology*. Oxford University Press, New York-Oxford.

UNAIDED: a PC System for Binary and Ternary Segmentation Analysis

Claudio Capiluppi, Luigi Fabbris, Michele Scarabello
Faculty of Statistics, University of Padua
Via San Francesco 33, 35121 Padova, Italy

Abstract: UNAIDED is a software program for segmentation analysis. The program implements several techniques and criteria for segmenting a set of units whatever the measurement scale of the criterion variable. At the present, the techniques available are: binary and ternary segmentation, monotone *vs.* free analysis, ranking of predictors, "look-ahead" search of the best split. The analytical criterion may be chosen among a large set of implemented criteria.

Key words: Statistical Data Analysis, Segmentation Analysis, Automatic Interaction Detector, Regression Trees, Classification Trees.

1. Segmentation analysis

Let **Y** denote a set of criterion, or dependent variables, and **X** a set of explanatory variables measured on a set of *N* units. Segmentation analysis is a statistical method for stepwise partitioning the set of units with reference to a univariate, bivariate, or multivariate distribution of the dependent variables.

The segmentation procedure partitions the set of units into hierarchical clusters by selecting in a stepwise fashion the predictor that minimises the within-cluster heterogeneity of the criterion variable(s). While performing a segmentation analysis, researchers can insert their prior information and substantive hypotheses for a targeted data processing.

UNAIDED - *UNivariate Automatic Interaction Detector of Empirical Data* - is a software program for the segmentation analysis with reference to a univariate dependent variable measured on any scale. Aims and features of the program, together with the statistical rationale of the choices presented to users, are discussed in the following.

2. UNAIDED

UNAIDED 1.0 is the prototype of a project which aims at offering several options of binary and ternary segmentation scattered in several software programs, such as AID (Sonquist *et al.*, 1973), ELISEE (Cellard *et al.*, 1967), THAID (Morgan & Messenger, 1973), CART (Breiman *et al.*, 1984), CHAID (Kass, 1980), C4.5 (Quinlan, 1993).

The prototype is designed for quantitative, ordinal and nominal dependent variables.

The options available[1] for the analysis are (Scarabello, 1997):

- binary, ternary, and best-between-binary-and-ternary segmentation to disclose efficiently non-monotone relations between the dependent variable and an ordinal predictor;
- "free" and "monotone" combination of the categories of candidate predictors on ordinal measurement scale;
- "ranking" of predictors, i.e. the partition of predictors in classes ranked according to causality, to control the order of predictor processing. Before the segmentation process is started, the user can assign predictors to (up to 4) ranked groups of predictors, remote causality groups being processed first and late causality ones being processed last;
- "look-ahead" evaluation of the predictive power of explanatory variables, i.e. the analysis of couples, terns, etc. of predictors at each step, in order to select the best predictor according to its main and interaction (with the coupled predictors) effects. This feature is implemented for just one step ahead, i.e. for the evaluation of first order interaction of predictors.

A single algorithm performs all the analytic options. In fact, the combination of each option with the measurement scale of the criterion variable represents a route inside the main segmentation engine. To evaluate the goodness-of-split, the algorithm follows either of the two implemented segmentation strategies (monotone *vs.* free analysis), and other criteria appropriate to the measurement scale of the dependent variable.

Multiple stopping rules are available for the user's choice: (a) minimum number of observations in the group which is about to be split, (b) minimum number of observations in groups candidate for splitting, (c) maximum number of terminal groups, (d) minimum (proportion of) variance/entropy explained by a split, (e) maximum (proportion of) within-group residual variance/entropy.

The program is user-friendly and flexible; i.e. the user may specify his/her preferences among the offered options. If the user is not able to choose among them, a default option is imposed.

3. Optimisation criteria

Each split is qualified by a reduction in the within-group heterogeneity of the criterion variable. Three basic functions are used in UNAIDED to evaluate the reduction in heterogeneity of the within-group[2] distribution of Y:

a) *the Minkowski distance, d_λ', between parameters of the parent and the resulting groups*. For a quantitative variable Y, the formula is:

$$d'_\alpha = \left[\sum_{g=1}^{G} \left| \overline{Y}_g - \overline{Y} \right|^\alpha w_g \right]^{1/\alpha} \qquad \alpha \geq 1 \qquad (1)$$

[1] Other analytical devices, such as the "pruning" of less important branches (Breiman *et al.*, 1984) and the "premium for symmetry" of the tree (Sonquist *et al.*, 1973), are in progress.

[2] For statistical analysis, groups are considered categories of a nominal variable.

where $\overline{Y}_g$ is the mean of group g, $\overline{Y}$ is the mean of the parent group and w_g is an appropriate weight of group g. Two known distances are derived from (1), the mean absolute distance (α=1) and the Euclidean distance (α=2). For an ordinal variable, the mean has to be substituted by the median, and, because of its optimal properties, the absolute distance (α=1) from the median should be considered.
Each distance may be normalised with its maximum value, which may be either the initial between-unit distance, or the between-unit distance of the parent group. Two normalised indexes are: (i) Fisher's η^2, which is an Euclidean distance with $w_g=N_g/[N\,\sigma^2]$, where σ^2 is the population variance and N_g is the size of group g; (ii) the relative group absolute distance from the median, which is the group absolute distance with $w_g=N_g/[N\,\sigma^*]$, where σ^* is the parental absolute deviation from the median. Both indexes vary between 0 and 1: $d.'$=0 if the centre of the conditional distributions is the same for all groups, $d.'$=1 if the conditional distributions maximally differ.
b) *The distance between the observed frequency distribution of variable Y and a reference distribution.* For a discrete variable Y with K categories, the Minkowski distance between observed and predicted values is:

$$d''_\alpha = \sum_{g=1}^{G}\sum_{k}^{K}\left|p_{gk} - p^*_{gk}\right|^\alpha w_{gk} \tag{2}$$

where $p_{gk}=n_{gk}/n$ is the observed and $p^*_{gk} = p_{g.}\, p_{.k}$ is the predicted frequency of category k of the resulting group g under the hypothesis of independence between Y and X. Two popular indexes are derived from (2) with α=2: (i) Pearson's χ^2, for which $w_{gk}=n/p^*_{gk}$; (ii) Goodman-Kruskal's τ_b, for which $w_{gk}=1/[p_{g.}(1-\sum_k p_{.k}^2)]$.
χ^2 may be standardised with Bonferroni's correction, to account for the degrees of freedom of the analysis[3], and its maximum, $Max(\chi^2)=n(m-1)$, where $m=min(G,K)$. While standardised, $d.''$ varies between 0 and 1: $d.''$=0 in the case of independence between Y and the variable whose categories are groups, $d.'$=1 if groups are maximally different;
c) *The entropy, or uncertainty, of the Y criterion variable distribution within the groups resulting after segmentation with variable X.* For the segmentation analysis with a variable X, Shannon's entropy is:

$$H_{y.x} = H(y) - H(y.x) = -\sum_{k}^{K} p_{.k}\ln(p_{.k}) + \sum_{g=1}^{G} p_{g.}\sum_{k}^{K} p_{k|g}\ln(p_{k|g}) \tag{3}$$

where $p_{k|g}=p_{gk}/p_{g.}$ is the relative frequency of category k conditional to group g. The coefficient is normalised with the entropy, $H(x)$, of the classification variable. While standardised, $H_{y/x}$ varies between 0 and 1: $H_{y/x}$=0 in the case of independence between Y and the variable whose categories are groups, $H_{y/x}$=1 if groups are maximally different.
Default criteria in UNAIDED are the Euclidean between mean distance for quantitative and entropy for nominal variables.

---

[3] The number of degrees of freedom differs according to the type of analysis, monotone *vs.* free.

4. Program features and algorithm

UNAIDED has a graphic user interface in Windows style. All program functions are accessible through menu items and dialog windows. Required user inputs, such as parameters and options for the segmentation analysis, are supplied through standard Windows input controls.

Data may be imported from many file formats; in particular, UNAIDED can access all Windows formats supported by the Microsoft Jet engine (i.e. MDB, DBF, XLS, Paradox). The program is also an ODBC (Open Data-Base Connectivity) client; this means it can access virtually to all data sources, and in practice to all database formats whose ODBC driver is installed on the host system. For instance, SAS data sets are directly usable, provided a SAS ODBC driver is available.

The tree obtained by the segmentation analysis is displayed by means of a Windows *Tree* control, which allows an easy examination of the node/group statistics. The analysis output, that is the tree structure and the node/group statistics, is saved in an output file, in MBD format, containing a record for each node/group of the tree. This allows for reviewing a saved tree output without running the analysis again.

UNAIDED performs segmentation analysis by means of an iterative algorithm, processing the segmentation tree for levels, and not for branches as recursive algorithms do. The *level* of the segmentation process is the number of splits occurred to isolate a group. Beginning from the root group of size n at the zero level, the first split separates two or three subgroups, which are the groups of the first level, and so on.

The algorithm examines and eventually splits each group at current level. To split a group, for each predictor, all the possible splits, according to the type of predictor, are considered and for each one, the split function is computed to evaluate the gain in within-group homogeneity following that split. The best splits of all predictors are then compared, and the very best one is found, among those considered. The procedure is repeated until no group of current level can be split.

The search of the best split is extended to the set of splits based on a single predictor at a time. Admissible splits are then determined according to the type of predictor: when a predictor is denoted as “monotone”, the order of its categories is kept fixed, that is only adjacent categories on the ordinal scale can be associated to identify a group; when the predictor is denoted “free”, any combination of its categories is plausible.

To make the search computationally practicable when the analysis is free, predictor categories are ordered according to a function consistent with the split criterion function; then the algorithm proceeds as in the case of monotone analysis. For instance, when the dependent variable is quantitative and η^2 split criterion is used, predictor categories are ordered according to conditional averages of the criterion variable (Fisher, 1958); in this case the solution found is the very best one. For binary segmentation with a nominal dependent variable and Entropy split criterion, predictor categories are firstly ordered according to the conditional entropy of the Y variable. In this case, however, the found solution is not necessarily the best.

The standard segmentation, based on a single predictor at a time, is a procedure locally optimal for the single step, but the overall solution, obtained by the stepwise procedure, is not the best partition of the data set according to the criterion. The algorithm of UNAIDED implements the Look-ahead option, which enhances the overall solution by noting interactions between the currently processed predictor and all other predictors at the next step.
No restriction is set about the number of observations and variables to be processed by the program; limitations are determined only by available memory and computing time.

5. Technical section

UNAIDED 1.0 is a program for personal computers based on Intel processors and runs under Windows 95 operative system. It should also run under Windows NT 4.0 although further testing is needed for the latter environment. The program needs a set-up to install some Windows shared components (DLL and OCX) coming with Visual C++ developing environment. To use UNAIDED, it is advisable to have a system with a 586 class processor and at least 16 Mb of RAM.
The program is coded in VISUAL C++ 4.0 and developed under Windows 95 environment. Program structure was designed according to specifications of the document-view architecture, the object oriented application framework settled in MFC 4.0 (Microsoft Foundation Classes) object class library. The development of the first prototype took about 10 months of man work, but the program is still in progress.

6. A sample application

An application, which shows how the program works with a typical problem of classification, is presented. The well known Iris data set is used to exemplify the comparison between binary and ternary segmentation.
Data are directly get from a SAS Dataset through the File Open menu function, by specifying the ODBC SAS data source (Fig. 1). From the list of on-line SAS datasets in the defined SAS libnames, the Iris dataset is picked up. Read data are browsed in a grid window, and a dialog window allows specifying the Iris species as nominal criterion variable and the four variables, petal and sepal height and length, as monotone predictors (Fig. 2).
Binary versus ternary segmentation is performed, using Entropy as split criterion and keeping fixed all the other options; in particular, stopping rules were set on minimum values to get the maximum growth of the tree. Figures 3 and 4 report the resulting trees in the output window: by clicking on a node, one gets the node statistics displayed on the left part of this window. The variable sepal height dominates both analyses, being selected repeatedly as the best split predictor. Ternary segmentation produces a tree simpler and compact, much easily readable than the binary tree. Multiple segmentation could do even better, separating in a single split the effect of this predictor. The Look-ahead option does not improve the results, giving the same tree of ternary segmentation without Look-ahead. This outcome suggests no interaction among predictors is present in the data.

Fig. 1 - Data loading from an ODBC data source (SAS dataset)

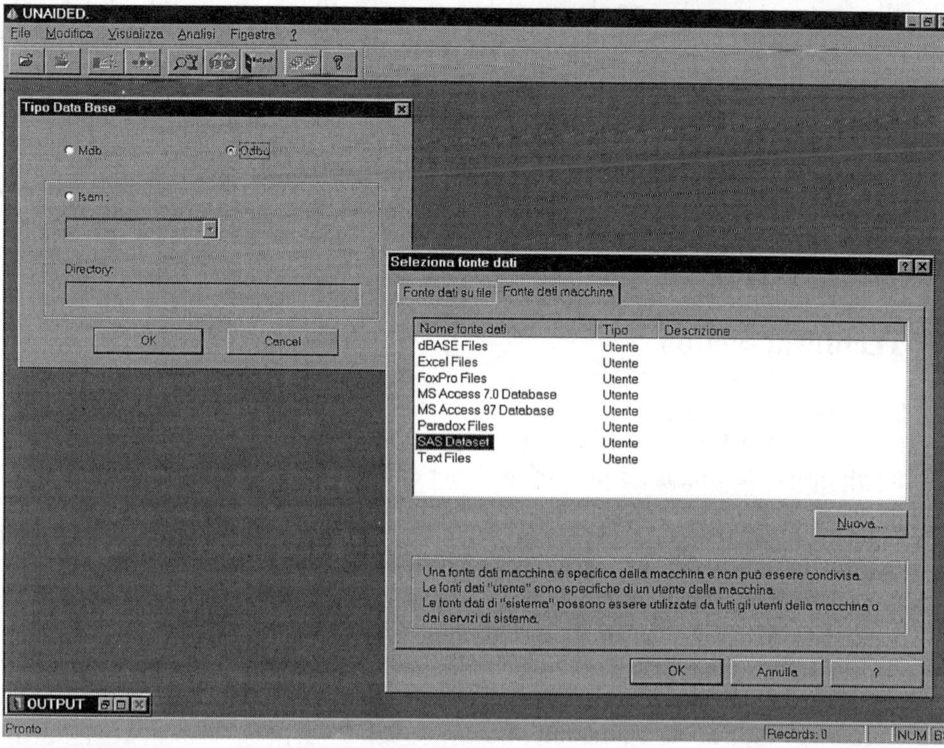

Fig. 2: The Data Browser Window and the Variable Selection Dialog Window

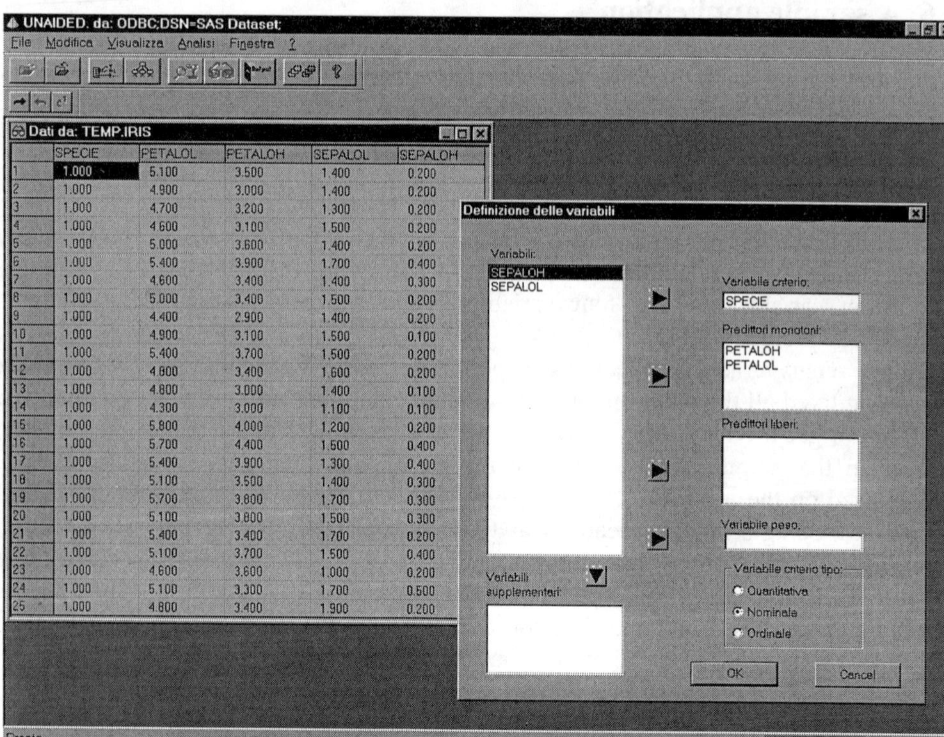

Fig 3: The Output Dialog Window: binary segmentation of Iris data set

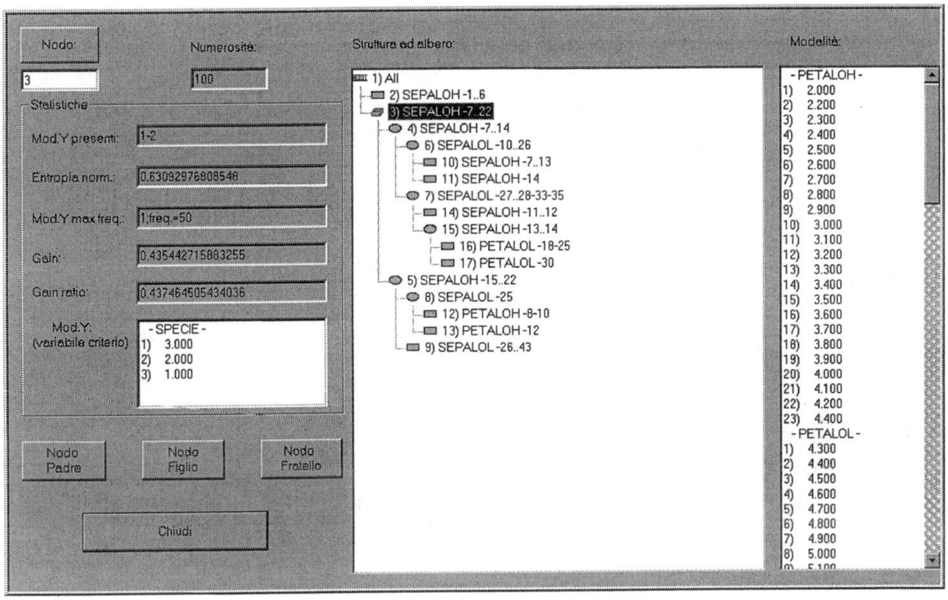

Fig 4: The Output Dialog Window: ternary segmentation of Iris data set

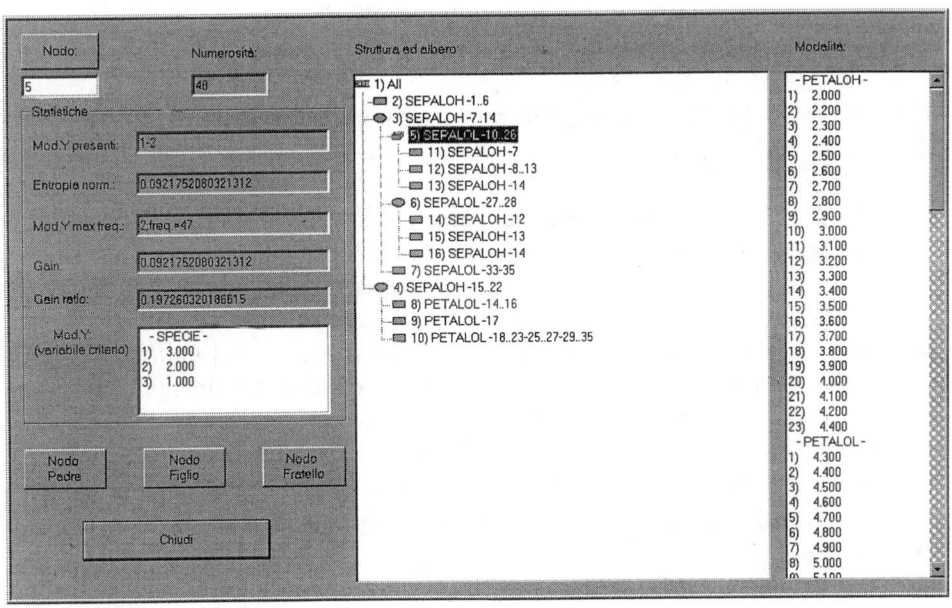

7. Conclusions

Program requires further development. Not all the considered splitting criteria are supported. Some important functionality is lacking, such as an output printing function.

The present version of the program is the first step of a project for the development of statistical methods based on segmentation. Some important developments are designing to extend the analysis capabilities of the program:

- larger set of split functions
- best split search extended to split based on combinations of predictors.

We plan to distribute the software on the FTP site ftp.stat.unipd.it, as soon as the necessary test phase will be completed.

Acknowledgements

The authors worked jointly to the paper. Nevertheless, C. Capiluppi wrote sections 3 and 4, L. Fabbris the sections 1 and 7 and M. Scarabello the sections 2, 5 and 6.

References

Breiman L., Friedman J.H., Olshen R.A. & Stone C.J. (1984) *Classification and Regression Trees*. Wadsworth Inc., Belmont California.

Cellard J.C., Labbé B. & Savitsky G. (1967) Le programme ELISEE. Presentation and application. *Metra*, **3**, 511-519.

Fisher W.D. (1958) On grouping for maximum homogeneity. *JASA*, **53**, 789-798.

Hays W.L. (1973), *Statistics for the Social Sciences*, II ed.. Holt, Rinehart & Winston, New York

Kass G.V. (1980) An Exploratory Technique for Investigating Large Quantities of Categorical Data. *Applied Statistics*, **29**, 119-127.

Magidson J. (1981) Qualitative variance, entropy, and correlation ratios for nominal dependent variables. *Social Science Research*, **10**, 177-194.

Morgan J.N. and Messenger R.C. (1973) *THAID A Sequential Analysis Program for the Analysis of Nominal Scale Dependent Variables*. Institute for Social Research, Ann Arbor, Michigan.Quinlan J.R. (1993) *C4.5: Programs for Machine Learning*. Morgan Kaufman, San Mateo, California.

Scarabello M. (1997) UNAIDED. Un programma per la segmentazione binaria e ternaria di campioni. Graduation Thesis, Statistics Faculty, University of Padua

Sonquist J.A., Baker E.L. & Morgan J.N. (1973) *Searching for Structure*. Institute for Social Research, Ann Arbor, Michigan.

AUTHOR INDEX

A

Alfò 73
Amenta 263

B

Baiocchi 237
Bellacicco 35
Bolasco 237
Borra 345
Bove 131

C

Caggese 81
Calò 97
Camiz 139
Capiluppi 367
Capobianchi 211
Cerbara 43
Cerioli 3
Chiodi 247
Coli 255
Corazziari 171
Cornillon 263

D

D'Esposito 279
D'Urso 11
Di Battista 19
Di Ciaccio 345
Di Marzio 287
Di Spalatro 19

E

Esposito F. 81
Esposito V. 179

F

Fabbris 367
Ferri 55

G

Gazzei 271
Giacalone 295
Giusti 221

I

Iacovacci 49
Iezzi 27
Ingrassia 89
Ippoliti 255

J

Jona-Lasinio 211

L

Lafratta 287
Laghi 303
Lemmi 271
Lizzani 303
Lombardo 187

M

Malerba 81
Mantovan 311
Maturo 55
Miglio 105
Milioli 63
Mineo 247;353
Mola 113
Montanari 97; 147
Morrone 237

N

Nissi 255

P

Pallini 319
Pastore 311
Petrucci 221
Pillati 105
Pittau 11
Plaia 353
Porzio 327
Postiglione 73

Q

Quintano 155

R

Ragozini 279
Rocci 131
Romagnoli 229
Roverato 335

S

Sabatier 263
Scarabello 367
Scepi 179
Semeraro 81
Siciliano 121
Soffritti 147

T

Tessitore 187
Tonnellato 311
Turrini 361

V

Verde 195
Vichi 27;163
Vittadini 203

KEY WORDS

A

adaptive cluster sampling ... 19
Akaike criterion ... 121
array of data or cubic matrices ... 171
asymmetry ... 131
automatic interaction detector ... 367

B

Bayes rule ... 121
bias ... 19
bivariate time series ... 105
bootstrap ... 19
B-spline function ... 195

C

classification trees ... 367
cluster ... 35
cluster analysis ... 27; 147
cluster stability ... 3
cluster validity ... 63
clusters analysis ... 221
co-inertia ... 179
combining ... 105
conditional independence ... 335
consensus classification ... 11
constrained principal component analysis ... 187
contiguity constraints ... 221
convex hull ... 279
covariance ... 319
cross-validation ... 319
crossvariogram ... 211
curse of dimensionality ... 287

D

data analysis ... 295
deletion diagnostics ... 3
dimensional scaling (DS) ... 155
discrimination and classification ... 81
dissimilarities ... 147
distance between time series ... 11
distance-based regression model ... 303
distances ... 139
dominant eigenvalue scores (DES) ... 155
dynamic linear models ... 311

E

eigenanalysis ... 139
evaluation urban projects ... 55
event ... 35
expected proportion of samples in kernel supports ... 287

F

factorial analysis ... 171
financial time series ... 345
firm performance ... 271
forecast ... 345
functional data analysis ... 319
fuzzy ... 35
fuzzy average linkage ... 43
fuzzy clustering ... 63
fuzzy c-means algorithm ... 49
fuzzy methods ... 55

G

Gauss-Markov ... 229
Geary coefficient ... 263
generalised canonical analysis ... 195
generalised additive models ... 113
graph ... 263
graphical displays ... 179
graphical methods ... 131
graphical models ... 335

H

hierarchical cluster analysis ... 43
hierarchical clustering time series ... 11
hyperstructures ... 55

I

identification ... 203
independence graph ... 335
indeterminacy ... 203
interpretation ... 35

K

Kalman filter ... 229; 255
kernel density estimation ... 319
Kronecker product ... 263

L

latent variable ... 121
local influence ... 327
loess ... 113
logistic discrimination ... 89

M

marginal model plots ... 327
matrix completion ... 335
MDS ... 353
measures ... 353
measures of fuzziness ... 63
method of moments estimators ... 311
missing data ... 255
mixed predictors ... 303
mixing parameter ... 121
MORALS ... 303
multidimensional scaling ... 147
multidimensional scaling(MDS) ... 155
multigraphs ... 147
multiple qualitative response variable ... 121
multiple sets ... 179
multivariate adaptive splines ... 187

N

Neighbourhood operator.......................................263
network.......................................35
neural.......................................35
neural networks.......................................105; 345
nonhierachical clustering.......................................3
nonparametric discriminant analysis.......................................97
non-parametric models.......................................345

O

orthogonal projections.......................................179
outliers.......................................3; 279

P

panel data.......................................27; 271
parallel Kalman filters.......................................311
parameter estimation.......................................89
partitioning.......................................27
pattern recognition.......................................81
posterior distribution.......................................335
pre-processing.......................................237
principal component analysis.......................................211
principal co-ordinate analysis.......................................303
principal surface.......................................187
profession market.......................................155
projection pursuit.......................................97
projection pursuit density estimation.......................................97
projection pursuit regression.......................................303
qualitative variables.......................................203

R

rand index.......................................3
random fiels.......................................229
regression and autoregressive models.......................................171
regression tree.......................................113; 367
regressogram.......................................113
resampling.......................................19
restricted regression component decomposition method.......................................203

S

segmentation analysis.......................................367
semantic.......................................35
semi-fuzzy classification.......................................49
separability measures.......................................89
Shewart's control chart.......................................295
sigma's estimate.......................................295
similarity.......................................43; 353
singular value decomposition.......................................211
smoothing.......................................319
soft clustering.......................................49
software.......................................237
spatial processes.......................................211
spline approximation.......................................187
spline smoothers.......................................113
stalactite plot.......................................3
state-space model.......................................255
STATIS method.......................................263
statistical data analysis.......................................367
stock location assignment.......................................353
structural model.......................................203
subpopolations.......................................327
supervised classifiers.......................................105
symbolic data analysis.......................................81
symbolic objects.......................................195
system parameter estimation.......................................311

T

technical efficiency.......................................271
telephone surveys.......................................221
textual variables.......................................237
three way data.......................................263
three way environmental data matrix.......................................255
time series.......................................171
three-way Asymmetric scaling.......................................131

U

unilateral representation.......................................229

V

variable selection.......................................63

W

Wilks statistic.......................................279